Electronic Properties of Materials

Third Edition

Rolf E. Hummel

Electronic Properties of Materials

Third Edition

With 252 Illustrations

 Springer

Rolf E. Hummel
Department of Materials Science
   and Engineering
University of Florida
Gainesville, FL 32611
USA

Library of Congress Cataloging-in-Publication Data
Hummel, Rolf E., 1934–
   Electronic properties of materials / Rolf E. Hummel.—3rd ed.
      p.   cm.
   Includes bibliographical references and index.
   ISBN 0-387-95144-X (alk. paper)
   1. Solid-state physics.   2. Energy-band theory of solids.   3. Electronics—
Materials.   4. Electrical engineering—Materials.   I. Title.
QC176.H86   2000
530.4′1—dc20   00-061866

ISBN 0-387-95144-X   Printed on acid-free paper.

Printed in the United States of America.   (AS/EB)

9 8 7 6 5 4 3 2   Corrected second printing, 2005.   SPIN 11005834

springeronline.com

Preface to the Third Edition

Books are seldom finished. At best, they are abandoned. The second edition of "Electronic Properties of Materials" has been in use now for about seven years. During this time my publisher gave me ample opportunities to update and improve the text whenever the book was reprinted. There were about six of these reprinting cycles. Eventually, however, it became clear that substantially more new material had to be added to account for the stormy developments which occurred in the field of electrical, optical, and magnetic materials. In particular, expanded sections on flat-panel displays (liquid crystals, electroluminescence devices, field emission displays, and plasma displays) were added. Further, the recent developments in blue- and green-emitting LED's and in photonics are included. Magnetic storage devices also underwent rapid development. Thus, magneto-optical memories, magneto-resistance devices, and new magnetic materials needed to be covered. The sections on dielectric properties, ferroelectricity, piezoelectricity, electrostriction, and thermoelectric properties have been expanded. Of course, the entire text was critically reviewed, updated, and improved. However, the most extensive change I undertook was the conversion of all equations to SI-units throughout. In most of the world and in virtually all of the international scientific journals use of this system of units is required. If today's students do not learn to utilize it, another generation is "lost" on this matter. In other words, it is important that students become comfortable with SI units.

If plagiarism is the highest form of flattery, then I have indeed been flattered. Substantial portions of the first edition have made up verbatim most of another text by a professor in Madras without giving credit to where it first appeared. In addition, pirated copies of the first and second editions have surfaced in Asian countries. Further, a translation into Korean appeared. Of course, I feel that one should respect the rights of the owner of intellectual property.

I am grateful for the many favorable comments and suggestions promulgated by professors and students from the University of Florida and other schools who helped to improve the text. Dr. H. Rüfer from Wacker Siltronic AG has again appraised me of many recent developments in wafer fabrication. Professor John Reynolds (University of Florida) educated me on the current trends in conducting polymers. Drs. Regina and Gerd Müller (Agilent Corporation) enlightened me on recent LED developments. Professor Paul Holloway (University of Florida) shared with me some insights in phosphors and flat-panel displays. Professor Volkmar Gerold (MPI Stuttgart) was always available when help was needed. My thanks go to all of them.

Gainesville, Florida   Rolf E. Hummel
October 2000

Preface to the Second Edition

It is quite satisfying for an author to learn that his brainchild has been favorably accepted by students as well as by professors and thus seems to serve some useful purpose. This horizontally integrated text on the electronic properties of metals, alloys, semiconductors, insulators, ceramics, and polymeric materials has been adopted by many universities in the United States as well as abroad, probably because of the relative ease with which the material can be understood. The book has now gone through several reprinting cycles (among them a few pirate prints in Asian countries). I am grateful to all readers for their acceptance and for the many encouraging comments which have been received.

I have thought very carefully about possible changes for the second edition. There is, of course, always room for improvement. Thus, some rewording, deletions, and additions have been made here and there. I withstood, however, the temptation to expand considerably the book by adding completely new subjects. Nevertheless, a few pages on recent developments needed to be inserted. Among them are, naturally, the discussion of ceramic (high-temperature) superconductors, and certain elements of the rapidly expanding field of optoelectronics. Further, I felt that the readers might be interested in learning some more practical applications which result from the physical concepts which have been treated here. Thus, the second edition describes common types of field-effect transistors (such as JFET, MOSFET, and MESFET), quantum semiconductor devices, electrical memories (such as D-RAM, S-RAM, and electrically erasable-programmable read-only memories), and logic circuits for computers. The reader will also find an expansion of the chapter on semiconductor device fabrication. The principal mechanisms behind some consumer devices, such as xerography, compact disc players, and optical computers, are also discussed.

Part III (Magnetic Properties of Materials) has been expanded to include more details on magnetic domains, as well as magnetostriction, amorphous ferromagnetics, the newest developments in permanent magnets, new magnetic recording materials, and magneto-optical memories.

Whenever appropriate, some economic facts pertaining to the manufacturing processes or sales figures have been given. Responding to occasional requests, the solutions for the numerical problems are now contained in the Appendix.

I am grateful for valuable expert advice from a number of colleagues, such as Professor Volkmar Gerold, Dr. Dieter Hagmann, Dr. H. Rüfer, Mr. David Malone, Professor Chris Batich, Professor Rolf Haase, Professor Robert Park, Professor Rajiv Singh, and Professor Ken Watson. Mrs. Angelika Hagmann and, to a lesser extent, my daughter, Sirka Hummel, have drawn the new figures. I thank them for their patience.

Gainesville, Florida   Rolf E. Hummel
1993

Preface to the First Edition

*Die meisten Grundideen der
Wissenschaft sind an sich einfach
und lassen sich in der Regel
in einer für jedermann
verständlichen Sprache
wiedergeben.*

—ALBERT EINSTEIN

The present book on electrical, optical, magnetic, and thermal properties of materials is, in many aspects, different from other introductory texts in solid state physics. First of all, this book is written for engineers, particularly materials and electrical engineers who want to gain a fundamental understanding of semiconductor devices, magnetic materials, lasers, alloys, etc. Second, it stresses concepts rather than mathematical formalism, which should make the presentation relatively easy to understand. Thus, this book provides a thorough preparation for advanced texts, monographs, or specialized journal articles. Third, this book is not an encyclopedia. The selection of topics is restricted to material which is considered to be essential and which can be covered in a 15-week semester course. For those professors who want to teach a two-semester course, supplemental topics can be found which deepen the understanding. (These sections are marked by an asterisk [*].) Fourth, the present text leaves the teaching of crystallography, X-ray diffraction, diffusion, lattice defects, etc., to those courses which specialize in these subjects. As a rule, engineering students learn this material at the beginning of their upper division curriculum. The reader is, however, reminded of some of these topics whenever the need arises. Fifth, this book is distinctly divided into five self-contained parts which may be read independently. All are based on the first part, entitled "Fundamentals of Electron Theory," because the electron theory of materials is a basic tool with which most material properties can be understood. The modern electron theory of solids is relatively involved. It is, however, not my intent to train a student to become proficient in the entire field of quantum theory. This should be left to more specialized texts. Instead, the essential quantum mechanical concepts are introduced only to the extent to which they are needed for the understanding of materials science. Sixth, plenty of practical applications are presented in the text, as well as in the problem sections, so that the students may gain an understanding of many devices that are used every day. In other words, I tried to bridge the gap between physics and engineering. Finally, I gave the treatment of the optical properties of materials about equal coverage to that of the electrical properties. This is partly due to my personal inclinations and partly because it is felt that a more detailed description of the optical properties is needed since most other texts on solid state physics devote relatively little space to this topic. It should be kept in mind that the optical properties have gained an increasing amount of attention in recent years, because of their potential application in communication devices as well as their contributions to the understanding of the electronic structure of materials.

The philosophy and substance of the present text emerged from lecture notes which I accumulated during more than twenty years of teaching. A preliminary version of Parts I and II appeared several years ago in *Journal of Educational Modules for Materials Science and Engineering* **4**, 1 (1982) and **4**, 781 (1982).

I sincerely hope that students who read and work with this book will enjoy, as much as I, the journey through the fascinating field of the physical properties of materials.

Each work benefits greatly from the interaction between author and colleagues or students. I am grateful in particular to Professor R.T. DeHoff, who read the entire manuscript and who helped with his inquisitive mind to clarify many points in the presentation. Professor Ken Watson read the part dealing with magnetism and made many helpful suggestions. Other colleagues to whom I am indebted are Professor Fred Lindholm, Professor Terry Orlando, and Dr. Siegfried Hofmann. My daughter, Sirka Hummel, contributed with her skills as an artist. Last, but not least, I am obliged to my family, to faculty, and to the chairman of the Department of Materials Science and Engineering at the University of Florida for providing the harmonious atmosphere which is of the utmost necessity for being creative.

Gainesville, Florida   Rolf E. Hummel
1985

Contents

PART II
Electrical Properties of Materials   75

PART IV
Magnetic Properties of Materials   303

CHAPTER 14
Foundations of Magnetism   305

CHAPTER 15
Magnetic Phenomena and Their Interpretation—Classical Approach   312

Note: Sections marked with an asterisk (*) are topics which are beyond a
      15-week semester course or may be treated in a graduate course.

PART I

FUNDAMENTALS OF ELECTRON THEORY

CHAPTER 1

Introduction

The understanding of the behavior of electrons in solids is one of the keys to understanding materials. The electron theory of solids is capable of explaining the optical, magnetic, thermal, as well as the electrical properties of materials. In other words, the electron theory provides important fundamentals for a technology which is often considered to be the basis for modern civilization. A few examples will illustrate this. Magnetic materials are used in electric generators, motors, loudspeakers, transformers, tape recorders, and tapes. Optical properties of materials are utilized in lasers, optical communication, windows, lenses, optical coatings, solar collectors, and reflectors. Thermal properties play a role in refrigeration and heating devices and in heat shields for spacecraft. Some materials are extremely good electrical conductors, such as silver and copper; others are good insulators, such as porcelain or quartz. Semiconductors are generally poor conductors at room temperature. However, if traces of certain elements are added, the electrical conductivity increases.

Since the invention of the transistor in the late 1940s, the electronics industry has grown to an annual sales level of about five trillion dollars. From the very beginning, materials and materials research have been the lifeblood of the electronics industry.

For the understanding of the electronic properties of materials, three approaches have been developed during the past hundred years or so which differ considerably in their philosophy and their level of sophistication. In the last century, a phenomenological description of the experimental observation was widely used. The laws which were eventually discovered were empirically derived. This "**continuum theory**" considered only macroscopic quantities and interrelated experimental data. No assumptions were made about the structure of matter when the equations were formulated. The

3

conclusions that can be drawn from the empirical laws still have validity, at least as long as no oversimplifications are made during their interpretation. Ohm's law, the Maxwell equations, Newton's law, and the Hagen–Rubens equation may serve as examples.

A refinement in understanding the properties of materials was accomplished at the turn to the twentieth century by introducing atomistic principles into the description of matter. The **"classical electron theory"** postulated that free electrons in metals drift as a response to an external force and interact with certain lattice atoms. Paul Drude was the principal proponent of this approach. He developed several fundamental equations that are still widely utilized today. We will make extensive use of the Drude equations in subsequent parts of this book.

A further refinement was accomplished at the beginning of the twentieth century by **quantum theory**. This approach was able to explain important experimental observations which could not be readily interpreted by classical means. It was realized that Newtonian mechanics become inaccurate when they are applied to systems with atomic dimensions, i.e., when attempts are made to explain the interactions of electrons with solids. Quantum theory, however, lacks vivid visualization of the phenomena which it describes. Thus, a considerable effort needs to be undertaken to comprehend its basic concepts; but mastering its principles leads to a much deeper understanding of the electronic properties of materials.

The first part of the present book introduces the reader to the fundamentals of quantum theory. Upon completion of this part the reader should be comfortable with terms such as Fermi energy, density of states, Fermi distribution function, band structure, Brillouin zones, effective mass of electrons, uncertainty principle, and quantization of energy levels. These concepts will be needed in the following parts of the book.

It is assumed that the reader has taken courses in freshman physics, chemistry, and differential equations. From these courses the reader should be familiar with the necessary mathematics and relevant equations and definitions, such as:

$$\text{Newton's law: force equals mass times acceleration } (F = ma); \quad (1.1)$$

$$\text{Kinetic energy: } E_{kin} = \tfrac{1}{2}mv^2 \text{ } (v \text{ is the particle velocity}); \quad (1.2)$$

$$\text{Momentum: } p = mv; \quad (1.3)$$

$$\text{Combining (1.2) and (1.3) yields } E_{kin} = \frac{p^2}{2m}; \quad (1.4)$$

$$\text{Speed of light: } c = v\lambda \text{ } (v = \text{frequency of the light wave, and}$$
$$\lambda \text{ its wavelength}); \quad (1.5)$$

$$\text{Velocity of a wave: } v = v\lambda; \quad (1.6)$$

$$\text{Angular frequency: } \omega = 2\pi v; \quad (1.7)$$

$$\text{Einstein's mass–energy equivalence: } E = mc^2. \quad (1.8)$$

It would be further helpful if the reader has taken an introductory course in materials science or a course in crystallography in order to be familiar with terms such as lattice constant, Miller's indices, X-ray diffraction, Bragg's law, etc. Regardless, these concepts are briefly summarized in this text whenever they are needed. In order to keep the book as self-contained as possible, some fundamentals in mathematics and physics are summarized in the Appendices.

CHAPTER 2

The Wave-Particle Duality

This book is mainly concerned with the interactions of electrons with matter. Thus, the question "What is an electron?" is quite in order. Now, to our knowledge, nobody has so far seen an electron, even by using the most sophisticated equipment. We experience merely the *actions* of electrons, e.g., on a television screen or in an electron microscope. In each of these instances, the electrons seem to manifest themselves in quite a different way, i.e., in the first case as a particle and in the latter case as an electron wave. Accordingly, we shall use, in this book, the terms "wave" and "particle" as convenient means to describe the different aspects of the properties of electrons. This "duality" of the manifestations of electrons should not overly concern us. The reader has probably been exposed to a similar discussion when the properties of light have been introduced.

We perceive **light** intuitively as a wave (specifically, an electromagnetic wave) which travels in undulations from a given source to a point of observation. The color of the light is related to its wavelength, λ, or to its frequency, v, i.e., its number of vibrations per second. Many crucial experiments, such as diffraction, interference, and dispersion clearly confirm the wavelike nature of light. Nevertheless, at least since the discovery of the photoelectric effect in 1887 by Hertz, and its interpretation in 1905 by Einstein, we do know that light also has a particle nature. (The photoelectric effect describes the emission of electrons from a metallic surface that has been illuminated by light of appropriately high energy, e.g., by blue light.) Interestingly enough, Newton, about 300 years ago, was a strong proponent of the particle concept of light. His original ideas, however, were in need of some refinement, which was eventually provided in 1901 by quantum theory. We know today (based on Planck's famous hypothesis) that a certain minimal energy of light, i.e., at least one **light quantum**, called a **photon**, with the energy

$$E = vh = \omega\hbar, \tag{2.1}$$

needs to impinge on a metal in order that a negatively charged electron may overcome its binding energy to its positively charged nucleus and escape into free space. (This is true regardless of the *intensity* of the light.) In (2.1) h is the Planck constant whose numerical value is given in Appendix 4. Frequently, the reduced Planck constant

$$\hbar = \frac{h}{2\pi} \tag{2.2}$$

is utilized in conjunction with the angular frequency, $\omega = 2\pi v$ (1.7). In short, the wave-particle duality of *light* (or more generally, of electromagnetic radiation) had been firmly established at the beginning of the twentieth century.

On the other hand, the wave-particle duality of **electrons** needed more time until it was fully recognized. The **particle** property of electrons, having a rest mass m_0 and charge e, was discovered in 1897 by the British physicist J.J. Thomson at the Cavendish Laboratory of Cambridge University in an experiment in which he observed the deviation of a cathode ray by electric and magnetic fields. These cathode rays were known to consist of an invisible radiation that emanated from a negative electrode (called a cathode) which was sealed through the walls of an evacuated glass tube that also contained at the opposite wall a second, positively charged electrode. It was likewise known at the end of the nineteenth century that cathode rays travel in straight lines and produce a glow when they strike glass or some other materials. J.J. Thomson noticed that cathode rays travel slower than light and transport negative electricity. In order to settle the lingering question of whether cathode rays were "vibrations of the ether" or instead "streams of particles," he promulgated a bold hypothesis, suggesting that cathode rays were "charged corpuscles which are miniscule constituents of the atom." This proposition—that an atom should consist of more than one particle— was startling for most people at that time. Indeed, atoms were considered since antiquity to be indivisible, that is, the most fundamental building blocks of matter.

The charge of these "corpuscles" was found to be the same as that carried by hydrogen ions during electrolysis (about 10^{-19}C). Further, the mass of these corpuscles turned out to be 1/2000th the mass of the hydrogen atom.

A second hypothesis brought forward by J.J. Thomson, suggesting that the "corpuscles of cathode rays are the *only* constituents of atoms," was eventually proven to be incorrect. Specifically, E. Rutherford, one of Thomson's former students, by using a different kind of particle beam, concluded in 1910 that the atom resembled a tiny solar system in which a few electrons orbited around a "massive" positively charged center. Today, one knows that the electron is the lightest stable elementary particle of matter and that it carries the basic charge of electricity.

Eventually, it was also discovered that the electrons in metals can move freely under certain circumstances. This critical experiment was performed by Tolman who observed inertia effects of the electrons when rotating metals.

In 1924, **de Broglie**, who believed in a unified creation of the universe, introduced the idea that electrons should also possess a wave-particle duality. In other words, he suggested, based on the hypothesis of a general reciprocity of physical laws, the **wave nature of electrons**. He connected the wavelength, λ, of an electron wave and the momentum, p, of the particle by the relation

$$\lambda p = h. \tag{2.3}$$

This equation can be "derived" by combining equivalents to the photonic equations $E = vh$ (2.1), $E = mc^2$ (1.8), $p = mc$ (1.3), and $c = \lambda v$ (1.5).

In 1926, **Schrödinger** gave this idea of de Broglie a mathematical form. In 1927, Davisson and Germer and, independently in 1928, G.P. Thomson (the son of J.J. Thomson; see above) discovered electron diffraction by a crystal, which finally proved the wave nature of electrons.

What is a wave? A **wave** is a "disturbance" which is periodic in position *and* time. (In contrast to this, a **vibration** is a disturbance which is only periodic in position *or* time.[1]) Waves are characterized by a velocity, v, a frequency, v, and a wavelength, λ, which are interrelated by

$$v = v\lambda. \tag{2.4}$$

Quite often, however, the wavelength is replaced by its inverse quantity (multiplied by 2π), i.e., λ is replaced by the **wave number**

$$k = \frac{2\pi}{\lambda}. \tag{2.5}$$

Concomitantly, the frequency, v, is replaced by the angular frequency $\omega = 2\pi v$ (1.7). Equation (2.4) then becomes

$$v = \frac{\omega}{k}. \tag{2.6}$$

One of the simplest waveforms is mathematically expressed by a sine (or a cosine) function. This simple disturbance is called a "**harmonic wave**." (We restrict our discussion below to harmonic waves since a mathematical manipulation, called a Fourier transformation, can substitute any odd type of waveform by a series of harmonic waves, each having a different frequency.)

The properties of electrons will be described in the following by a harmonic wave, i.e., by a wave function Ψ (which contains, as outlined above, a

---

[1] A summary of the equations which govern waves and vibrations is given in Appendix 1.

time- and a space-dependent component):

$$\Psi = \sin(kx - \omega t). \tag{2.7}$$

This wave function does not represent, as far as we know, any physical waves or other physical quantities. It should be understood merely as a mathematical description of a particle (the electron) which enables us to calculate its actual behavior in a convenient way. This thought probably sounds unfamiliar to a beginner in quantum physics. However, by repeated exposure, one can become accustomed to this kind of thought.

The wave-particle duality may be better understood by realizing that the electron can be represented by a *combination* of several wave trains having slightly different frequencies, for example, ω and $\omega + \Delta\omega$, and different wave numbers, k and $k + \Delta k$. Let us study this, assuming at first only *two* waves, which will be written as above:

$$\Psi_1 = \sin[kx - \omega t] \tag{2.7}$$

and

$$\Psi_2 = \sin[(k + \Delta k)x - (\omega + \Delta\omega)t]. \tag{2.8}$$

Superposition of Ψ_1 and Ψ_2 yields a new wave Ψ. With $\sin\alpha + \sin\beta = 2\cos\frac{1}{2}(\alpha - \beta) \cdot \sin\frac{1}{2}(\alpha + \beta)$ we obtain

$$\Psi_1 + \Psi_2 = \Psi = \underbrace{2\cos\left(\frac{\Delta\omega}{2}t - \frac{\Delta k}{2}x\right)}_{\substack{\text{Modulated} \\ \text{amplitude}}} \cdot \underbrace{\sin\left[\left(k + \frac{\Delta k}{2}\right)x - \left(\omega + \frac{\Delta\omega}{2}\right)t\right]}_{\text{Sine wave}}. \tag{2.9}$$

Equation (2.9) describes a sine wave (having a frequency intermediate between ω and $\omega + \Delta\omega$) whose amplitude is slowly modulated by a cosine function. (This familiar effect in acoustics can be heard in the form of "beats" when two strings of a piano have a slightly different pitch. The beats become less rapid the smaller the difference in frequency, $\Delta\omega$, between the two strings until they finally cease once both strings have the same pitch, (2.9).) Each of the "beats" represents a **"wave packet"** (Fig. 2.1). The wave

Figure 2.1. Combination of two waves of slightly different frequencies. ΔX is the distance over which the particle can be found.

Figure 2.2. Monochromatic matter wave ($\Delta\omega$ and $\Delta k = 0$). The wave has constant amplitude. The matter wave travels with the phase velocity, v.

packet becomes "longer" the slower the beats, i.e., the smaller $\Delta\omega$. The extreme conditions are as follows:

(a) No variation in ω and k (i.e., $\Delta\omega = 0$ and $\Delta k = 0$). This yields an "infinitely long" wave packet, i.e., a **monochromatic wave**, which corresponds to the wave picture of an electron (see Fig. 2.2).

(b) Alternately, $\Delta\omega$ and Δk could be assumed to be very large. This yields short wave packets. Moreover, if a large number of different waves are combined (rather than only two waves Ψ_1 and Ψ_2), having frequencies $\omega + n\Delta\omega$ (where $n = 1, 2, 3, 4 \ldots$), then the string of wave packets shown in Fig. 2.1 reduces to *one* wave packet only. The electron is then represented as a particle. This is shown in Fig. 2.3, in which a number of Ψ-waves have been superimposed on each other, as just outlined. It is evident from Fig. 2.3 that a superposition of, say, 300 Ψ-waves yields essentially one wave packet only.

Different velocities need to be distinguished:

(a) The velocity of the *matter wave* is called the wave velocity or "**phase velocity**," v. As we saw above, the matter wave is a monochromatic wave (or a stream of particles of equal velocity) whose frequency, ω, wavelength, λ, momentum, p, or energy, E, can be exactly determined (Fig. 2.2). The location of the particles, however, is undetermined. From the second part of (2.9) (marked "sine wave"), we deduce

$$v = \frac{x}{t} = \frac{\omega + \Delta\omega/2}{k + \Delta k/2} = \frac{\omega'}{k'}, \tag{2.6a}$$

which is a restatement of (2.6). We obtain the velocity of a matter wave that has a frequency $\omega + \Delta\omega/2$ and a wave number $k + \Delta k/2$. The phase velocity varies for different wavelengths (a phenomenon which is called "dispersion," and which the reader knows from the rainbow colors that emerge from a prism when white light impinges on it).

(b) We mentioned above that a *particle* can be understood to be "composed of" a group of waves or a "wave packet." Each individual wave has a slightly different frequency. Appropriately, the velocity of a particle is

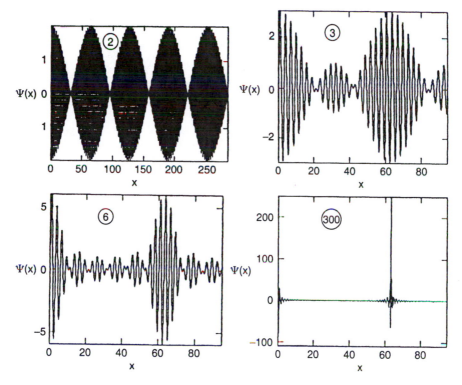

Figure 2.3. Superposition of Ψ-waves. The number of Ψ-waves is given in the graphs. (See also Fig. 2.1 and Problem 2.8.)

called "**group velocity**," v_g. The "envelope" in Fig. 2.1 propagates with the group velocity, v_g. From the left part of (2.9) (marked "modulated amplitude") we obtain this group velocity

$$v_g = \frac{x}{t} = \frac{\Delta\omega}{\Delta k} = \frac{d\omega}{dk}. \qquad (2.10)$$

Equation (2.10) is the velocity of a "**pulse wave**," i.e., of a moving particle.

The location X of a particle is known precisely, whereas the frequency is not. This is due to the fact that a wave packet can be thought to "consist" of several wave functions $\Psi_1, \Psi_2, \ldots, \Psi_n$, with slightly different frequencies. Another way of looking at it is to perform a Fourier analysis of a pulse wave (Fig. 2.4) which results in a series of sine and cosine functions (waves) which have different wavelengths. The better the location, ΔX, of a particle can be determined, the wider is the frequency range, $\Delta\omega$, of its waves. This is one form of **Heisenberg's uncertainty principle**,

$$\Delta p \cdot \Delta X \geqq h, \qquad (2.11)$$

Figure 2.4. Particle (pulse wave) moving with a group velocity v_g ($\Delta\omega$ is large).

stating that the product of the distance over which there is a finite probability of finding an electron, ΔX, and the range of momenta, Δp (or wavelengths (2.3)), of the electron wave is greater than or equal to a constant. This means that both the location *and* frequency of an electron cannot be accurately determined at the same time.

A word of encouragement should be added at this point for those readers who (quite legitimately) might ask the question: What can I do with wave functions which supposedly have no equivalent in real life? For the interpretation of the wave functions, we will use in future chapters Born's postulate, which states that the square of the wave function (or because Ψ is generally a complex function, the quantity $\Psi\Psi^*$) is the probability of finding a particle at a certain location. (Ψ^* is the complex conjugate quantity of Ψ.) In other words,

$$\Psi\Psi^* \, dx \, dy \, dz = \Psi\Psi^* \, d\tau \qquad (2.12)$$

is the probability of finding an electron in the volume element $d\tau$. This makes it clear that in wave mechanics probability statements are often obtained, whereas in classical mechanics the location of a particle can be determined exactly. We will see in future chapters, however, that this does not affect the usefulness of our results.

Finally, the reader may ask the question: Is an electron wave the same as an electromagnetic wave? Most definitely not! *Electromagnetic waves* (radio waves, infrared radiation (heat), visible light, ultraviolet (UV) light, X-rays, or γ-rays) propagate by an interaction of electrical and magnetic disturbances. Detection devices for electromagnetic waves include the human eye, photomultiplier tubes, photographic films, heat-sensitive devices, such as the skin, and antennas in conjunction with electrical circuits. For the detection of *electrons* (e.g., in an electron microscope or on a television screen) certain chemical compounds called "phosphors" are utilized. Materials which possess "phosphorescence" (see Section 13.8) include zinc sulfide, zinc–cadmium sulfide, tungstates, molybdates, salts of the rare earths, uranium compounds, and organic compounds. They vary in color and strength and in the length in time during which visible light is emitted.

At the end of this chapter, let us revisit the fundamental question that stood at the outset of our discussion concerning the wave-particle duality: Are particles and waves really two completely unrelated phenomena? Seen conceptually, they probably are. But consider (2.9) and its discussion. Both waves and particles are *mathematically* described essentially by the same equation, i.e., the former by setting $\Delta\omega$ and $\Delta k = 0$ and the latter by making $\Delta\omega$ and Δk large. Thus, waves and particles appear to be interrelated in a certain way. It is left to the reader to contemplate further on this idea.

Problems

1. Calculate the wavelength of an electron which has a kinetic energy of 4 eV.

2. What should be the energy of an electron so that the associated electron waves have a wavelength of 600 nm?

3. Since the visible region spans between approximately 400 nm and 700 nm, why can the electron wave mentioned in Problem 2 not be seen by the human eye? What kind of device is necessary to detect electron waves?

4. What is the energy of a light quantum (photon) which has a wavelength of 600 nm? Compare the energy with the electron wave energy calculated in Problem 2 and discuss the difference.

5. A tennis ball, having a mass of 50 g, travels with a velocity of 200 km/h. What is the equivalent wavelength of this "particle"? Compare your result with that obtained in Problem 1 above and discuss the difference.

6. Derive (2.9) by adding (2.7) and (2.8).

7. "Derive" (2.3) by combining (1.3), (1.5), (1.8), and (2.1).

*8. *Computer problem.*
   (a) Insert numerical values of your choice into (2.9) and plot the result. For example, set a constant time (e.g. $t = 0$) and vary Δk.
   (b) Add more than two equations of the type of (2.7) and (2.8) by using different values of $\Delta\omega$ and plot the result. Does this indeed reduce the number of wave packets, as stated in the text? Compare to Fig. 2.3.

CHAPTER 3

The Schrödinger Equation

We shall now make use of the conceptual ideas which we introduced in the previous chapter, i.e., we shall cast, in mathematical form, the description of an electron as a wave, as suggested by Schrödinger in 1926. All "derivations" of the Schrödinger equation start in one way or another from certain assumptions, which cause the uninitiated reader to ask the legitimate question, "Why just in this way?" The answer to this question can naturally be given, but these explanations are relatively involved. In addition, the "derivations" of the Schrödinger equation do not further our understanding of quantum mechanics. It is, therefore, not intended to "derive" here the Schrödinger equation. We consider this relation as a fundamental equation for the description of wave properties of electrons, just as the Newton equations describe the matter properties of large particles.

3.1. The Time-Independent Schrödinger Equation

The *time-independent* Schrödinger equation will always be applied when the properties of atomic systems have to be calculated in stationary conditions, i.e., when the property of the surroundings of the electron does not change with time. This is the case for most of the applications which will be discussed in this text. Thus, we introduce, at first, this simpler form of the Schrödinger equation in which the potential energy (or potential barrier), V, depends only on the location (and not, in addition, on the time). Therefore, the time-independent Schrödinger equation is an equation of a *vibration*. It has the following form:

$$\nabla^2\psi + \frac{2m}{\hbar^2}(E - V)\psi = 0, \qquad (3.1)$$

14

where

$$\nabla^2 \psi = \frac{\partial^2 \psi}{\partial x^2} + \frac{\partial^2 \psi}{\partial y^2} + \frac{\partial^2 \psi}{\partial z^2}, \tag{3.2}$$

and m is the (rest) mass of the electron,[2] and

$$E = E_{\text{kin}} + V \tag{3.3}$$

is the total energy of the system. E provides values for allowed energies once ψ and V are given, as we shall see later on.

In (3.1) we wrote for the wave function a lowercase ψ, which we will use from now on when we want to state explicitly that the wave function is only space dependent. Thus, we split from Ψ a time-dependent part:

$$\Psi(x, y, z, t) = \psi(x, y, z) \cdot e^{i\omega t}. \tag{3.4}$$

*3.2. The Time-Dependent Schrödinger Equation

The *time-dependent* Schrödinger equation is a *wave* equation, because it contains derivatives of Ψ with respect to space *and* time (see below, (3.8)). One obtains this equation from (3.1) by eliminating the total energy,

$$E = vh = \omega h, \tag{2.1}$$

where ω is obtained by differentiating (3.4) with respect to time:

$$\frac{\partial \Psi}{\partial t} = \psi i\omega e^{i\omega t} = \Psi i\omega. \tag{3.5}$$

This yields

$$\omega = -\frac{i}{\Psi} \frac{\partial \Psi}{\partial t}. \tag{3.6}$$

Combining (2.1) with (3.6) provides

$$E = -\frac{\hbar i}{\Psi} \frac{\partial \Psi}{\partial t}. \tag{3.7}$$

Finally, combining (3.1) with (3.7) yields

$$\boxed{\nabla^2 \Psi - \frac{2mV}{\hbar^2} \Psi - \frac{2mi}{\hbar} \frac{\partial \Psi}{\partial t} = 0.} \tag{3.8}$$

It should be noted here that quantum mechanical equations can be obtained from classical equations by applying differential operators to the wave func-

---

[2] In most cases we shall denote the rest mass by m instead of m_0.

tion Ψ (Hamiltonian operators). They are

$$E \triangleq -\hbar i \frac{\partial}{\partial t} \tag{3.9}$$

and

$$\mathbf{p} \triangleq -\hbar i \nabla. \tag{3.10}$$

When these operators are applied to

$$E_{\text{total}} = E_{\text{kin}} + E_{\text{pot}} = \frac{p^2}{2m} + V \tag{3.11}$$

we obtain

$$-\hbar i \frac{\partial \Psi}{\partial t} = \frac{\hbar^2 i^2}{2m} \nabla^2 \Psi + V \Psi, \tag{3.12}$$

which yields, after rearranging, the time-dependent Schrödinger equation (3.8).

*3.3. Special Properties of Vibrational Problems

The solution to an equation for a vibration is determined, except for certain constants. These constants are calculated by using boundary or starting conditions

$$(\text{e.g., } \psi = 0 \text{ at } x = 0). \tag{3.13}$$

As we will see in Section 4.2, only certain vibrational forms are possible when boundary conditions are imposed. This is similar to the vibrational forms of a vibrating string, where the fixed ends cannot undergo vibrations. Vibrational problems that are determined by boundary conditions are called **boundary or eigenvalue problems**. It is a peculiarity of vibrational problems with boundary conditions that not all frequency values are possible and, therefore, because of

$$E = v h, \tag{3.14}$$

not all values for the energy are allowed (see next chapter). One calls the allowed values eigenvalues. The functions ψ, which belong to the eigenvalues and which are a solution of the vibration equation and, in addition, satisfy the boundary conditions, are called **eigenfunctions** of the differential equation.

In Section 2 we related the product $\psi \psi^*$ (which is called the "**norm**") to the probability of finding a particle at a given location. The probability of finding a particle somewhere in space is *one*, or

$$\int \psi \psi^* \, d\tau = \int |\psi|^2 \, d\tau = 1. \tag{3.15}$$

Equation (3.15) is called the normalized eigenfunction.

Problems

1. Write a mathematical expression for a vibration (vibrating string, for example) and for a wave. (See Appendix 1.) Familiarize yourself with the way these differential equations are solved. What is a "trial solution?" What is a boundary condition?

2. Define the terms "vibration" and "wave."

3. What is the difference between a damped and an undamped vibration? Write the appropriate equations.

4. What is the complex conjugate function of:
   (a) $\hat{x} = a + bi$; and
   (b) $\Psi = 2Ai\sin\alpha x$.

CHAPTER 4

Solution of the Schrödinger Equation for Four Specific Problems

4.1. Free Electrons

At first we solve the Schrödinger equation for a simple but, nevertheless, very important case. We consider electrons which propagate freely, i.e., in a potential-free space in the positive x-direction. In other words, it is assumed that no "wall," i.e., no potential barrier (V), restricts the propagation of the electron wave. The potential energy V is then zero and the Schrödinger equation (3.1) assumes the following form:

$$\frac{d^2\psi}{dx^2} + \frac{2m}{\hbar^2} E\psi = 0. \tag{4.1}$$

This is a differential equation for an undamped vibration[3] with spatial periodicity whose solution is known to be[3]

$$\psi(x) = Ae^{i\alpha x}, \tag{4.2}$$

where

$$\alpha = \sqrt{\frac{2m}{\hbar^2} E}. \tag{4.3}$$

(For our special case we do not write the second term in (A.5)[3],

$$u = Ae^{i\alpha x} + Be^{-i\alpha x}, \tag{4.4}$$

---

[3] See Appendix 1.

Figure 4.1. Energy continuum of a free electron (compare with Fig. 4.3).

because we stipulated above that the electron wave[3]

$$\Psi(x) = A e^{i\alpha x} \cdot e^{i\omega t} \tag{4.5}$$

propagates only in the positive x-direction and not, in addition, in the nega-tive x-direction.)

From (4.3), it follows that

$$E = \frac{\hbar^2}{2m}\alpha^2. \tag{4.6}$$

Since no boundary condition had to be considered for the calculation of the free-flying electron, all values of the energy are "allowed," i.e., one obtains an **energy continuum** (Fig. 4.1). This statement seems to be trivial at this point. The difference to the bound electron case will become, however, evi-dent in the next section.

Before we move ahead, let us combine equations (4.3), (2.3), and (1.4), i.e.,

$$\alpha = \sqrt{\frac{2mE}{\hbar^2}} = \frac{p}{\hbar} = \frac{2\pi}{\lambda} = k, \tag{4.7}$$

which yields

$$E = \frac{\hbar^2}{2m}k^2. \tag{4.8}$$

The term $2\pi/\lambda$ was defined in (2.5) to be the wave number, k. Thus, α is here identical with k. We see from (4.7) that the quantity k is proportional to the momentum p and, because of $\mathbf{p} = m\mathbf{v}$, also proportional to the velocity of the electrons. Since both momentum and velocity are vectors, it follows that k is a vector, too. Therefore, we actually should write k as a vector which has the components k_x, k_y, and k_z:

$$|\mathbf{k}| = \frac{2\pi}{\lambda}. \tag{4.9}$$

Since $\mathbf{k}$ is inversely proportional to the wavelength, λ, it is also called the "wave vector." We shall use the wave vector in the following sections frequently. The k-vector describes the wave properties of an electron, just as one describes in classical mechanics the particle property of an electron with the momentum. As mentioned above, **k** and **p** are mutually proportional, as one can see from (4.7). The proportionality factor is $1/\hbar$.

4.2. Electron in a Potential Well (Bound Electron)

We now consider an electron that is bound to its atomic nucleus. For simplicity, we assume that the electron can move freely between two infinitely high potential barriers (Fig. 4.2). The potential barriers do not allow the electron to escape from this potential well, which means that $\psi = 0$ for $x \leq 0$ and $x \geq a$. We first treat the one-dimensional case just as in Section 4.1, i.e., we assume that the electron propagates only along the x-axis. However, because the electron is reflected on the walls of the well, it can now propagate in the positive, as well as in the negative, x-direction. In this respect, the present problem is different from the preceding one. The potential energy inside the well is zero, as before, so that the Schrödinger equation for an electron in this region can be written, as before,

$$\frac{d^2\psi}{dx^2} + \frac{2m}{\hbar^2} E\psi = 0. \tag{4.10}$$

Because of the two propagation directions of the electron, the solution of (4.10) is

$$\psi = Ae^{i\alpha x} + Be^{-i\alpha x} \tag{4.11}$$

(see Appendix 1), where

$$\alpha = \sqrt{\frac{2m}{\hbar^2} E}. \tag{4.12}$$

We now determine the constants A and B by means of boundary conditions. We just mentioned that at $x \leq 0$ and $x \geq a$ the ψ function is zero.

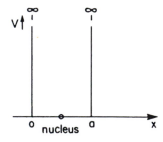

Figure 4.2. One-dimensional potential well. The walls consist of infinitely high potential barriers.

This boundary condition is similar to that known for a vibrating string, which does not vibrate at the two points where it is clamped down. (See also Fig. 4.4(a).) Thus, for $x = 0$ we stipulate $\psi = 0$. Then we obtain from (4.11)

$$B = -A. \tag{4.13}$$

Similarly, we stipulate $\psi = 0$ for $x = a$. Using this boundary condition and (4.13), equation (4.11) becomes

$$0 = Ae^{i\alpha a} + Be^{-i\alpha a} = A(e^{i\alpha a} - e^{-i\alpha a}). \tag{4.14}$$

With the Euler equation

$$\sin\rho = \frac{1}{2i}(e^{i\rho} - e^{-i\rho}) \tag{4.15}$$

(see Appendix 2), we rewrite Equation (4.14)

$$A[e^{i\alpha a} - e^{-i\alpha a}] = 2Ai \cdot \sin\alpha a = 0. \tag{4.16}$$

Equation (4.16) is only valid if $\sin\alpha a = 0$, i.e., if

$$\alpha a = n\pi, \qquad n = 0, 1, 2, 3, \ldots. \tag{4.17}$$

(because 2, A, and i cannot be zero).

Substituting the value of α from (4.12) into (4.17) provides

$$\boxed{E_n = \frac{\hbar^2}{2m}\alpha^2 = \frac{\hbar^2\pi^2}{2ma^2}n^2,} \qquad n = 1, 2, 3, \ldots. \tag{4.18}$$

(We exclude $n = 0$, which would yield $\psi = 0$, that is, no electron wave.) We notice immediately a striking difference from the case in Section 4.1. Because of the boundary conditions, only certain solutions of the Schrödinger equation exist, namely those for which n is an integer. In the present case the energy assumes only those values which are determined by (4.18). All other energies are not allowed. The allowed values are called "**energy levels.**" They

Figure 4.3. Allowed energy values of an electron that is bound to its atomic nucleus. E is the excitation energy in the present case. $C = \hbar^2\pi^2/2ma^2$, see (4.18). (E_1 is the zero-point energy.)

are shown in Fig. 4.3 for a one-dimensional case. Because of the fact that an electron of an isolated atom can assume only certain energy levels, it follows that the energies which are excited or absorbed also possess only discrete values. The result is called an "**energy quantization**." The lowest energy that an electron may assume is called the "**zero-point energy**". It can be calculated from (4.18) for $n = 1$. In other words, the lowest energy of the electron is not that of the bottom of the potential well, but rather a slightly higher value.

We discuss now the wave function, ψ, and the probability $\psi\psi^*$ for finding an electron within the potential well (see Chapter 2). According to (4.11), (4.13), and the Euler equation (4.15), we obtain within the well

$$\psi = 2Ai \cdot \sin \alpha x, \tag{4.19}$$

and the complex conjugate of ψ

$$\psi^* = -2Ai \sin \alpha x. \tag{4.20}$$

The product $\psi\psi^*$ is then

$$\psi\psi^* = 4A^2 \sin^2 \alpha x. \tag{4.21}$$

Equations (4.19) and (4.21) are plotted for various n-values in Fig. 4.4. From Fig. 4.4(a), we see that standing electron waves are created between the walls of the potential well. Note that integer multiples of half a wavelength are equal to the length, a, of the potential well. The present case, in its mathematical treatment, as well as in its result, is analogous to that of a vibrating string.

Of special interest is the behavior of the function $\psi\psi^*$, i.e., the probability of finding the electron at a certain place within the well (Fig. 4.4(b)). In the classical case the electron would travel back and forth between the walls. Its probability function would therefore be equally distributed along the whole length of the well. In wave mechanics the deviation from the classical case is

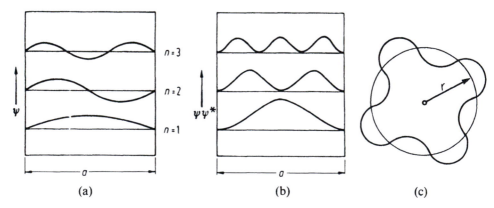

Figure 4.4. (a) ψ function and (b) probability function $\psi\psi^*$ for an electron in a potential well for different n-values. (c) Allowed electron orbit of an atom.

most pronounced for $n = 1$. In this case, $\psi\psi^*$ is largest in the middle of the well and vanishes at the boundaries. For $n = 2$, the probability of finding an electron at the center of the well (and at the boundaries) is zero, whereas the largest $\psi\psi^*$ is found at $\frac{1}{4}a$ and $\frac{3}{4}a$. For successively higher n-values, i.e., for higher energies, the wave mechanical values for $\psi\psi^*$ are eventually approaching the classical value.

In order to deepen the understanding of the behavior of bound electrons, the reader is reminded of the **Rutherford model** (Chapter 2), in which the electrons are described to move in distinct orbits about a positively charged nucleus. Similarly, as shown in Fig. 4.4(a), the electron waves associated with an orbiting electron have to be standing waves. If this were not the case, the wave would be out of phase with itself after one orbit. After a large number of orbits, all possible phases would be obtained and the wave would be annihilated by destructive interference. This can only be avoided if a radius is chosen so that the wave joins on itself (Fig. 4.4(c)). In this case the circumference, $2\pi r$, of the orbit is an integer multiple, n, of the wavelength, λ, or

$$2\pi r = n\lambda, \tag{4.22}$$

which yields

$$r = \frac{\lambda}{2\pi}n. \tag{4.23}$$

This means that only certain distinct orbits are allowed, which brings us back to the allowed energy levels which we discussed above. Actually, this model was proposed in 1913 by Niels Bohr.

*For the above discussions, we did not need to evaluate the constant 'A'. Those readers who are interested in this detail may simply rewrite (4.21) in conjunction with (3.15):

$$\int_0^a \psi\psi^* \, dx = 4A^2 \int_0^a \sin^2(\alpha x) \, dx = \frac{4A^2}{\alpha}\left[-\frac{1}{2}\sin\alpha x \cos\alpha x + \frac{\alpha x}{2}\right]_0^a = 1. \tag{4.24}$$

Inserting the boundaries in (4.24) and using (4.17) provides

$$A = \sqrt{\frac{1}{2a}}. \tag{4.25}$$

*The results that are obtained by considering an electron in a square well are similar to the ones which one receives when the wave mechanical properties of a **hydrogen atom** are calculated. As above, one considers an electron with charge $-e$ to be bound to its nucleus. The potential, V, in which the electron propagates is taken as the Coulombic potential $V = -e^2/(4\pi\varepsilon_0 r)$. Since V is a function of the radius, r, the Schrödinger equation is more conveniently expressed in polar coordinates. Of main interest are, again, the conditions under which solutions to this Schrödinger equation exist. The

Figure 4.5. Energy levels of atomic hydrogen. E is the binding energy.

treatment leads, similarly as above, to discrete energy levels:

$$E = \frac{me^4}{2(4\pi\varepsilon_0\hbar)^2}\frac{1}{n^2} = -13.6\cdot\frac{1}{n^2}\,(\text{eV}). \tag{4.18a}$$

The main difference compared to the square well model is, however, that the energy is now proportional to $-1/n^2$ (and not to n^2 as in (4.18)). This results in a "crowding" of energy levels at higher energies. The energy at the lowest level is called the ionization energy, which has to be supplied to remove an electron from its nucleus. Energy diagrams, as in Fig. 4.5, are common in spectroscopy. The origin of the energy scale is arbitrarily set at $n = \infty$ and the ionization energies are counted negative. Since we are mainly concerned with the solid state, the detailed calculation of the hydrogen atom is not treated here.

  *So far, we have considered the electron to be confined to a one-dimensional well. A similar calculation for a three-dimensional potential well ("**electron in a box**") leads to an equation which is analogous to (4.18):

$$E_n = \frac{\hbar^2\pi^2}{2ma^2}(n_x^2 + n_y^2 + n_z^2). \tag{4.26}$$

The smallest allowed energy in a three-dimensional potential well is occupied by an electron if $n_x = n_y = n_z = 1$. For the next higher energy there are three different possibilities for combining the n-values; namely, $(n_x, n_y, n_z) = (1,1,2)$, $(1,2,1)$, or $(2,1,1)$. One calls the states which have the same energy but different quantum numbers "**degenerate**" states. The example just given describes a threefold degenerate energy state.

4.3. Finite Potential Barrier (Tunnel Effect)

Let us assume that a free electron, propagating in the positive x-direction, encounters a potential barrier whose potential energy, V_0, ("height" of the barrier) is larger than the total energy, E, of the electron, but is still finite (Fig. 4.6). For this case we have to write two Schrödinger equations, which

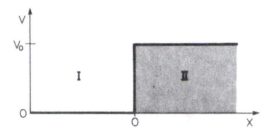

Figure 4.6. Finite potential barrier.

take into account the two different areas. In region I ($x < 0$) the electron is assumed to be free, and we can write

$$\text{(I)} \quad \frac{d^2\psi}{dx^2} + \frac{2m}{\hbar^2} E\psi = 0. \tag{4.27}$$

Inside the potential barrier ($x > 0$) the Schrödinger equation reads

$$\text{(II)} \quad \frac{d^2\psi}{dx^2} + \frac{2m}{\hbar^2}(E - V_0)\psi = 0. \tag{4.28}$$

The solutions to these equations are as before (see Appendix 1):

$$\text{(I)} \quad \psi_{\mathrm{I}} = Ae^{i\alpha x} + Be^{-i\alpha x}, \tag{4.29}$$

where

$$\alpha = \sqrt{\frac{2mE}{\hbar^2}}, \tag{4.30}$$

and

$$\text{(II)} \quad \psi_{\mathrm{II}} = Ce^{i\beta x} + De^{-i\beta x}, \tag{4.31}$$

with

$$\beta = \sqrt{\frac{2m}{\hbar^2}(E - V_0)}. \tag{4.32}$$

A word of caution has to be inserted here. We stipulated above that V_0 is larger than E. As a consequence of this $(E - V_0)$ is negative and β becomes imaginary. To prevent this, we define a new parameter:

$$\gamma = i\beta. \tag{4.33}$$

This yields, for (4.32),

$$\gamma = \sqrt{\frac{2m}{\hbar^2}(V_0 - E)}. \tag{4.34}$$

The parameter γ is now prevented under the stated conditions from becoming imaginary. Rearranging (4.33) to obtain

$$\beta = \frac{\gamma}{i} \tag{4.35}$$

and inserting (4.35) into (4.31) yields

$$\psi_{II} = Ce^{\gamma x} + De^{-\gamma x}. \tag{4.36}$$

Next, one of the constants C or D needs to be determined by means of a boundary condition:

For $x \to \infty$ it follows from (4.36) that

$$\psi_{II} = C \cdot \infty + D \cdot 0. \tag{4.37}$$

The consequence of (4.37) could be that ψ_{II} and therefore $\psi_{II}\psi_{II}^*$ are infinity.

Since the probability $\psi\psi^*$ can never be larger than one (certainty), $\psi_{II} \to \infty$ is no solution. To avoid this, C has to go to zero:

$$C \to 0. \tag{4.38}$$

Then, (4.36) reduces to

$$\psi_{II} = De^{-\gamma x}, \tag{4.39}$$

which reveals that the ψ-function decreases in Region II exponentially, as shown in Fig. 4.7 (solid line). The decrease is stronger the larger γ is chosen, i.e., for a large potential barrier, V_0.

The electron wave $\Psi(x, t)$ is then given, using (A.27) and (4.39), by

$$\Psi = De^{-\gamma x} \cdot e^{i(\omega t - kx)}$$

(damped wave) as shown by the dashed curve in Fig. 4.7. In other words, (4.39) provides the envelope (or decreasing amplitude) for the electron wave that propagates in the finite potential barrier. If the potential barrier is only moderately high and relatively narrow, the electron wave may continue on the opposite side of the barrier. This behavior is analogous to that for a light wave, which likewise penetrates to a certain degree into a material and whose amplitude also decreases exponentially, as we shall see in the optics part of this book, specifically in Fig. 10.4. The penetration of a potential barrier by an electron wave is called "**tunneling**" and has important applications in solid state physics (tunnel diode, tunnel electron microscope, field ion microscope). Tunneling is a quantum mechanical effect. In classical

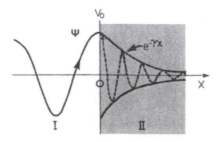

Figure 4.7. ψ-function (solid line) and electron wave (dashed line) meeting a finite potential barrier.

physics, the electron (particle) would be described to be entirely reflected back from the barrier (at $x = 0$) if its kinetic energy is smaller than V_0.

*For the complete solution of the behavior of an electron wave that penetrates a finite potential barrier (Fig. 4.6), some additional boundary conditions need to be taken into consideration:

(1) The functions ψ_I and ψ_{II} are continuous at $x = 0$. As a consequence, $\psi_I \equiv \psi_{II}$ at $x = 0$. This yields, with (4.29), (4.36), and (4.38),

$$Ae^{i\alpha x} + Be^{-i\alpha x} = De^{-\gamma x}.$$

With $x = 0$, we obtain

$$A + B = D. \tag{4.40}$$

(2) The slopes of the wave functions in Regions I and II are continuous at $x = 0$, i.e., $(d\psi_I/dx) \equiv (d\psi_{II}/dx)$. This yields

$$Ai\alpha e^{i\alpha x} - Bi\alpha e^{-i\alpha x} = -\gamma De^{-\gamma x}. \tag{4.41}$$

With $x = 0$, one obtains

$$Ai\alpha - Bi\alpha = -\gamma D. \tag{4.42}$$

Inserting (4.40) into (4.42) yields

$$A = \frac{D}{2}\left(1 + i\frac{\gamma}{\alpha}\right) \tag{4.43}$$

and

$$B = \frac{D}{2}\left(1 - i\frac{\gamma}{\alpha}\right). \tag{4.43a}$$

From this, the ψ-functions can be expressed in terms of a constant D. Figure 4.8 illustrates the modification of Fig. 4.4(a) when tunneling is taken into consideration. A penetration of the ψ-function into the potential barriers is depicted.

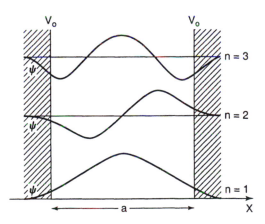

Figure 4.8. Square well with finite potential barriers. (The zero points on the vertical axis have been shifted for clarity.)

4.4. Electron in a Periodic Field of a Crystal (the Solid State)

In the preceding sections we became acquainted with some special cases, namely, the completely free electron and the electron which is confined to a potential well. The goal of this section is to study the behavior of an electron in a crystal. We will see eventually that the extreme cases which we treated previously can be derived from this general case.

Our first task is to find a potential distribution that is suitable for a solid. From high resolution transmission electron microscopy and from X-ray diffraction investigations, it is known that the atoms in a crystal are arranged periodically. Thus, for the treatment of our problem a periodic repetition of the potential well of Fig. 4.2, i.e., a periodic arrangement of potential wells and potential barriers, is probably close to reality and is also best suited for a calculation. Such a periodic potential is shown in Fig. 4.9 for the one-dimensional case.[4]

The potential distribution shows potential wells of length a, which we call Region I. These wells are separated by potential barriers of height V_0 and width b (Region II), where V_0 is assumed to be larger than the energy E of the electron.

This model is certainly a coarse simplification of the actual potential distribution in a crystal. It does not take into consideration that the inner electrons are more strongly bound to the core, i.e., that the potential function of a point charge varies as $1/r$. It also does not consider that the individual potentials from each lattice site overlap. A potential distribution which takes these features into consideration is shown in Fig. 4.10. It is immediately evident, however, that the latter model is less suitable for a simple calculation than the one which is shown in Fig. 4.9. Thus, we utilize the model shown in Fig. 4.9.

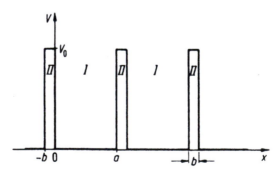

Figure 4.9. One-dimensional periodic potential distribution (simplified) (**Kronig-Penney model**).

---

[4] R. De. L. Kronig and W.G. Penney, *Proc. Roy. Soc. London*, **130**, 499 (1931).

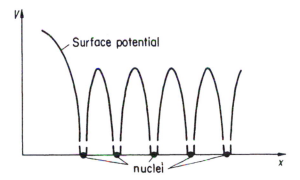

Figure 4.10. One-dimensional periodic potential distribution for a crystal (**muffin tin potential**).

We now write the Schrödinger equation for Regions I and II:

$$\text{(I)} \qquad \frac{d^2\psi}{dx^2} + \frac{2m}{\hbar^2} E\psi = 0, \qquad (4.44)$$

$$\text{(II)} \quad \frac{d^2\psi}{dx^2} + \frac{2m}{\hbar^2}(E - V_0)\psi = 0. \qquad (4.45)$$

For abbreviation we write, as before,

$$\alpha^2 = \frac{2m}{\hbar^2} E, \qquad (4.46)$$

and

$$\gamma^2 = \frac{2m}{\hbar^2}(V_0 - E). \qquad (4.47)$$

(γ^2 is chosen in a way to keep it from becoming imaginary, see Section 4.3.) Equations (4.44) and (4.45) need to be solved simultaneously, a task which can be achieved only with considerable mathematical effort. Bloch[5] showed that the solution of this type of equation has the following form:

$$\psi(x) = u(x) \cdot e^{ikx} \qquad (4.48)$$

(**Bloch function**), where $u(x)$ is a periodic function which possesses the periodicity of the lattice in the x-direction. Therefore, $u(x)$ is no longer a constant (amplitude A) as in (4.2), but changes periodically with increasing x (modulated amplitude). Of course, $u(x)$ is different for various directions in the crystal lattice.

The reader who is basically interested in the results, and their implications for the electronic structure of crystals, may skip the mathematical treatment given below and refer directly to (4.67).

---

[5] F. Bloch, Z. Phys. **52**, 555 (1928); **59**, 208 (1930).

Differentiating the Bloch function (4.48) twice with respect to x provides

$$\frac{d^2\psi}{dx^2} = \left(\frac{d^2u}{dx^2} + \frac{du}{dx}2ik - k^2u\right)e^{ikx}. \tag{4.49}$$

We insert (4.49) into (4.44) and (4.45) and take into account the abbreviations (4.46) and (4.47):

$$(I) \quad \frac{d^2u}{dx^2} + 2ik\frac{du}{dx} - (k^2 - \alpha^2)u = 0, \tag{4.50}$$

$$(II) \quad \frac{d^2u}{dx^2} + 2ik\frac{du}{dx} - (k^2 + \gamma^2)u = 0. \tag{4.51}$$

Equations (4.50) and (4.51) have the form of an equation of a damped vibration. The solution[6] to (4.50) and (4.51) is

$$(I) \quad u = e^{-ikx}(Ae^{i\alpha x} + Be^{-i\alpha x}), \tag{4.55}$$

$$(II) \quad u = e^{-ikx}(Ce^{-\gamma x} + De^{\gamma x}). \tag{4.56}$$

We have four constants A, B, C, and D which we need to dispose of by means of four boundary conditions: The functions ψ and $d\psi/dx$ pass over continuously from Region I into Region II at the point $x = 0$. Equation I = Equation II for $x = 0$ yields

$$A + B = C + D. \tag{4.57}$$

(du/dx) for I $= (du/dx)$ for II at $x = 0$ provides

$$A(i\alpha - ik) + B(-i\alpha - ik) = C(-\gamma - ik) + D(\gamma - ik). \tag{4.58}$$

Further, ψ, and therefore u, is continuous at the distance $(a + b)$. This means that Equation I at $x = 0$ must be equal to Equation II at $x = a + b$, or, more simply, Equation I at $x = a$ is equal to Equation II at $x = -b$ (see Fig. 4.9). This yields

$$Ae^{(i\alpha - ik)a} + Be^{(-i\alpha - ik)a} = Ce^{(ik+\gamma)b} + De^{(ik-\gamma)b}. \tag{4.59}$$

---

[6] Differential equation of a damped vibration for spatial periodicity (see Appendix 1)

$$\frac{d^2u}{dx^2} + D\frac{du}{dx} + Cu = 0. \tag{4.52}$$

Solution:

$$u = e^{-(D/2)x}(Ae^{i\delta x} + Be^{-i\delta x}), \tag{4.53}$$

where

$$\delta = \sqrt{C - \frac{D^2}{4}}. \tag{4.54}$$

Finally, (du/dx) is periodic in $a + b$:

$$Ai(\alpha - k)e^{ia(\alpha-k)} - Bi(\alpha + k)e^{-ia(\alpha+k)}$$

$$= -C(\gamma + ik)e^{(ik+\gamma)b} + D(\gamma - ik)e^{(ik-\gamma)b}. \tag{4.60}$$

The constants A, B, C, and D can be determined by means of these four equations which, when inserted in (4.55) and (4.56), provide values for u. This also means that solutions for the function ψ can be given by using (4.48). However, as in the preceding sections, the knowledge of the ψ function is not of primary interest. We are searching instead for a condition which tells us where solutions to the Schrödinger equations (4.44) and (4.45) exist. We recall that these limiting conditions were leading to the energy levels in Section 4.2. We proceed here in the same manner. We use the four equations (4.57)–(4.60) and eliminate the four constants A–D. (This can be done by simple algebraic manipulation or by forming the determinant out of the coefficients A–D and equating this determinant to zero). The lengthy calculation provides, using some Euler equations,[7]

$$\frac{\gamma^2 - \alpha^2}{2\alpha\gamma} \sinh(\gamma b) \cdot \sin(\alpha a) + \cosh(\gamma b) \cos(\alpha a) = \cos k(a + b). \tag{4.61}$$

For simplification of the discussion of this equation we make the following stipulation. The potential barriers in Fig. 4.9 will be of the kind such that b is very small and V_0 is very large. It is further assumed that the product $V_0 b$, i.e., the area of this potential barrier, remains finite. In other words, if V_0 grows, b diminishes accordingly. The product $V_0 b$ is called the potential barrier strength.

If V_0 is very large, then E in (4.47) can be considered to be small compared to V_0 and can therefore be neglected so that

$$\gamma = \sqrt{\frac{2m}{\hbar^2}} \sqrt{V_0}. \tag{4.62}$$

Multiplication of (4.62) by b yields

$$\gamma b = \sqrt{\frac{2m}{\hbar^2}} \sqrt{(V_0 b)b}. \tag{4.63}$$

Since $V_0 b$ has to remain finite (see above) and $b \to 0$ it follows that γb becomes very small. For a small γb we obtain (see tables of the hyperbolic functions)

$$\cosh(\gamma b) \approx 1 \quad \text{and} \quad \sinh(\gamma b) \approx \gamma b. \tag{4.64}$$

---

[7] See Appendix 2.

Finally, one can neglect α^2 compared to γ^2 and b compared to a (see (4.46), (4.47), and Fig. 4.9) so that (4.61) reads as follows:

$$\frac{m}{\alpha \hbar^2} V_0 b \sin \alpha a + \cos \alpha a = \cos ka. \qquad (4.65)$$

With the abbreviation

$$P = \frac{m a V_0 b}{\hbar^2} \qquad (4.66)$$

we finally get from (4.65)

$$\boxed{P\frac{\sin \alpha a}{\alpha a} + \cos \alpha a = \cos ka.} \qquad (4.67)$$

This is the desired relation which provides the allowed solutions to the Schrödinger equations (4.44) and (4.45). We notice that the boundary conditions lead to an equation with trigonometric function similarly as in Section 4.2. Therefore, only certain values of α are possible. This in turn means, because of (4.46), that only certain values for the energy E are defined. One can assess the situation best if one plots the function $P(\sin \alpha a/\alpha a) + \cos \alpha a$ versus αa, which is done in Fig. 4.11 for $P = (3/2)\pi$. It is of particular significance that the right-hand side of (4.67) allows only certain values of this function because $\cos ka$ is only defined between $+1$ and -1 (except for imaginary k-values). This is shown in Fig. 4.11, in which the allowed values of the function $P(\sin \alpha a/\alpha a) + \cos \alpha a$ are marked by heavy lines on the αa-axis.

We arrive herewith at the following very important result: Because αa is a function of the energy, the above-mentioned limitation means that an electron that moves in a periodically varying potential field can only occupy

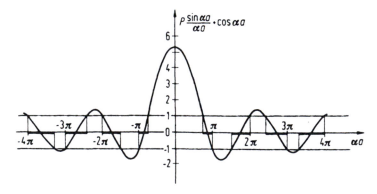

Figure 4.11. Function $P(\sin \alpha a/\alpha a) + \cos \alpha a$ versus αa. P was arbitrarily set to be $(3/2)\pi$.

Figure 4.12. Function $P(\sin \alpha a/\alpha a) + \cos \alpha a$ with $P = \pi/10$.

certain allowed energy zones. Energies outside of these allowed zones or "bands" are prohibited. One sees from Fig. 4.11 that with increasing values of αa (i.e., with increasing energy), the disallowed (or forbidden) bands become narrower. The size of the allowed and forbidden energy bands varies with the variation of P. Below, four special cases will be discussed.

(a) If the "**potential barrier strength**" $V_0 b$ (see Fig. 4.9) is large, then, according to (4.66), P is also large and the curve in Fig. 4.11 proceeds more steeply. The allowed bands are narrow.

(b) If the potential barrier strength, and therefore P, is small, the allowed bands become wider (see Fig. 4.12).

(c) If the potential barrier strength becomes smaller and smaller and finally disappears completely, P goes toward zero, and one obtains from (4.67)

$$\cos \alpha a = \cos k a \qquad (4.68)$$

or $\alpha = k$. From this it follows, with (4.46), that

$$E = \frac{\hbar^2 k^2}{2m}.$$

This is the well-known equation (4.8) for free electrons which we derived in Section 4.1.

(d) If the potential barrier strength is very large, P approaches infinity. However, because the left-hand side of (4.67) has to stay within the limits ± 1, i.e., it has to remain finite, it follows that

$$\frac{\sin \alpha a}{\alpha a} \to 0,$$

i.e., $\sin \alpha a \to 0$. This is only possible if $\alpha a = n\pi$ or

$$\alpha^2 = \frac{n^2 \pi^2}{a^2} \qquad \text{for} \quad n = 1, 2, 3, \ldots. \qquad (4.69)$$

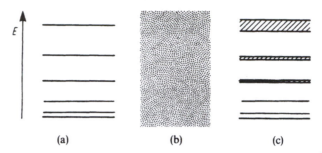

Figure 4.13. Allowed energy levels for (a) bound electrons, (b) free electrons, and (c) electrons in a solid.

Combining (4.46) with (4.69) yields

$$E = \frac{\pi^2 \hbar^2}{2ma^2} \cdot n^2,$$

which is the result of Section 4.2, equation (4.18).

We summarize (Fig. 4.13): If the electrons are strongly bound, i.e., if the potential barrier is very large, one obtains sharp energy levels (electron in the potential field of *one* ion). If the electron is not bound, one obtains a continuous energy region (free electrons). If the electron moves in a periodic potential field, one receives energy bands (solid).

The widening of the energy levels into energy bands and the transition into a quasi-continuous energy region is shown in Fig. 4.14. This widening occurs because the atoms increasingly interact as their separation distance decreases. The arrows *a*, *b*, and *c* refer to the three sketches of Fig. 4.13.

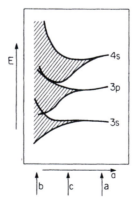

Figure 4.14. Widening of the sharp energy levels into bands and finally into a quasi-continuous energy region with decreasing interatomic distance, *a*, for a metal (after calculations of Slater). The quantum numbers are explained in Appendix 3.

Problems

1. Describe the energy for:
   (a) a free electron;
   (b) a strongly bound electron; and
   (c) an electron in a periodic potential.
   Why do we get these different band schemes?

2. *Computer problem.* Plot $\psi\psi^*$ for an electron in a potential well. Vary n from 1 to ~100. What conclusions can be drawn from these graphs? (*Hint*: If for large values for n you see strange periodic structures, then you need to choose more data points!)

3. State the two Schrödinger equations for electrons in a periodic potential field (Kronig–Penney model). Use for their solutions, instead of the Bloch function, the trial solution

$$\psi(x) = Ae^{ikx}.$$

   Discuss the result. (*Hint*: For free electrons $V_0 = 0$.)

*4. When treating the Kronig–Penney model, we arrived at four equations for the constants A, B, C, and D. Confirm (4.61).

5. The differential equation for an undamped vibration is

$$a\frac{d^2u}{dx^2} + bu = 0, \tag{1}$$

   whose solution is

$$u = Ae^{ikx} + Be^{-ikx}, \tag{2}$$

   where

$$k = \sqrt{b/a}. \tag{3}$$

   Prove that (2) is indeed a solution of (1).

6. Calculate the "ionization energy" for atomic hydrogen.

7. Derive (4.18a) in a semiclassical way by assuming that the centripetal force of an electron, mv^2/r, is counterbalanced by the Coulombic attraction force, $-e^2/4\pi\varepsilon_0 r^2$, between the nucleus and the orbiting electron. Use Bohr's postulate which states that the angular momentum $L = mvr$ (v = linear electron velocity and r = radius of the orbiting electron) is a multiple integer of Planck's constant (i.e., $n \cdot \hbar$). (*Hint*: The kinetic energy of the electron is $E = \frac{1}{2}mv^2$.)

8. *Computer problem.* Plot equation (4.67) and vary values for P.

9. *Computer problem.* Plot equation (4.39) for various values for D and γ.

10. The width of the potential well (Fig. 4.2) of an electron can be assumed to be about 2 Å. Calculate the energy of an electron (in Joules and in eV) from this information for various values of n. Give the zero-point energy.

CHAPTER 5

Energy Bands in Crystals

5.1. One-Dimensional Zone Schemes

We are now in a position to make additional important statements which contribute considerably to the understanding of the properties of crystals. For this we plot the energy versus the momentum of the electrons, or, because of (4.8), versus the wave vector, **k**. As before, we first discuss the one-dimensional case.

The relation between E and k_x is particularly simple in the case of free electrons, as can be seen from (4.8),

$$k_x = \text{const. } E^{1/2}. \tag{5.1}$$

The plot of E versus k_x is a parabola (Fig. 5.1).

We return now to (4.68), which we obtained from (4.67) for $P = 0$ (free electrons). Because the cosine function is periodic in 2π, (4.68) should be written in the more general form

$$\cos \alpha a = \cos k_x a \equiv \cos(k_x a + n2\pi), \tag{5.2}$$

where $n = 0, \pm 1, \pm 2, \dots$. This gives

$$\alpha a = k_x a + n2\pi. \tag{5.3}$$

Combining (4.8),

$$\alpha = \sqrt{\frac{2m}{\hbar^2}} E^{1/2},$$

with (5.3) yields

$$k_x + n\frac{2\pi}{a} = \sqrt{\frac{2m}{\hbar^2}} E^{1/2}. \tag{5.4}$$

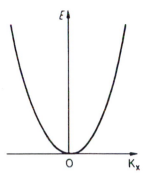

Figure 5.1. Electron energy E versus the wave vector k_x for free electrons.

We see from (5.4) that in the general case the parabola, shown in Fig. 5.1, is repeated periodically in intervals of $n \cdot 2\pi/a$ (Fig. 5.2). The energy is thus a periodic function of k_x with the periodicity $2\pi/a$.

We noted, when discussing Fig. 4.11, that if an electron propagates in a periodic potential we always observe discontinuities of the energies when $\cos k_x a$ has a maximum or a minimum, i.e., when $\cos k_x a = \pm 1$. This is only the case when

$$k_x a = n\pi, \quad n = \pm 1, \pm 2, \pm 3, \ldots, \tag{5.5}$$

or

$$k_x = n \cdot \frac{\pi}{a}. \tag{5.6}$$

At these singularities a deviation from the parabolic E versus k_x curve

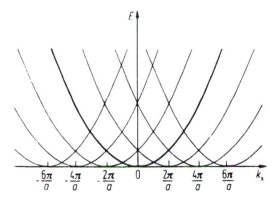

Figure 5.2. Periodic repetition of Fig. 5.1 at the points $k_x = n \cdot 2\pi/a$. The figure depicts a family of free electron parabolas having a periodicity of $\pm 2\pi/a$.

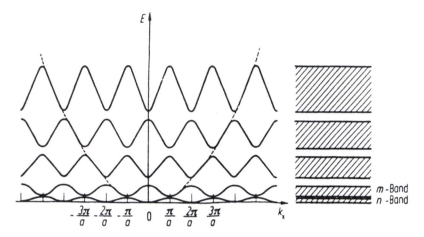

Figure 5.3. Periodic zone scheme.

occurs and the branches of the individual parabolas merge into the neighboring ones.[8] This is shown in Fig. 5.3.

The aforementioned consideration leads to a very important result. The electrons in a crystal behave, for most k_x values, like free electrons, except when k_x approaches the value $n \cdot \pi/a$.

Besides this "**periodic zone scheme**" (Fig. 5.3), two further zone schemes are common. In the future we will use mostly the "**reduced zone scheme**" (Fig. 5.4), which is a section of Fig. 5.3 between the limits $\pm \pi/a$. In the

---

[8] If two energy functions with equal symmetry cross, the quantum mechanical "noncrossing rule" requires that the eigenfunctions be split, so that they do not cross.

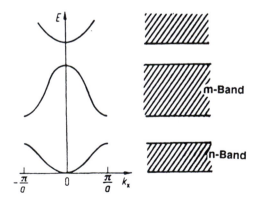

Figure 5.4. Reduced zone scheme. (This is a section of Fig. 5.3 between $-\pi/a$ and $+\pi/a$.)

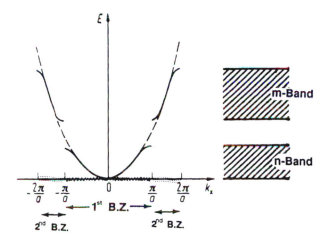

Figure 5.5. Extended zone scheme. The first and second Brillouin zones (BZ) are shown, see Section 5.2.

"**extended zone scheme**" (Fig. 5.5), the deviations from the free electron parabola at the critical points $k_x = n \cdot \pi/a$ are particularly easy to identify.

Occasionally, it is useful to plot *free* electrons in a reduced zone scheme. In doing so, one considers the width of the forbidden bands to be reduced until the energy gap between the individual branches disappears completely. This leads to the "**free electron bands**" which are shown in Fig. 5.6 for a special case. The well-known band character disappears for free electrons, however, and one obtains a continuous energy region as explained in Section 4.1. As before, the shape of the individual branches in Fig. 5.6 is due to the $2\pi/a$ periodicity, as a comparison with Fig. 5.2 shows. From (5.4), it follows that

$$E = \frac{\hbar^2}{2m}\left(k_x + n\frac{2\pi}{a}\right)^2, \quad n = 0, \pm 1, \pm 2, \ldots. \quad (5.7)$$

By inserting different n-values in (5.7), one can calculate the shape of the branches of the free electron bands. A few examples might illustrate this:

n = 0 yields $E = \dfrac{\hbar^2}{2m}k_x^2$ (parabola with 0 as origin);

n = −1 yields $E = \dfrac{\hbar^2}{2m}\left(k_x - \dfrac{2\pi}{a}\right)^2$ $\left(\text{parabola with } \dfrac{2\pi}{a} \text{ as origin}\right)$;

specifically, for $k_x = 0$ follows $E = 4\dfrac{\pi^2\hbar^2}{2ma^2}$;

and for $k_x = \dfrac{\pi}{a}$ follows $E = 1\dfrac{\pi^2\hbar^2}{2ma^2}$.

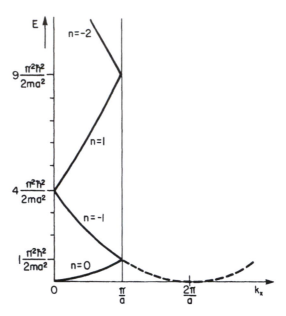

Figure 5.6. "Free electron bands" plotted in the reduced zone scheme (cubic primitive crystal structure). Compare this figure with the central portion of Fig. 5.2, that is, with the region from zero to π/a. Note the sameness of the individual bands.

The calculated data are depicted in Fig. 5.6. (The calculation of the remaining branches (bands) is left to the reader, see Problem 5.)

One important question has remained essentially unanswered: What do these E versus $|k|$ curves really mean? Simply stated, they relate the energy of an electron to its k-vector, i.e., with its momentum. They provide in principle quite similar information as, for example, a distance versus time diagram for a moving car, or a "stress-strain diagram" in mechanical metallurgy, or a "phase diagram" in materials science. All these diagrams relate in graphic form one parameter with another variable in order to provide an easier interpretation of data. We shall eventually learn to appreciate complete band diagrams in later chapters, from which we will draw important conclusions about the electronic properties of materials.

In Figs. 5.3, 5.4, and 5.5 the individual allowed energy regions and the disallowed energy regions, called **band gaps**, are clearly seen. We call the allowed bands, for the time being, the n-band, or the m-band, and so forth. In later sections and particularly in semiconductor physics (see Chapter 8) we will call one of these bands the **valence band** (because it contains the valence electrons) and the next higher one the **conduction band**.

An additional item needs to be mentioned: It is quite common to use the word "band" for both the allowed energy regions, such as the n-band or the m-band, as well as for the individual branches within a band as seen, for example, in Fig. 5.6. As a rule this does not cause any confusion.

Finally, we need to stress one more point: The wave vector $\mathbf{k}$ is inversely proportional to the wavelength of the electrons (see equation (4.9)). Thus, $\mathbf{k}$ has the unit of a reciprocal length and is therefore defined in "**reciprocal space**." The reader might recall from a course in crystallography that each crystal structure has two lattices associated with it, one of them being the crystal (or real) lattice and the other the reciprocal lattice. We will show in Section 5.5 how these two lattices are related. The following may suffice for the moment: each lattice plane in real space can be represented by a vector which is normal to this plane and whose length is made proportional to the reciprocal of the interplanar distance. The tips of all such vectors from sets of parallel lattice planes form the points in a reciprocal lattice. An X-ray diffraction pattern is a map of such a reciprocal lattice.

5.2. One- and Two-Dimensional Brillouin Zones

Let us again inspect Fig. 5.5. We noticed there that the energy versus k_x curve, between the boundaries $-\pi/a$ and $+\pi/a$, corresponds to the first electron band, which we arbitrarily labeled as *n-band*. This region in *k*-space between $-\pi/a$ and $+\pi/a$ is called the first **Brillouin zone (BZ)**. Accordingly, the area between π/a and $2\pi/a$, and also between $-\pi/a$ and $-2\pi/a$, which corresponds to the *m*-band, is called the second Brillouin zone. In other words, the lowest band shown in Fig. 5.5 corresponds to the first Brillouin zone, the next higher band corresponds to the second Brillouin zone, and so on. Now, we learned above that the individual branches in an extended zone scheme (Fig. 5.5) are $2\pi/a$ periodic, i.e., they can be shifted by $2\pi/a$ to the left or to the right. We make use of this concept and shift the branch of the second Brillouin zone on the positive side of the $E - (k_x)$ diagram in Fig. 5.5 by $2\pi/a$ to the left, and likewise the left band of the second Brillouin zone by $2\pi/a$ to the right. A reduced zone scheme as shown in Fig. 5.4 is the result. Actually, we projected the second Brillouin zone into the first Brillouin zone. The same can be done with the third Brillouin zone, etc. This has very important implications: we do not need to plot E versus *k*-curves for *all* Brillouin zones; the relevant information is, because of the $2\pi/a$ periodicity, already contained in the first Brillouin zone, i.e., in a reduced zone scheme.

We now consider the behavior of an electron in the potential of a two-dimensional lattice. The electron movement in two dimensions can be described as before by the wave vector $\mathbf{k}$ that has the components k_x and k_y, which are parallel to the *x*- and *y*-axes in reciprocal space. Points in the $k_x - k_y$ coordinate system form a two-dimensional reciprocal lattice (see Fig. 5.7). One obtains, in the two-dimensional case, a two-dimensional field of allowed energy regions which corresponds to the allowed energy bands, i.e., one obtains two-dimensional Brillouin zones.

We shall illustrate the construction of the Brillouin zones for a two-dimensional reciprocal lattice (Fig. 5.7). For the first zone one constructs the

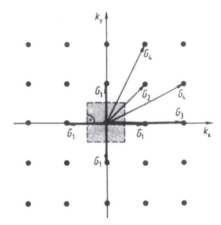

Figure 5.7. Four shortest lattice vectors in a $k_x - k_y$ coordinate system and the first Brillouin zone in a two-dimensional reciprocal lattice. (Cubic primitive crystal structure.)

perpendicular bisectors on the shortest lattice vectors, G_1. The area that is enclosed by these four "Bragg planes" is the first Brillouin zone. For the following zones the bisectors of the next shortest lattice vectors are constructed. It is essential that for the zones of higher order the extended limiting lines of the zones of lower order are used as additional limiting lines. The first four Brillouin zones are shown in Fig. 5.8. Note that all the zones have

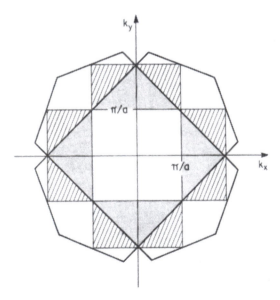

Figure 5.8. The first four Brillouin zones of a two-dimensional, cubic primitive reciprocal lattice.

the same area. The first four shortest lattice vectors G_1 through G_4 are drawn in Fig. 5.7.

The significance of the Brillouin zones will become evident in later sections, when the energy bands of solids are discussed. A few words of explanation will be given here, nevertheless. The Brillouin zones are useful if one wants to calculate the behavior of an electron which may travel in a specific direction in reciprocal space. For example, if in a two-dimensional lattice an electron travels at 45° to the k_x-axis, then the boundary of the Brillouin zone is reached, according to Fig. 5.8, for

$$k_{\text{crit}} = \frac{\pi}{a}\sqrt{2}. \tag{5.8}$$

This yields with (4.8) a maximal attainable energy of

$$E_{\text{max}} = \frac{\hbar^2}{2m}k_{\text{crit}}^2 = \frac{\pi^2\hbar^2}{a^2m}. \tag{5.8a}$$

On the other hand, the boundary of a Brillouin zone is reached at

$$k_{\text{crit}} = \frac{\pi}{a} \tag{5.9}$$

when an electron moves parallel to the k_x- or k_y-axes. The largest energy that electrons can assume in this second case is only

$$E_{\text{max}} = \frac{1}{2}\left(\frac{\pi^2\hbar^2}{a^2m}\right). \tag{5.9a}$$

Once the maximal energy has been reached, the electron waves form standing waves (or equivalently, the electrons are reflected back into the Brillouin zone).

The consequence of (5.8) and (5.9) is an overlapping of energy bands which can be seen when the bands are drawn in different directions in k-space (Fig. 5.9). We will learn later that these considerations can be utilized to determine the difference between metals, semiconductors, and insulators.

*The occurrence of critical energies at which a reflection of the electron wave takes place can also be illustrated in a completely different way. This will be done briefly here because of its immediate intuitive power. We consider an electron wave that propagates in a lattice at an angle θ to a set of parallel lattice planes (Fig. 5.10). The corresponding rays are diffracted on the lattice atoms. At a certain angle of incidence, constructive interference between rays $1'$ and $2'$ occurs. It has been shown by Bragg that each ray which is diffracted in this way can be considered as being reflected by a mirror parallel to the lattice planes. In other words, at a critical angle the "reflected" rays will be enhanced considerably. This is always the case when the path difference $2a\sin\theta$ is an integer multiple of the electron wavelength

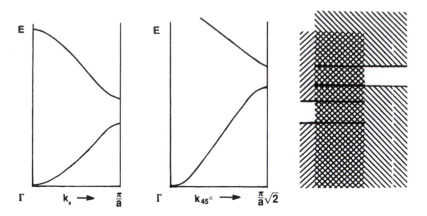

Figure 5.9. Overlapping of allowed energy bands.

λ, i.e., when

$$2a \sin \theta = n\lambda, \qquad n = 1, 2, 3, \ldots \tag{5.10}$$

(**Bragg relation**). With (4.9) one obtains, from (5.10),

$$2a \sin \theta = n \frac{2\pi}{k}$$

and therefore

$$k_{\text{crit}} = n \frac{\pi}{a \sin \theta}. \tag{5.11}$$

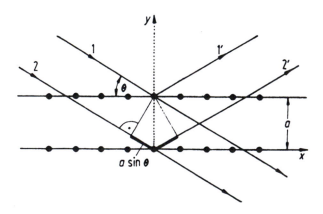

Figure 5.10. Bragg reflection of an electron wave in a lattice. The angle of incidence is θ.

For perpendicular incidence ($\theta = 90°$) equation (5.11) becomes (5.9). On the other hand, if $\theta = 45°$, one obtains (5.8).

Equation (5.11) leads to the result that for increasing electron energies a critical **k**-value is finally reached for which "reflection" of the electron wave at the lattice planes occurs. At this critical **k**-value the transmission of an electron beam through the lattice is prevented. Then, the incident and the Bragg-reflected electron wave form a standing wave.

*5.3. Three-Dimensional Brillouin Zones

In the previous section, the physical significance of the Brillouin zones was discussed. It was shown that at the boundaries of these zones the electron waves are Bragg-reflected by the crystal. The wave vector $|\mathbf{k}| = 2\pi/\lambda$ was seen to have the unit of a reciprocal length and is therefore defined in the reciprocal lattice. We will now attempt to construct three-dimensional Brillouin zones for two important crystal structures, namely, the face-centered cubic (fcc) and the body-centered cubic (bcc) crystals. Since the Brillouin zones for these structures have some important features in common with the so-called Wigner–Seitz cells, it is appropriate to discuss, at first, the Wigner–Seitz cells and also certain features of the reciprocal lattice before we return to the Brillouin zones at the end of Section 5.5.

*5.4. Wigner–Seitz Cells

Crystals have symmetrical properties. Therefore, a crystal can be described as an accumulation of "**unit cells**." In general, the smaller such a unit cell, i.e., the fewer atoms it contains, the simpler its description. The smallest possible cell is called a "**primitive unit cell**." Frequently, however, a larger, nonprimitive unit cell is used, which might have the advantage that the symmetry can be better recognized. Body-centered cubic and face-centered cubic are characteristic representatives of such "**conventional**" unit cells.[9]

The Wigner–Seitz cell is a special type of primitive unit cell that shows the cubic symmetry of the cubic cells. For its construction, one bisects the vectors from a given atom to its nearest neighbors and places a plane per-

---

[9] A lattice is a regular periodic arrangement of points in space; it is, consequently, a mathematical abstraction. All crystal structures can be traced to one of the 14 types of **Bravais lattices** (see textbooks on crystallography).

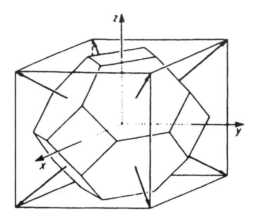

Figure 5.11. Wigner–Seitz cell for the body-centered cubic (bcc) structure.

pendicular to these vectors at the bisecting points. This is shown in Fig. 5.11 for the bcc lattice.

In the fcc lattice, the atoms are arranged on the corners and faces of a cube, which is equivalent to the center points of the edges and the center of the cell (Fig. 5.12). The Wigner–Seitz cell for this structure is shown in Fig. 5.13.

Figure 5.12. Conventional unit cell of the fcc structure. In the cell which is marked black, the atoms are situated on the corners and faces of the cubes. In the white cell, the atoms are at the centers of the edges and the center of the cell.

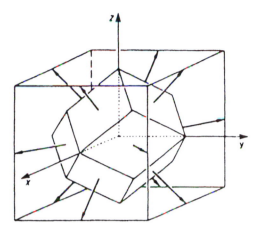

Figure 5.13. Wigner–Seitz cell for the fcc structure. It is constructed from the white cell which is marked in Fig. 5.12.

*5.5. Translation Vectors and the Reciprocal Lattice

In Fig. 5.14(a) the fundamental vectors t_1, t_2, t_3 are inserted in a unit cell of a cubic primitive lattice. By combination of these "**primitive vectors**" a **translation vector**,

$$\mathbf{R} = n_1 \mathbf{t}_1 + n_2 \mathbf{t}_2 + n_3 \mathbf{t}_3, \tag{5.12}$$

can be defined. Using this translation vector it is possible to reach, from a given lattice point, any other equivalent lattice point. For this, the factors n_1, n_2, n_3 have to be integers. In Fig. 5.14(b) the fundamental vectors t_1, t_2, t_3 are shown in a conventional unit cell of a bcc lattice.

Similarly, as above, we now introduce for the reciprocal lattice three vectors, $\mathbf{b}_1, \mathbf{b}_2, \mathbf{b}_3$, and a translation vector

$$\mathbf{G} = 2\pi(h_1 \mathbf{b}_1 + h_2 \mathbf{b}_2 + h_3 \mathbf{b}_3), \tag{5.13}$$

where h_1, h_2, and h_3 are, again, integers. (The factor 2π is introduced for convenience. In X-ray crystallography, this factor is omitted.)

The real and reciprocal lattices are related by a definition which states that the scalar product of the vectors t_1 and $\mathbf{b}_1$ should be unity, whereas the scalar products of $\mathbf{b}_1$ and t_2 or $\mathbf{b}_1$ and t_3 are zero:

$$\mathbf{b}_1 \cdot \mathbf{t}_1 = 1, \tag{5.14}$$

$$\mathbf{b}_1 \cdot \mathbf{t}_2 = 0, \tag{5.15}$$

$$\mathbf{b}_1 \cdot \mathbf{t}_3 = 0. \tag{5.16}$$

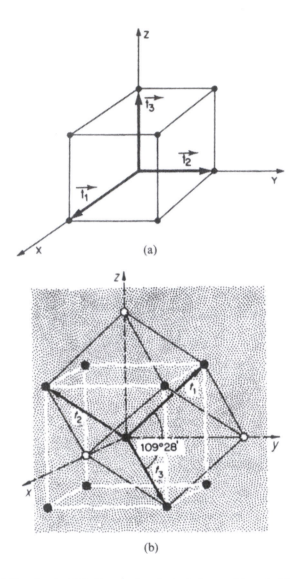

Figure 5.14. (a) Fundamental lattice vectors t_1, t_2, t_3 in a cubic primitive lattice. (b) Fundamental lattice vectors in a conventional (white) and primitive, noncubic unit cell (black) of a bcc lattice. The axes of the primitive (noncubic) unit cell form angles of $109°\ 28'$.

Equivalent equations are defined for $\mathbf{b}_2$ and $\mathbf{b}_3$. These nine equations can be combined by using the **Kronecker-Delta symbol,**

$$\mathbf{b}_n \cdot \mathbf{t}_m = \delta_{nm}, \tag{5.17}$$

where $\delta_{nm} = 1$ for $n = m$ and $\delta_{nm} = 0$ for $n \neq m$. Equation (5.17) is from now on our *definition* for the three vectors $\mathbf{b}_n$, which are reciprocal to the vectors

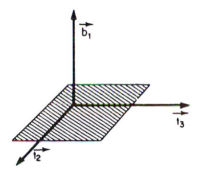

Figure 5.15. Plane formed by t_2 and t_3 with perpendicular vector b_1.

t_m. From (5.15) and (5.16) it follows[10] that b_1 is perpendicular to t_2 and to t_3, which means that t_2 and t_3 form a plane perpendicular to the vector b_1 (Fig. 5.15). We therefore write[11]

$$b_1 = \text{const.}\, t_2 \times t_3. \tag{5.18}$$

To evaluate the constant, we form the scalar product of t_1 and b_1 (5.18) and make use of (5.14):

$$b_1 \cdot t_1 = \text{const.}\, t_1 \cdot t_2 \times t_3 = 1. \tag{5.19}$$

This yields

$$\text{const.} = \frac{1}{t_1 \cdot t_2 \times t_3}. \tag{5.20}$$

Combining (5.18) with (5.20) gives

$$b_1 = \frac{t_2 \times t_3}{t_1 \cdot t_2 \times t_3}. \tag{5.21}$$

Equivalent equations can be obtained for b_2 and b_3:

$$b_2 = \frac{t_3 \times t_1}{t_1 \cdot t_2 \times t_3}, \tag{5.22}$$

$$b_3 = \frac{t_1 \times t_2}{t_1 \cdot t_2 \times t_3}. \tag{5.23}$$

Equations (5.21)–(5.23) are the transformation equations which express the fundamental vectors b_1, b_2, and b_3 of the reciprocal lattice in terms of real lattice vectors.

---

[10] The *scalar product* of two vectors a and b is $a \cdot b = ab\cos(ab)$. If i, j, and l are mutually perpendicular unit vectors, then we can write $i \cdot j = j \cdot l = l \cdot i = 0$ and $i \cdot i = j \cdot j = l \cdot l = 1$.

[11] The *vector product* of two vectors a and b is a vector which stands perpendicular to the plane formed by a and b. It is $i \times i = j \times j = l \times l = 0$ and $i \times j = l$ and $j \times i = -l$.

As an example of how these transformations are performed, we calculate now the reciprocal lattice of a bcc crystal. The real crystal may have the lattice constant "a." We express the lattice vectors t_1, t_2, t_3 in terms of the unit vectors, i, j, l in the x, y, z coordinate system (see Fig. 5.14(b)):

$$t_1 = \frac{a}{2}(-i + j + l), \tag{5.24}$$

or, abbreviated,

$$t_1 = \frac{a}{2}(\bar{1}11) \tag{5.25}$$

and

$$t_2 = \frac{a}{2}(1\bar{1}1), \tag{5.26}$$

$$t_3 = \frac{a}{2}(11\bar{1}). \tag{5.27}$$

To calculate b_1, using (5.21), we form at first the vector product[12]

$$t_2 \times t_3 = \frac{a^2}{4} \begin{vmatrix} i & j & l \\ 1 & -1 & 1 \\ 1 & 1 & -1 \end{vmatrix} = \frac{a^2}{4}(i + j + l + l - i + j) = \frac{a^2}{4}(2j + 2l)$$

$$= \frac{a^2}{2}(j + l) \tag{5.28}$$

and the scalar[13] product

$$t_1 \cdot t_2 \times t_3 = \frac{a^3}{4}(-i + j + l) \cdot (0 + j + l) = \frac{a^3}{4}(0 + 1 + 1) = \frac{a^3}{2}. \tag{5.29}$$

Combining (5.21) with (5.28) and (5.29) yields

$$b_1 = \frac{\dfrac{a^2}{2}(j + l)}{\dfrac{a^3}{2}} = \frac{1}{a}(j + l), \tag{5.30}$$

or

$$b_1 = \frac{1}{a}(011). \tag{5.31}$$

---

[12] $a \times b = \begin{vmatrix} i & j & l \\ a_x & a_y & a_z \\ b_x & b_y & b_z \end{vmatrix}.$

[13] $a \cdot b = a_x b_x + a_y b_y + a_z b_z.$

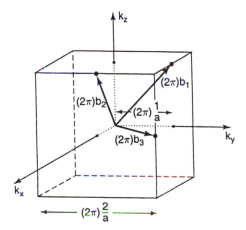

Figure 5.16. Lattice vectors in reciprocal space of a bcc crystal. The primitive vectors in the reciprocal lattice are (because of (5.13)) larger by a factor of 2π. The lattice constant of the cube then becomes $2\pi \cdot 2/a$.

Similar calculations yield

$$\mathbf{b}_2 = \frac{1}{a}(101), \tag{5.32}$$

$$\mathbf{b}_3 = \frac{1}{a}(110). \tag{5.33}$$

In Fig. 5.16, the vectors $\mathbf{b}_1, \mathbf{b}_2, \mathbf{b}_3$ are inserted into a cube of length $2/a$. We note immediately an important result. The end points of the reciprocal lattice vectors of a bcc crystal are at the center of the edges of a cube. This means that points of the reciprocal lattice of the bcc structure are identical to the lattice points in a real lattice of the fcc structure, see Fig. 5.12. Conversely, the reciprocal lattice points of the fcc structure and the real lattice points of the bcc structure are identical.

In Section 5.2, we constructed two-dimensional Brillouin zones by drawing perpendicular bisectors on the shortest lattice vectors. Similarly, a three-dimensional Brillouin zone can be obtained by bisecting all lattice vectors $\mathbf{b}$ and placing planes perpendicular on these points. As has been shown in Section 5.4, this construction is identical for a Wigner–Seitz cell. A comparison of the fundamental lattice vectors $\mathbf{b}$ and $\mathbf{t}$ gives the striking result that the Wigner–Seitz cell for an fcc crystal (Fig. 5.13) and the first Brillouin zone for a bcc crystal (Fig. 5.17) are identical in shape. The same is true for the Wigner–Seitz cell for bcc and the first Brillouin zone for fcc. Thus, a Brillouin zone can be defined as a Wigner–Seitz cell in the reciprocal lattice.

From (5.31) it can again be seen that the reciprocal lattice vector has the unit of a reciprocal length.

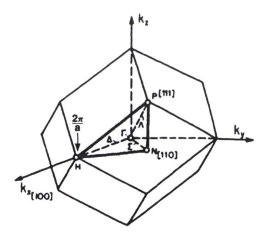

Figure 5.17. First Brillouin zone of the bcc crystal structure.

*5.6. Free Electron Bands

We mentioned in Section 5.1 that, because of the $E(\mathbf{k})$ periodicity, all information pertaining to the electronic properties of materials is contained in the first Brillouin zone. In other words, the energy $E_{\mathbf{k}'}$ for $\mathbf{k}'$ outside the first zone is identical to the energy $E_{\mathbf{k}}$ within the first zone if a suitable translation vector $\mathbf{G}$ can be found so that a wave vector $\mathbf{k}'$ becomes

$$\mathbf{k}' = \mathbf{k} + \mathbf{G}. \qquad (5.34)$$

We have already used this feature in Section 5.1, where we plotted one-dimensional energy bands in the form of a reduced zone scheme. We proceed now to three-dimensional zone pictures. We might correctly expect that the energy bands are not alike in different directions in $\mathbf{k}$-space. This can be demonstrated by using the "free electron bands" which we introduced in Fig. 5.6. We explain the details using the bcc crystal structure as an example.

In three dimensions the equation analogous to (5.7) reads

$$E_{\mathbf{k}'} = \frac{\hbar^2}{2m}(\mathbf{k} + \mathbf{G})^2. \qquad (5.35)$$

In Fig. 5.17 three important directions in $\mathbf{k}$-space are inserted into the first Brillouin zone of a bcc lattice. They are the [100] direction from the origin (Γ) to point H, the [110] direction from Γ to N, and the [111] direction from Γ to P.[14] These directions are commonly labeled by the symbols Δ, Σ, and Λ, respectively. Figure 5.18 depicts the bands, calculated by using (5.35), for

---

[14] Directions in unit cells are identified by subtracting the coordinates of the tail from the coordinates of the tip of a distance vector. The set of numbers thus gained is inserted into square brackets; see textbooks on materials science.

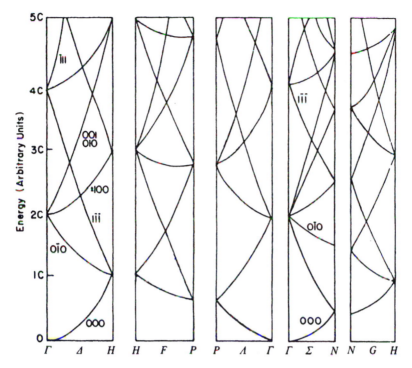

Figure 5.18. Energy bands of the free electrons for the bcc structure. The numbers given on the branches are the respective h_i values (see the calculation in the text). Compare to Fig. 5.6. $C = \hbar^2 2\pi^2 / ma^2$, see (5.38).

these distinct directions in **k**-space. The sequence of the individual subgraphs is established by convention and can be followed using Fig. 5.17.

We now show how some of these bands are calculated for a simple case. We select the $\Gamma - H$ direction as an example. We vary the modulus of the vector $\mathbf{k}_{\Gamma H} \equiv \mathbf{k}_x$ between 0 and $2\pi/a$, the latter being the boundary of the Brillouin zone (see Fig. 5.16).[15] For this direction, (5.35) becomes

$$E = \frac{\hbar^2}{2m}\left(\frac{2\pi}{a}x\mathbf{i} + \mathbf{G}\right)^2, \tag{5.36}$$

where x may take values between 0 and 1. To start with, let $\mathbf{G}$ be 0. Then (5.36) reads

$$E = \frac{\hbar^2}{2m}\left(\frac{2\pi}{a}\right)^2 (x\mathbf{i})^2 \equiv Cx^2 \tag{5.37}$$

---

[15] The attentive reader may have noticed that the boundary of the first Brillouin zone in the k_x direction for the bcc lattice is $2\pi/a$, and not π/a as for the cubic primitive unit cell (Fig. 5.6). This can be convincingly seen by comparing Figures 5.13, 5.16, and 5.17.

(see footnote 10), where

$$C = \frac{\hbar^2}{2m}\left(\frac{2\pi}{a}\right)^2 = \frac{2\hbar^2\pi^2}{ma^2}. \tag{5.38}$$

This yields the well-known parabolic $E(\mathbf{k})$-dependence. The curve which represents (5.37) is labeled (000) in Fig. 5.18, because h_1, h_2, and h_3 in (5.13) are all zero for $\mathbf{G} = 0$.

Now we let $h_1 = 0$, $h_2 = -1$, and $h_3 = 0$. Then we obtain, by using (5.13) and (5.32),

$$\mathbf{G} = -\frac{2\pi}{a}(\mathbf{i} + \mathbf{l}). \tag{5.39}$$

Combining (5.36) with (5.38) and (5.39) provides

$$E = \frac{\hbar^2}{2m}\left[\frac{2\pi x}{a}\mathbf{i} - \frac{2\pi}{a}(\mathbf{i} + \mathbf{l})\right]^2 = C[\mathbf{i}(x-1) - \mathbf{l}]^2$$

$$= C[(x-1)^2 + 1] = C(x^2 - 2x + 2) \tag{5.40}$$

(see footnote 10), which yields for

$$x = 0 \rightarrow E = 2C$$

and for

$$x = 1 \rightarrow E = 1C.$$

We obtain the band labeled $(0\bar{1}0)$ in Fig. 5.18. Similarly, all bands in Fig. 5.18 can be calculated by variation of the h values and $\mathbf{k}$-directions and by using (5.35).

The free electron bands are very useful for the following reason: by comparing them with the band structures of actual materials, an assessment is possible if and to what degree the electrons in that material can be considered to be free.

In Figs. 5.19 and 5.20 the first Brillouin zone and the free electron bands of the fcc structure are shown.

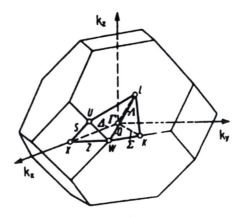

Figure 5.19. First Brillouin zone of the fcc structure.

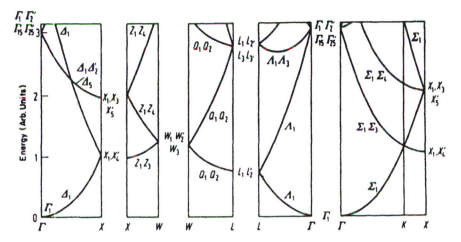

Figure 5.20. Free electron bands of the fcc structure. The letters on the bottom of the graphs correspond to letters in Fig. 5.19 and indicate specific symmetry points in k-space.

5.7. Band Structures for Some Metals and Semiconductors

Those readers who have skipped Sections 5.3 through 5.6 need to familiarize themselves with the (three-dimensional) first Brillouin zone for the face centered cubic (fcc) crystal structure (Fig. 5.19). The [100], the [110], and the [111] directions in k-space are indicated by the letters $\Gamma - X$, $\Gamma - K$, and $\Gamma - L$, respectively. Other directions in **k**-space are likewise seen. These specific symmetry points and directions are selected by convention from a much larger number of possible directions. They sufficiently characterize the properties of materials, as we will see below.

We inspect now some calculated energy-band structures. They should resemble the one shown in Fig. 5.4. In the present case, however, they are depicted for more than one direction in **k**-space. Additionally, they are displayed in the *positive* **k**-direction only, similarly as in Figs. 5.6 or 5.20.

We start with the band diagram for **aluminum**, Fig. 5.21. We recognize immediately the characteristic parabola-shaped bands in the $k_x(\Gamma - X)$ direction as seen before in Fig. 5.4. Similar parabolic bands can be detected in the $\Gamma - K$ and the $\Gamma - L$ directions. The band diagram for aluminum looks quite similar to the free electron bands shown in Fig. 5.20. This suggests that the electrons in aluminum behave essentially free-electronlike (which is indeed the case).

We also detect in Fig. 5.21 some band gaps, for example, between the X_4' and X_1 symmetry points, or between W_3 and W_2'. Note, however, that the individual energy bands overlap in different directions in k-space, so that as a whole no band gap exists. (This is in marked difference to the band dia-

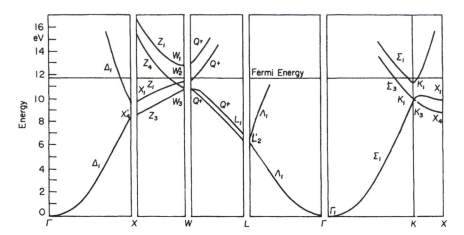

Figure 5.21. Energy bands for aluminum. Adapted from B. Segal, *Phys. Rev.* **124**, 1797 (1961). (The meaning of the Fermi energy will be explained in Section 6.1.)

gram of a semiconductor, as we shall see in a moment.) The lower, parabola-shaped bands are associated with the aluminum 3s electrons (see Appendix 3). These bands are therefore called "3s bands". The origin of the energy scale is positioned for convenience in the lower end of this s-band.

Next, we discuss the band structure for **copper**, Fig. 5.22. We notice in the lower half of this diagram closely spaced and flat running bands. Calculations show that these can be attributed to the 3d-bands of copper (see Appendix 3). They superimpose the 4s-bands (which are heavily marked in Fig. 5.22). The band which starts at Γ is, at first, s-electronlike, and becomes d-electronlike while approaching point X. The first half of this band is continued at higher energies. It is likewise heavily marked. It can be seen, therefore, that the d-bands overlap the s-bands. Again, as for aluminum, no band gap exists if one takes all directions in **k**-space into consideration.

As a third example, the band structure of **silicon** is shown (Fig. 5.23). Of particular interest is the area between 0 and approximately 1 eV in which no energy bands are shown. This "energy gap," which is responsible for the well-known semiconductor properties, will be the subject of detailed discussion in a later chapter. For semiconductors, the zero point of the energy scale is placed at the bottom of this energy gap, even though other conventions are possible and in use.

Finally, the band structure of **gallium arsenide** is shown in Fig. 5.24. The so-called III–V semiconductor compounds, such as GaAs, are of great technical importance for optoelectronic devices, as we will discuss in later sections. They have essentially the same crystal structure and the same total number of valence electrons as the element silicon. Again, a band gap is clearly seen.

Figure 5.22. Band structure of copper (fcc). Adapted from B. Segal, *Phys. Rev.* **125**, 109 (1962). The calculation was made using the *l*-dependent potential. (For the definition of the Fermi energy, see Section 6.1.)

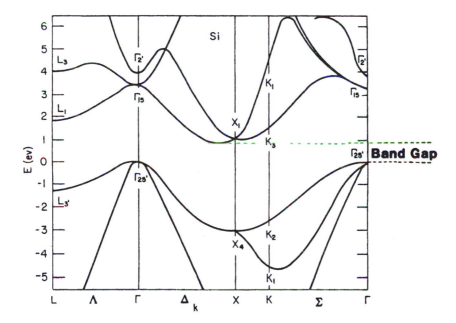

Figure 5.23. Calculated energy band structure of silicon (diamond-cubic crystal structure). Adapted from M.L. Cohen and T.K. Bergstresser, *Phys. Rev.* **14**, 789 (1966). See also J.R. Chelikowsky and M.L. Cohen, *Phys. Rev.* **B14**, 556 (1976).

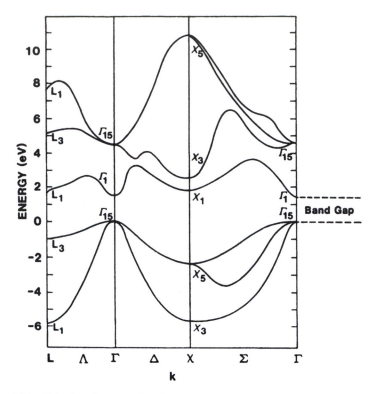

Figure 5.24. Calculated energy band structure of GaAs. Adapted from F. Herman and W.E. Spicer, *Phys. Rev.* **174**, 906 (1968).

It should be mentioned, in closing, that the band structures of actual solids, as shown in Figs. 5.21–5.24, are the result of extensive, computer-aided calculations, and that various investigators using different starting potentials arrive at slightly different band structures. Experimental investigations, such as measurements of the frequency dependence of the optical properties, can help determine which of the various calculated band structures are closest to reality.

5.8. Curves and Planes of Equal Energy

We conclude this chapter by discussing another interesting aspect of the energy versus wave vector relationship.

In one-dimensional **k**-"space" there is only *one* (positive) **k**-value which is connected with a given energy (see Fig. 5.1). In the two-dimensional case, i.e., when we plot the electron energy over a $k_x - k_y$ plane, more than one **k**-value can be assigned to a given energy. This leads to curves of equal energy,

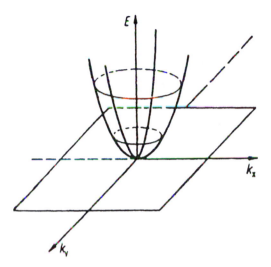

Figure 5.25. Electron energy E versus wave vector $\mathbf{k}$ (two-dimensional). This figure demonstrates various curves of equal energy for free electrons.

as shown in Fig. 5.25. For a two-dimensional square lattice and for small electron energies, the curves of equal energy are circles. However, if the energy of the electrons is approaching the energy of the boundary of a Brillouin zone, then a deviation from the circular form is known to occur. This is shown in Fig. 5.26, where curves of equal energy for a two-dimensional square lattice are inserted into the first Brillouin zone. It is of particular interest that the energy which belongs to point K in Fig. 5.26 is larger than the energy which belongs to point X (see (5.8a) and (5.9a)). Consequently, the curves of equal energy for the first Brillouin zone may extend into the second zone. This leads to an overlapping of energy bands as schematically

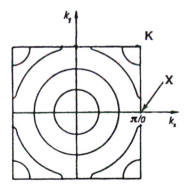

Figure 5.26. Curves of equal energy inserted into the first Brillouin zone for a two-dimensional square lattice.

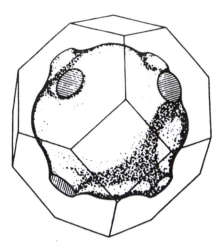

Figure 5.27. A particular surface of equal energy (Fermi surface, see Section 6.1) and the first Brillouin zone for copper. Adapted from A.B. Pippard, *Phil. Trans. Roy. Soc. London*, **A 250**, 325 (1957).

shown in Fig. 5.9, and in the band structures of Figs. 5.21–5.24. For copper and aluminum the band overlapping leads to quasi-continuous allowed energies (in different directions of **k**-space). For semiconductors the band overlapping is not complete, which results in the already-mentioned energy gap (Figs. 5.23 and 5.24).

In three-dimensional **k**-space one obtains *surfaces* of equal energy. For the free electron case and for a cubic lattice they are spheres. For a nonparabolic E-($\mathbf{k}$) behavior these surfaces become more involved. This is demonstrated in Fig. 5.27 for a special case.

Problems

1. What is the energy difference between the points L_2' and L_1 (upper) in the band diagram for copper?

2. How large is the "gap energy" for silicon? (*Hint*: Consult the band diagram for silicon.)

3. Calculate how much the kinetic energy of a free electron at the corner of the first Brillouin zone of a simple cubic lattice (three dimensions!) is larger than that of an electron at the midpoint of the face.

4. Construct the first four Brillouin zones for a simple cubic lattice in two dimensions.

5. Calculate the shape of the free electron bands for the cubic primitive crystal structure for $n = 1$ and $n = -2$ (see Fig. 5.6).

6. Calculate the free energy bands for a bcc structure in the k_x-direction having the following values for $h_1/h_2/h_3$: (a) $1\bar{1}\bar{1}$; (b) 001; and (c) 010. Plot the bands in k-space. Compare with Fig. 5.18.

7. Calculate the main lattice vectors in reciprocal space of an fcc crystal.

8. Calculate the bands for the bcc structure in the 110 $[\Gamma - N]$ direction for: (a) (000); (b) $(0\bar{1}0)$; and (c) $(1\bar{1}\bar{1})$.

9. If $\mathbf{b}_1 \cdot \mathbf{t}_1 = 1$ is given (see equation (5.14)), does this mean that $\mathbf{b}_1$ is parallel to $\mathbf{t}_1$?

CHAPTER 6

Electrons in a Crystal

In the preceding chapters we considered essentially only *one* electron, which was confined to the field of the atoms of a solid. This electron was in most cases an outer, i.e., a valence, electron. However, in a solid of one cubic centimeter at least 10^{22} valence electrons can be found. In this section we shall describe how these electrons are distributed among the available energy levels. It is impossible to calculate the exact place and the kinetic energy of each individual electron. We will see, however, that probability statements nevertheless give meaningful results.

6.1. Fermi Energy and Fermi Surface

The Fermi energy, E_F, is an important part of an electron band diagram. Many of the electronic properties of materials, such as optical, electrical, or magnetic properties, are related to the location of E_F within a band.

The Fermi energy is often defined as the "highest energy that the electrons assume at $T = 0$ K." This can be compared to a vessel, like a cup, (the electron band) into which a certain amount of water (electrons) is poured. The top surface of the water contained in this vessel can be compared to the Fermi energy. The more electrons are "poured" into the vessel, the higher the Fermi energy. The Fermi energies for aluminum and copper are shown in Figs. 5.21 and 5.22. Numerical values for the Fermi energies for some materials are given in Appendix 4. They range typically from 2 eV to 12 eV.

The above-stated definition, even though convenient, can occasionally be misleading, particularly when dealing with semiconductors. Therefore, a more accurate definition of the Fermi energy will be given in Section 6.2. We will see there that at the Fermi energy the Fermi function, F(E), equals $\frac{1}{2}$. An equation for the Fermi energy is given in (6.11).

In three-dimensional **k**-space the one-dimensional Fermi energy is replaced by a Fermi surface. The energy surface shown in Fig. 5.27 is the Fermi surface for copper.

6.2. Fermi Distribution Function

The distribution of the energies of a large number of particles and its change with temperature can be calculated by means of statistical considerations. The kinetic energy of an electron gas is governed by Fermi–Dirac statistics, which states that the probability that a certain energy level is occupied by electrons is given by the **Fermi function, F(E)**,

$$F(E) = \frac{1}{\exp\left(\dfrac{E - E_F}{k_B T}\right) + 1}. \tag{6.1}$$

If an energy level E is completely occupied by electrons, the Fermi distribution function F(E) equals 1 (certainty); for an empty energy level one obtains F(E) = 0. E_F is the Fermi energy which we introduced in Section 6.1, k_B is the Boltzmann constant, and T is the absolute temperature. In Fig. 6.1, the Fermi function is plotted versus the energy for $T \to 0$ by using (6.1). One sees from this figure that at $T = 0$ all levels that have an energy smaller than E_F are completely filled with electrons, whereas higher energy states are empty.

The Fermi distribution function for higher temperatures ($T \neq 0$) is shown in Fig. 6.2. It is noticed there that F(E) varies around E_F in a gradual manner and not by a step as for $T = 0$. To characterize this behavior, one says that F(E) is "smeared out," i.e., it is extended to an energy interval

Figure 6.1. Fermi distribution function, F(E), versus energy, E, for $T = 0$. (For $E > E_F$ and $T \to 0$ (6.1) yields $F(E) \to 0$.)

Figure 6.2. Fermi distribution function for $T \neq 0$.

$2\Delta E$. This decrease in $F(E)$ with increasing energy is heavily exaggerated in Fig. 6.2. ΔE at room temperature is in reality only about 1% of E_F.

At high energies $(E \gg E_F)$ the upper end of the Fermi distribution function can be approximated by the classical (Boltzmann) distribution function. This is best seen from (6.1) in which for large energies the exponential factor becomes significantly larger than 1. Then, $F(E)$ is approximately

$$F(E) \approx \exp\left[-\left(\frac{E - E_F}{k_B T}\right)\right]. \tag{6.1a}$$

Equation (6.1a) is known to be the Boltzmann factor, which gives, in classical thermodynamics, the probability that a given energy state is occupied. The $F(E)$ curve for high energies is thus referred to as the "**Boltzmann tail**" of the Fermi distribution function.

Of particular interest is the value of the Fermi function $F(E)$ at $E = E_F$ and $T \neq 0$. As can be seen from (6.1) and Fig. 6.2, $F(E)$ is in this particular case $\frac{1}{2}$. This serves as a definition for the Fermi energy, as outlined in Section 6.1.

6.3. Density of States

We are now interested in the question of how energy levels are distributed over a band. We restrict our discussion for the moment to the lower part of the valence band (the $3s$-band in aluminum, for example) because there the electrons can be considered to be essentially free due to their weak binding force to the nucleus. We assume that the free electrons (or the "electron gas") are confined in a square potential well (box) from which they cannot escape. The dimensions of this potential well are thought to be identical to the dimensions of the crystal under consideration. Then our problem is similar to the case of *one* electron in a potential well of size a, which we treated in

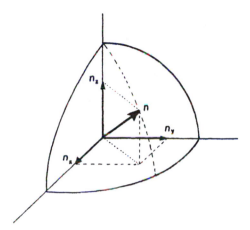

Figure 6.3. Representation of an energy state in quantum number space.

Section 4.2. By using the appropriate boundary conditions, the solution of the Schrödinger equation yields an equation that has the same form as (4.18) or (4.26),

$$E_n = \frac{\pi^2 \hbar^2}{2ma^2}(n_x^2 + n_y^2 + n_z^2), \tag{6.2}$$

where n_x, n_y, and n_z are the principal quantum numbers and a is now the *length*, etc., of the crystal. Now we pick an arbitrary set of quantum numbers n_x, n_y, n_z. To each such set we can find a specific energy level E_n, frequently called "**energy state.**" An energy state can therefore be represented by a point in quantum number space (Fig. 6.3). In this space, n is the radius from the origin of the coordinate system to a point (n_x, n_y, n_z) where

$$n^2 = n_x^2 + n_y^2 + n_z^2. \tag{6.3}$$

Equal values of the energy E_n lie on the surface of a sphere with radius n. All points within the sphere therefore represent quantum states with energies smaller than E_n. The number of quantum states, η, with an energy equal to or smaller than E_n is proportional to the volume of the sphere. Since the quantum numbers are positive integers, the n-values can only be defined in the positive octant of the n-space. One-eighth of the volume of the sphere with radius n therefore gives the number of energy states, η, the energy of which is equal to or smaller than E_n. Thus, with (6.2) and (6.3), we obtain

$$\eta = \frac{1}{8} \cdot \frac{4}{3}\pi n^3 = \frac{\pi}{6}\left(\frac{2ma^2}{\pi^2 \hbar^2}\right)^{3/2} E^{3/2}. \tag{6.4}$$

Differentiation of η with respect to the energy E provides the **number of energy states per unit energy** in the energy interval dE, i.e., the density of the

Figure 6.4. Density of states $Z(E)$ within a band. The electrons in this band are considered to be free.

energy states, briefly called **density of states**, $Z(E)$:

$$\frac{d\eta}{dE} = Z(E) = \frac{\pi}{4}\left(\frac{2ma^2}{\pi^2\hbar^2}\right)^{3/2} E^{1/2} = \frac{V}{4\pi^2}\left(\frac{2m}{\hbar^2}\right)^{3/2} E^{1/2} \qquad (6.5)$$

(a^3 is the volume, V, that the electrons can occupy).

The density of states plotted versus the energy gives, according to (6.5), a parabola. Figure 6.4 shows that at the lower end of the band considerably fewer energy levels (per unit energy) are available than at higher energies. One can compare the density of states concept with a high-rise apartment building in which the number of apartments per unit height (e.g., 8 feet) is counted. To stay within this analogy, only a very few apartments are thought to be available on the ground level. However, with increasing height of the building, the number of apartments per unit height becomes larger.

The area within the curve in Fig. 6.4 is, by definition, the number of states that have an energy equal to or smaller than E_n. Therefore, one obtains, for an area element $d\eta$,

$$d\eta = Z(E) \cdot dE, \qquad (6.6)$$

as can be seen from (6.5) and Fig. 6.4.

6.4. Population Density

The number of electrons per unit energy, $N(E)$, within an energy interval dE can be calculated by multiplying the number of possible energy levels, $Z(E)$, by the probability for the occupation of these energy levels. We have to note, however, that because of the **Pauli principle**, each energy state can be occupied by one electron of positive spin and one of negative spin,[16] i.e., each

---

[16] See Appendix 3

energy state can be occupied by two electrons. Therefore,

$$N(E) = 2 \cdot Z(E) \cdot F(E) \tag{6.7}$$

or, with (6.1) and (6.5),

$$N(E) = \frac{V}{2\pi^2} \left(\frac{2m}{\hbar^2}\right)^{3/2} E^{1/2} \frac{1}{\exp\left(\dfrac{E - E_F}{k_B T}\right) + 1}. \tag{6.8}$$

$N(E)$ is called the (electron) **population density**. We see immediately that for $T \to 0$ and $E < E_F$, the function $N(E)$ equals $2 \cdot Z(E)$ because F(E) is unity in this case. For $T \neq 0$ and $E \simeq E_F$, the Fermi distribution function (6.1) causes a smearing out of $N(E)$ (Fig. 6.5).

The area within the curve in Fig. 6.5 represents the number of electrons, N^*, that have an energy equal to or smaller than the energy E_n. For an energy interval between E and $E + dE$, one obtains

$$dN^* = N(E)\,dE. \tag{6.9}$$

We are now in a position to calculate the Fermi energy by making use of (6.8) and (6.9). We consider the simple case $T \to 0$ and $E < E_F$, which yields $F(E) = 1$. Integration from the lower end of the band to the Fermi energy, E_F, provides

$$N^* = \int_0^{E_F} N(E)\,dE = \int_0^{E_F} \frac{V}{2\pi^2} \left(\frac{2m}{\hbar^2}\right)^{3/2} E^{1/2}\,dE = \frac{V}{3\pi^2} \left(\frac{2m}{\hbar^2}\right)^{3/2} E_F^{3/2}. \tag{6.10}$$

Rearranging (6.10) yields

$$E_F = \left(3\pi^2 \frac{N^*}{V}\right)^{2/3} \frac{\hbar^2}{2m}. \tag{6.11}$$

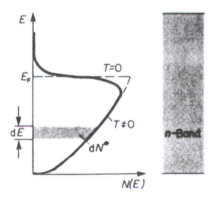

Figure 6.5. Population density $N(E)$ within a band for free electrons. dN^* is the number of electrons in the energy interval dE.

We define $N' = N^*/V$ as the **number of electrons per unit volume**. Then we obtain

$$E_F = (3\pi^2 N')^{2/3} \frac{\hbar^2}{2m}. \tag{6.11a}$$

It should be noted that N^* was calculated for simplicity for $T \to 0$ and $E < E_F$. This does not limit the applicability of (6.11), however, since the number of electrons does not change when the temperature is increased. In other words, integrating from zero to infinity and using $T \neq 0$ would yield essentially the same result as above.

6.5. Complete Density of States Function Within a Band

We have seen in Section 6.3 that for the free electron case the density of states has a parabolic E versus $Z(E)$ relationship. In actual crystals, however, the density of states is modified by the energy conditions within the first Brillouin zone. Let us consider, for example, the curves of equal energy depicted in Fig. 5.26. For low energies, the equal energy curves are circles. Thus, the electrons behave free-electronlike for these low energies. The density of states curve is then, as before, a parabola. For larger energies, however, fewer energy states are available, as is seen in Fig. 5.26. Thus, $Z(E)$ decreases with increasing E, until eventually the corners of the Brillouin zones are filled. At this point $Z(E)$ has dropped to zero. The largest number of energy states is thus found near the center of a band, as shown schematically in Fig. 6.6.

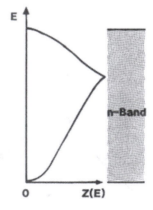

Figure 6.6. Schematic representation of the complete density of states function within a band.

6.6. Consequences of the Band Model

We mentioned in Section 6.4 that, because of the **Pauli principle**, each s-band of a crystal, consisting of N atoms, has space for $2N$ electrons, i.e., for two electrons per atom. If the highest filled s-band of a crystal is occupied by two electrons per atom, i.e., if the band is completely filled, we would expect that the electrons cannot drift through the crystal when an external electric field is applied (as it is similarly impossible to move a car in a completely occupied parking lot). An electron has to absorb energy in order to move. Keep in mind that for a completely occupied band higher energy states are not allowed. (We exclude the possibility of electron jumps into higher bands.) Solids in which the highest filled band is completely occupied by electrons are, therefore, **insulators** (Fig. 6.7(a)).

In solids with one valence electron per atom (e.g., **alkali metals**) the valence band is essentially half-filled. An electron drift upon application of an external field is possible; the crystal shows metallic behavior (Fig. 6.7(b)).

Bivalent metals should be insulators according to this consideration, which is not the case. The reason for this lies in the fact that the upper bands partially overlap, which occurs due to the weak binding forces of the valence electrons on their atomic nuclei (see Fig. 5.9). If such an overlapping of bands occurs, the valence electrons flow in the lower portion of the next higher band, because the electrons tend to assume the lowest potential energy (Fig. 6.7(c)). As a result, bivalent solids may also possess partially filled bands. Thus, they are also conductors.

We shall see in Chapter 8 that the valence as well as the conduction bands of **semiconductors** can accommodate $4N$ electrons. Because germanium and silicon possess four valence electrons, the valence band is completely filled with electrons. Intrinsic semiconductors have a relatively narrow forbidden energy zone (Fig. 6.7(d)). A sufficiently large energy can, therefore, excite

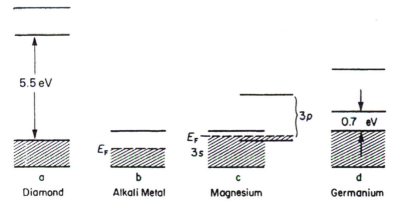

Figure 6.7. Simplified representation for energy bands for (a) insulators, (b) alkali metals, (c) bivalent metals, and (d) intrinsic semiconductors.

electrons from the completely filled valence band into the empty conduction band and thus provide some electron conduction.

This preliminary and very qualitative discussion on electronic conduction will be expanded substantially and the understanding will be deepened in Part II of this book.

6.7. Effective Mass

We implied in the previous sections that the mass of an electron in a solid is the same as the mass of a free electron. Experimentally determined physical properties of solids, such as optical, thermal, or electrical properties, indicate, however, that for some solids the mass is larger while for others it is slightly smaller than the free electron mass. This experimentally determined electron mass is usually called the effective mass, m^*. The deviation of m^* from the free electron mass[17] m_0 can be easily appreciated by stating the ratio m^*/m_0, which has values slightly above or below 1 (see Appendix 4). The cause for the deviation of the effective mass from the free electron mass is usually attributed to interactions between the drifting electrons and the atoms in a crystal. For example, an electron which is accelerated in an electric field might be slowed down slightly because of "collisions" with some atoms. The ratio m^*/m_0 is then larger than 1. On the other hand, the electron wave in another crystal might have just the right phase in order that the response to an external electric field is enhanced. In this case, m^*/m_0 is smaller than 1.

We shall now attempt to find an expression for the effective mass. For this, we shall compare the acceleration of an electron in an electric field calculated by classical as well as by quantum mechanical means. At first, we write an expression for the velocity of an electron in an energy band. We introduced in Chapter 2 the group velocity, i.e., the velocity with which a wave packet moves. Let ω be the angular frequency and $|\mathbf{k}| = 2\pi/\lambda$ the wave number of the electron wave. Then, the group velocity is, according to (2.10),

$$v_g = \frac{d\omega}{dk} = \frac{d(2\pi v)}{dk} = \frac{d(2\pi E/h)}{dk} = \frac{1}{\hbar}\frac{dE}{dk}.$$

(6.12)

From this we calculate the acceleration

$$a = \frac{dv_g}{dt} = \frac{1}{\hbar}\frac{d^2E}{dk^2}\frac{dk}{dt}.$$

(6.13)

The relation between the energy E and the wave number $|\mathbf{k}|$ is known from the preceding sections. We now want to determine the factor dk/dt. Forming

---

[17] We shall use the symbol m_0 only when we need to distinguish the free electron (rest) mass from the effective mass.

the first derivative of (4.7) ($p = \hbar k$) with respect to time yields

$$\frac{dp}{dt} = \hbar \frac{dk}{dt}.$$ (6.14)

Combining (6.14) with (6.13) yields

$$a = \frac{1}{\hbar^2} \frac{d^2 E}{dk^2} \frac{dp}{dt} = \frac{1}{\hbar^2} \cdot \frac{d^2 E}{dk^2} \cdot \frac{d(mv)}{dt} = \frac{1}{\hbar^2} \frac{d^2 E}{dk^2} F,$$ (6.15)

where F is the force on the electron. The classical acceleration can be calculated from Newton's law (1.1)

$$a = \frac{F}{m}.$$ (6.16)

Comparing (6.15) with (6.16) yields the effective mass

$$m^* = \hbar^2 \left(\frac{d^2 E}{dk^2} \right)^{-1}.$$ (6.17)

We see from (6.17) that the effective mass is inversely proportional to the curvature of an electron band. Specifically, if the curvature of $E = f(k)$ at a given point in **k**-space is large, then the effective mass is small (and vice versa). When inspecting band structures (Fig. 5.4 or Figs. 5.21–5.24) we notice some regions of high curvature. These regions might be found, particularly, near the center or near the boundary of a Brillouin zone. At these places, the effective mass is substantially reduced and may be as low as 1% of the free electron mass m_0. At points in **k**-space for which more than one electron band is found (Γ-point in Fig. 5.23, for example) more than one effective mass needs to be defined.

We shall demonstrate the **k**-dependence of the effective mass for a simple case and defer discussions about actual cases to Section 8.4. In Fig. 6.8(a) an ideal electron band within the first Brillouin zone is depicted. From this curve, both the first derivative and the reciprocal function of the second derivative, i.e., m^*, have been calculated. These functions are shown in Fig. 6.8(b) and (c). We notice in Fig. 6.8(c) that the effective mass of the electrons is small and positive near the center of the Brillouin zone and eventually increases for larger values of k_x. We likewise observe in Fig. 6.8(c) that electrons in the upper part of the given band have a negative effective mass. A negative mass means that the "particle" under consideration travels in the opposite direction to an applied electric force (and opposite to an electron.) An electron with a negative effective mass is called a "defect electron" or an "**electron hole**." (It is, however, common to ascribe to the hole a *positive effective mass and a positive charge* instead of a negative mass and a negative charge.) Electron holes play an important role in crystals whose valence bands are almost filled, e.g., in semiconductors. Solids which possess different properties in various directions (anisotropy) have a different m^* in each

Figure 6.8. (a) Simple band structure, as shown in Fig. 5.4. (b) First derivative and (c) inverse function of the second derivative of the curve shown in (a).

direction. The effective mass is a tensor in this case. An electron/hole pair is called an "**exciton.**"

It should be noted here for clarification that an electron hole is not identical with a **positron**. The latter is a subatomic particle like the electron, however with a positive charge. Positrons are emitted in the β-decay or are found in cosmic radiation. When positrons and electrons react with each other they are both annihilated under emission of energy.

6.8. Conclusion

The first part of this book is intended to provide the reader with the necessary tools for a better understanding of the electronic properties of materials. We started our discussion by solving the Schrödinger equation for the free electron case, the bound electron case, and for electrons in a crystal. We learned that the distinct energy levels which are characteristic for isolated atoms widen into energy bands when the atoms are moved closer together and eventually form a solid. We also learned that the electron bands have "fine structure," i.e., they consist of individual "branches" in an energy versus momentum (actually **k**) diagram. We further learned that some of these energy bands are filled by electrons, and that the degree of this filling depends upon whether we consider a metal, a semiconductor, or an insulator. Finally, the degree to which electron energy levels are available within a band was found to be nonuniform. We discovered that the density of states is largest near the center of an electron band. All these relatively unfamiliar concepts will become more transparent to the reader when we apply them in the chapters to come.

Problems

1. What velocity has an electron near the Fermi surface of silver? ($E_F = 5.5$ eV).

2. Are there more electrons on the bottom or in the middle of the valence band of a metal? Explain.

3. At what temperature can we expect a 10% probability that electrons in silver have an energy which is 1% above the Fermi energy? ($E_F = 5.5$ eV).

4. Calculate the Fermi energy for silver assuming 6.1×10^{22} free electrons per cubic centimeter. (Assume the effective mass equals the free electron mass.)

5. Calculate the density of states of 1 m³ of copper at the Fermi level ($m^* = m_0, E_F = 7$ eV). *Note*: Take 1 eV as energy interval. (Why?)

6. The density of states at the Fermi level (7 eV) was calculated for 1 cm³ of a certain metal to be about 10^{21} energy states per electron volt. Someone is asked to calculate the number of electrons for this metal using the Fermi energy as the maximum kinetic energy which the electrons have. He argues that because of the Pauli principle, each energy state is occupied by two electrons. Consequently, there are 2×10^{21} electrons in that band.
   (a) What is wrong with that argument?
   (b) Why is the answer, after all, not too far from the correct numerical value?

7. Assuming the electrons to be free, calculate the total number of states below $E = 5$ eV in a volume of 10^{-5} m³.

8. (a) Calculate the number of free electrons per cubic centimeter in copper, assuming that the maximum energy of these electrons equals the Fermi energy ($m^* = m_0$).
   (b) How does this result compare with that determined directly from the density and the atomic mass of copper? Hint: Consider equation (7.5)
   (c) How can we correct for the discrepancy?
   (d) Does using the effective mass decrease the discrepancy?

9. What fraction of the 3s-electrons of sodium is found within an energy k_BT below the Fermi level? (Take room temperature, i.e., $T = 300$ K.)

10. Calculate the Fermi distribution function for a metal at the Fermi level for $T \neq 0$.

11. Explain why, in a simple model, a bivalent material could be considered to be an insulator. Also explain why this simple argument is not true.

12. We stated in the text that the Fermi distribution function can be approximated by classical Boltzmann statistics if the exponential factor in the Fermi distribution function is significantly larger than one.
    (a) Calculate $E - E_F = nk_B T$ for various values of n and state at which value for n,

    $$\exp\left(\frac{E - E_F}{k_B T}\right)$$

    can be considered to be "significantly larger" than 1 (assume $T = 300$ K). (*Hint*: Calculate the error in $F(E)$ for neglecting "1" in the denominator.)
    (b) For what *energy* can we use Boltzmann statistics? (Assume $E_F = 5$ eV and $E - E_F = 4k_B T$.)

Suggestions for Further Reading (Part I)

N.W. Ashcroft and N.D. Mermin, *Solid State Physics*, Holt, Rinehart and Winston, New York (1976).

L.V. Azároff and J.J. Brophy, *Electronic Processes in Materials*, McGraw-Hill, New York (1963).

J.S. Blakemore, *Solid State Physics*, W.B. Saunders, Philadelphia, PA (1969).

R.H. Bube, *Electronic Properties of Crystalline Solids*, Academic Press, New York (1974).

R.H. Bube, *Electrons in Solids*, 3rd edn., Academic Press, New York (1992).

S. Datta, *Quantum Phenomena*, Vol. VIII of Modular Series on Solid State Devices, Addison-Wesley, Reading, MA (1989).

D. Jiles, *Electronic Properties of Materials*, Chapman and Hall, London (1994).

J.D. Livingston, *Electronic Properties of Engineering Materials*, Wiley, New York (1999).

C. Kittel, *Solid State Physics*, 7th Edition, Wiley, New York (1996).

N.F. Mott and H. Jones, *The Theory of the Properties of Metals and Alloys*, Dover, New York (1936).

M.A. Omar, *Elementary Solid State Physics*, Addison-Wesley, Reading, MA (1978).

R.F. Pierret, *Advanced Semiconductor Fundamentals*, Vol. VI of Modular Series on Solid State Devices, Addison-Wesley, Reading, MA (1987).

A.B. Pippard, *The Dynamics of Conduction Electrons*, Gordon and Breach, New York (1965).

H.A. Pohl, *Quantum Mechanics for Science and Engineering*, Prentice-Hall Series in Materials Science, Prentice-Hall, Englewood Cliffs, NJ.

R.M. Rose, L.A. Shepard, and J. Wulff, *The Structure and Properties of Materials*, Vol. IV, Electronic Properties, Wiley, New York (1966).

L. Solymar and D. Walsh, *Electrical Properties of Materials* 6th ed., Oxford Univ Press, Oxford (1998)

C.A. Wert and R.M. Thomson, *Physics of Solids*, McGraw-Hill, New York (1964).

P. Wikes, *Solid State Theory in Metallurgy*, Cambridge University Press, Cambridge (1973).

PART II

ELECTRICAL PROPERTIES
OF MATERIALS

CHAPTER 7

Electrical Conduction in Metals and Alloys

7.1. Introduction

The first observations involving electrical phenomena probably began with the study of static electricity. Thales of Miletus, a Greek philosopher, discovered around 600 BC that a piece of amber, having been rubbed with a piece of cloth, attracted feathers and other light particles. Very appropriately, the word *electricity* was later coined by incorporating the Greek word *elektron*, which means *amber*.

It was apparently not before 2300 years later that man became again interested in electrical phenomena. Stephen Gray found in the early 1700s that some substances conduct electricity whereas others do not. In 1733 DuFay postulated the existence of two types of electricity, which he termed *glass electricity* and *amber electricity* dependent on which material was rubbed. From then on a constant stream of well-known scientists contributed to our knowledge of electrical phenomena. Names such as Coulomb, Galvani, Volta, Oersted, Ampère, Ohm, Seebeck, Faraday, Henry, Maxwell, Thomson, and others, come to mind. What started 2600 years ago as a mysterious effect has been applied quite recently in an impressive technology that culminated in large-scale integration of electronic devices.

A satisfactory understanding of electrical phenomena on an atomistic basis was achieved by Drude at the turn of the twentieth century. A few decades later quantum mechanics refined our understanding. Both, the classical as well as the quantum concepts of electrical phenomena will be covered in the chapters to come. Special emphasis is placed on the description of important applications.

7.2. Survey

One of the principal characteristics of materials is their ability (or lack of ability) to conduct electrical current. Indeed, materials are classified by this property, that is, they are divided into conductors, semiconductors, and nonconductors. (The latter are often called insulators or dielectrics.) The **conductivity**, σ, of different materials at room temperature spans more than 25 orders of magnitude, as depicted in Figure 7.1. Moreover, if one takes the conductivity of superconductors, measured at low temperatures, into consideration, this span extends to 40 orders of magnitude (using an estimated conductivity for superconductors of about 10^{20} $1/\Omega$ cm). This is the largest known variation in a physical property and is only comparable to the ratio between the diameter of the universe (about 10^{26} m) and the radius of an electron (10^{-14} m).

It is generally accepted that in metals and alloys the electrons, particularly the outer or valence electrons, play an important role in electrical conduction. Therefore, it seems most appropriate to make use of the electron theory that has been developed in the foregoing chapters. Before doing so, the reader is reminded of some fundamental equations of physics pertaining to electrical conduction. These laws have been extracted from experimental observations. **Ohm's law**,

$$V = RI, \tag{7.1}$$

relates the **potential difference**, V (in volts), with the electrical **resistance**, R (in ohms i.e. Ω), and the **electrical current**, I (in amps). Another form of Ohm's law,

$$j = \sigma\mathscr{E}, \tag{7.2}$$

links **current density**,

$$j = \frac{I}{A}, \tag{7.2a}$$

Figure 7.1. Room-temperature conductivity of various materials. (Superconductors, having conductivities many orders of magnitude larger than copper, near 0 K, are not shown. The conductivity of semiconductors varies substantially with temperature and purity.) It is customary in engineering to use the centimeter as unit of length rather than the meter. We follow this practice.

i.e., the current per unit area (A/cm^2), with conductivity, σ ($1/\Omega$ cm), and **electric field strength**[1],

$$\mathscr{E} = \frac{V}{L} \tag{7.3}$$

(V/cm). (In general, $\mathscr{E}$ and j are vectors. For our purpose, however, we need only their moduli.) The current density is frequently expressed by

$$j = Nve, \tag{7.4}$$

where N is the **number of electrons** (per unit volume), v their **velocity**, and e their **charge**. The resistance of a conductor can be calculated from its physical dimensions by

$$R = \frac{L\rho}{A}, \tag{7.4a}$$

where L is the length of the conductor, A is its cross-sectional area, and ρ is the **specific resistance**, or **resistivity** (Ω cm). We define

$$\rho = \frac{1}{\sigma}. \tag{7.4b}$$

The reciprocal of the ohm (Ω) is defined to be 1 siemens (S); see Appendix 4.

We discussed in Chapter 2 the existence of two alternatives to describe an electron. First, we may consider the electrons to have a particle nature. If this model is utilized, one can explain the resistance by means of collisions of the drifting electrons with certain lattice atoms. The more collisions are encountered, the higher is the resistance. This concept qualitatively describes the increase in resistance with an increasing amount of lattice imperfections. It also explains the observed increase in resistance with increasing temperature: the thermal energy causes the lattice atoms to oscillate about their equilibrium positions (see Part V), thus increasing the probability for collisions with the drifting electrons.

Second, one may consider the electrons to have a wave nature. The matter waves may be thought to be scattered by lattice atoms. Scattering is the dissipation of radiation on small particles in all directions. The atoms absorb the energy of an incoming wave and thus become oscillators. These oscillators in turn re-emit the energy in the form of spherical waves. If two or more atoms are involved, the phase relationship between the individual re-emitted waves has to be taken into consideration. A calculation[2] shows that for a *periodic crystal structure* the individual waves in the forward direction are *in phase*, and thus interfere constructively. As a result, a wave which propagates through an ideal crystal (having periodically arranged atoms) does

---

[1] We use for the electric field strength a script $\mathscr{E}$ to distinguish it from the energy.

[2] L. Brillouin, *Wave Propagation in Periodic Structures*, Dover, New York (1953).

not suffer any change in intensity or direction. In other words, the electron wave passes without hindrance through an ideal crystal. (Only its velocity is modified.) This mechanism is called **coherent scattering**.

If, however, the scattering centers are not periodically arranged (impurity atoms, vacancies, grain boundaries, thermal vibration of atoms, etc.) the scattered waves have no set phase relationship and the wave is said to be **incoherently scattered**. The energy of incoherently scattered waves is smaller in the forward direction, that is, the matter wave loses energy. This energy loss qualitatively explains the resistance. The wave picture provides, therefore, a deeper understanding of the electrical resistance in metals and alloys. In the following two sections we shall calculate the resistance or equivalently, the electrical conduction, using, at first, the particle and then the wave concept.

7.3. Conductivity—Classical Electron Theory

Our first approach towards an understanding of electrical conduction is to postulate, as Drude did, a free "electron gas" or "plasma," consisting of the valence electrons of the individual atoms in a crystal. We assume that in a monovalent metal, such as sodium, each atom contributes *one* electron to this plasma. The **number of atoms**, N_a, per cubic centimeter (and therefore the number of free electrons in a monovalent metal) can be obtained by applying

$$N_a = \frac{N_0 \delta}{M}, \tag{7.5}$$

where N_0 is the **Avogadro constant**, δ the density, and M the atomic mass of the element. One calculates about 10^{22} to 10^{23} atoms per cubic centimeter, i.e., 10^{22} to 10^{23} free electrons per cm^3 for a monovalent metal.

The electrons move randomly (in all possible directions) so that their individual velocities in the absence of an electric field cancel and no net velocity results. This situation changes when an electric field is applied. The electrons are then accelerated with a force $e\mathscr{E}$ towards the anode and a net **drift of the electrons** results, which can be expressed by a form of Newton's law $(F = ma)$

$$m\frac{dv}{dt} = e\mathscr{E}, \tag{7.6}$$

where e is the charge of the electrons and m is their mass. Equation (7.6) implies that as long as an electric field persists, the electrons are constantly accelerated. Equation (7.6) also suggests that after the field has been removed, the electrons keep drifting with constant velocity through the crystal. This is generally not observed, however, except for some materials at very

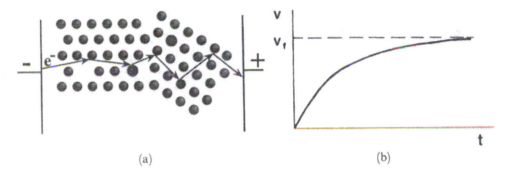

(a)   (b)

Figure 7.2. (a) Schematic representation of an electron path through a conductor (containing vacancies, impurity atoms, and a grain boundary) under the influence of an electric field. This classical model does not completely describe the resistance in materials. (b) Velocity distribution of electrons due to an electrostatic force and a counteracting friction force. The electron eventually reaches the final velocity v_f.

low temperatures (superconductors). The **free electron model** needs, therefore, an adjustment to take into account the electrical resistance.

An electron, accelerated by an electric field, may be described to increase its **drift velocity** until it encounters a collision. At this time, the electron has acquired the drift velocity v_{max} which it may lose, all or in part, at the collision (Fig. 7.2(a)). Alternatively, and more appropriately, one may describe an electron motion to be counteracted by a "friction" force γv which opposes the electrostatic force $e\mathscr{E}$. We postulate that the resistance in metals and alloys is due to interactions of the drifting electrons with some lattice atoms, i.e., essentially with the imperfections in the crystal lattice (such as impurity atoms, vacancies, grain boundaries, dislocations, etc.). Thus, (7.6) is modified as follows:

$$m\frac{dv}{dt} + \gamma v = e\mathscr{E}, \qquad (7.7)$$

where γ is a constant. The second term in (7.7) is a damping or *friction* force which contains the drift velocity, v, of the electrons. The electrons are thought to be accelerated until a final drift velocity v_f is reached (see Fig. 7.2(b)). At that time the electric field force and the friction force are equal in magnitude. In other words, the electrons are thought to move in a "viscous" medium.

For the steady state case ($v = v_f$) we obtain $dv/dt = 0$. Then (7.7) reduces to

$$\gamma v_f = e\mathscr{E}, \qquad (7.8)$$

which yields

$$\gamma = \frac{e\mathscr{E}}{v_f}. \qquad (7.9)$$

We insert (7.9) into (7.7) and obtain the complete equation for the drifting electrons under the influence of an electric field force and a friction force:

$$m\frac{dv}{dt} + \frac{e\mathscr{E}}{v_f}v = e\mathscr{E}. \tag{7.10}$$

The solution to this equation[3] is

$$v = v_f\left[1 - \exp\left(-\left(\frac{e\mathscr{E}}{mv_f}t\right)\right)\right]. \tag{7.11}$$

We note that the factor $mv_f/e\mathscr{E}$ in (7.11) has the unit of a time. It is customary to define this quantity

$$\tau = \frac{mv_f}{e\mathscr{E}}, \tag{7.12}$$

as a **relaxation time** (which can be interpreted as the average time between two consecutive collisions). Rearranging (7.12) yields

$$v_f = \frac{\tau e\mathscr{E}}{m}. \tag{7.13}$$

We make use of (7.4), which states that the current density, j, is proportional to the velocity of the drifting electrons and proportional to the **number of free electrons**, N_f (per cm^3). This yields, with (7.2),

$$j = N_f v_f e = \sigma\mathscr{E}. \tag{7.14}$$

Combining (7.13) with (7.14) finally provides the sought-for equation for the conductivity,

$$\boxed{\sigma = \frac{N_f e^2 \tau}{m}.} \tag{7.15}$$

Equation (7.15) teaches us that the conductivity is large for a large number of free electrons and for a large relaxation time. The latter is proportional to the mean free path between two consecutive collisions. The **mean free path** is defined to be

$$l = v\tau. \tag{7.15a}$$

---

[3] The reader may convince himself of the correctness of this solution by inserting (7.11) and its first derivative by time into (7.10). Further, inserting $t \to \infty$ into (7.11) yields correctly $v = v_f$ (Fig. 7.2(b)). See also Problem 8.

7.4. Conductivity—Quantum Mechanical Considerations

It was stated above that the valence electrons perform, when in equilibrium, random motions with no preferential velocity in any direction. One can visualize this fact conveniently by plotting the velocities of the electrons in velocity space (Fig. 7.3(a)). The points inside a sphere (or inside a circle when considering two dimensions) correspond to the endpoints of velocity vectors. The maximum velocity that the electrons are able to assume at $T = 0$ is the Fermi velocity, v_F (i.e., the velocity of the electrons at the Fermi energy). The sphere having v_F as a radius represents, therefore, the Fermi surface. All points inside the Fermi sphere are occupied. As a consequence the velocity vectors cancel each other pairwise at equilibrium and no net velocity of the electrons results.

If an electric field is applied, the Fermi sphere is displaced opposite to the field direction, i.e., towards the positive end of the electric field, due to the net velocity gain of the electrons (Fig. 7.3(b) dashed circle). The great majority of the electron velocities still cancel each other pairwise (shaded area). However, some electrons remain uncompensated; their velocities are shown cross hatched in Fig. 7.3(b). These electrons cause the observed current. The Drude description of conduction thus needs a modification. In the *classical* picture one would assume that *all* electrons drift, under the influence of an electric field, with a modest velocity. Quantum mechanics, instead, teaches us that *only specific electrons* participate in conduction and that these electrons drift with a high velocity which is approximately the Fermi velocity v_F.

An additional point needs to be discussed and leads to an even deeper understanding. The largest energy which the electrons can assume in a metal at $T = 0$ is the Fermi energy E_F (Chapter 6). A large number of electrons actually possess this very energy since the density of states and thus the

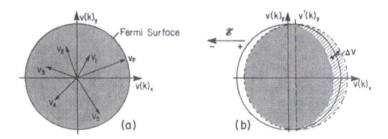

Figure 7.3. Velocity of electrons in two-dimensional velocity space. (a) Equilibrium and (b) when an electric field is applied. The shaded areas to the left and right of the $v(k)_y$-axis are of equal size. They cancel each other. The cross-hatched area remains uncompensated.

population density is highest around E_F (Fig. 7.4). Thus, only a little extra-energy ΔE is needed to raise a substantial number of electrons from the Fermi level into slightly higher states. As a consequence, the energy (or the velocity) of electrons accelerated by the electric field $\mathscr{E}$ is only slightly larger than the Fermi energy E_F (or the Fermi velocity v_F) so that for all practical purposes the mean velocity can be approximated by the Fermi velocity, v_F. We implied this fact already in our previous discussions.

We now calculate the conductivity by quantum mechanical means and apply, as before, Ohm's law $j = \sigma \mathscr{E}$, (7.2). The current density j is, as stated in (7.4), the product of the number of electrons, the electron velocity, and the electron charge. In our present case, we know that the velocity of the electrons which are responsible for the electron conduction is essentially the Fermi velocity, v_F. Further, the number of electrons which need to be considered here is N', i.e., the number of displaced electrons per unit volume, as shown in Fig. 7.4. Thus, (7.4) needs to be modified to read

$$j = v_F e N'. \tag{7.16}$$

The number of electrons displaced by the electric field $\mathscr{E}$ is

$$N' = N(E_F)\Delta E \tag{7.17}$$

(see Fig. 7.4), which yields for the current density

$$j = v_F e N(E_F)\Delta E = v_F e N(E_F)\frac{dE}{dk}\Delta k. \tag{7.18}$$

The factor dE/dk is calculated by using the E versus $|\mathbf{k}|$ relationship known for free electrons (4.8), i.e.,

$$E = \frac{\hbar^2}{2m}k^2. \tag{7.19}$$

Figure 7.4. Population density $N(E)$ versus energy for free electrons (see Fig. 6.5) and displacement ΔE by an electric field (see Fig. 7.3(b)). N' is the number of displaced electrons per unit volume (see (6.11a)) in the energy interval ΔE. $N(E)$ is defined per unit energy and, in the present case, also per unit volume, see (6.8).

Taking the first derivative of (7.19) yields, with $k = p/\hbar$ (4.7),

$$\frac{dE}{dk} = \frac{\hbar^2}{m}k = \frac{\hbar^2 p}{m\hbar} = \frac{\hbar m v_F}{m} = \hbar v_F. \tag{7.20}$$

Inserting (7.20) into (7.18) yields

$$j = v_F^2 e N(E_F)\hbar\Delta k. \tag{7.21}$$

The displacement, Δk, of the Fermi sphere in k-space under the influence of an electric field can be calculated by using (7.6) and $p = \hbar k$ (4.7):

$$F = m\frac{dv}{dt} = \frac{d(mv)}{dt} = \frac{dp}{dt} = \hbar\frac{dk}{dt} = e\mathscr{E}, \tag{7.22}$$

which yields

$$dk = \frac{e\mathscr{E}}{\hbar}\,dt,$$

or

$$\Delta k = \frac{e\mathscr{E}}{\hbar}\Delta t = \frac{e\mathscr{E}}{\hbar}\tau, \tag{7.23}$$

where τ is the time interval Δt between two "collisions" or the relaxation time (see Section 7.3). Inserting (7.23) into (7.21) yields

$$j = v_F^2 e^2 N(E_F)\mathscr{E}\tau. \tag{7.24}$$

One more consideration needs to be made. If the electric field vector points in the negative $v(k)_x$ direction, then only the components of those velocities that are parallel to the positive $v(k)_x$ direction contribute to the electric current (Fig. 7.5). The $v(k)_y$ components cancel each other pairwise. In other words, only the projections of the velocities v_F on the positive $v(k)_x$-axis ($v_{Fx} = v_F \cos\theta$) contribute to the current. Thus, we have to sum up all contributions of the velocities in the first and fourth quadrants in Fig. 7.5, which

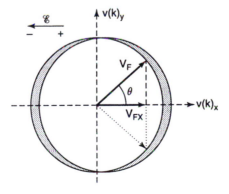

Figure 7.5. Two-dimensional velocity space.

yields

$$j = e^2 N(E_F) \mathscr{E} \tau \int_{-\pi/2}^{+\pi/2} (v_F \cos \theta)^2 \frac{d\theta}{\pi}$$

$$= e^2 N(E_F) \mathscr{E} \tau \frac{v_F^2}{\pi} \int_{-\pi/2}^{+\pi/2} \cos^2 \theta \, d\theta$$

$$= e^2 N(E_F) \mathscr{E} \tau \frac{v_F^2}{\pi} \left[\frac{1}{4} \sin 2\theta + \frac{\theta}{2} \right]_{-\pi/2}^{+\pi/2},$$

$$j = \tfrac{1}{2} e^2 N(E_F) \mathscr{E} \tau v_F^2.$$

A similar calculation for a spherical Fermi surface yields

$$j = \tfrac{1}{3} e^2 N(E_F) \mathscr{E} \tau v_F^2. \tag{7.25}$$

Thus, the conductivity finally becomes, with $\sigma = j/\mathscr{E}$ (7.2),

$$\boxed{\; \sigma = \tfrac{1}{3} e^2 v_F^2 \tau N(E_F). \;} \tag{7.26}$$

This quantum mechanical equation reveals that the conductivity depends on the Fermi velocity, the relaxation time, and the population density (per unit volume). The latter is, as we know, proportional to the density of states. Equation (7.26) is more meaningful than the expression derived from the classical electron theory (7.15). Specifically, (7.26) contains the information that not *all* free electrons N_f are responsible for conduction, i.e., the conductivity in metals depends to a large extent on the population density of the electrons near the Fermi surface. For example, monovalent metals (such as copper, silver, or gold) have partially filled valence bands, as shown in Figs. 5.22 or 6.7. Their electron population densities near their Fermi energy are high (Fig. 7.6), which results in a large conductivity according to (7.26). Bi-

Figure 7.6. Schematic representation of the density of states (Fig. 6.6) and thus, with minor modifications, also the population density (6.7). Examples for highest electron energies for a monovalent metal (E_M), for a bivalent metal (E_B), and for an insulator (E_I) are indicated.

valent metals, on the other hand, are distinguished by an overlapping of the upper bands and by a small electron concentration near the bottom of the valence band, as shown in Fig. 6.7(c). As a consequence, the electron population near the Fermi energy is small (Fig. 7.6), which leads to a comparatively low conductivity. Finally, insulators and semiconductors have, under certain conditions, completely filled electron bands, which results in a virtually zero population density near the top of the valence band (Fig. 7.6). Thus, the conductivity in these materials is extremely small.

7.5. Experimental Results and Their Interpretation

7.5.1. Pure Metals

The resistivity of a metal, such as copper, decreases linearly with decreasing temperature until it reaches a finite value (Fig. 7.7) according to the empirical equation

$$\rho_2 = \rho_1[1 + \alpha(T_2 - T_1)], \tag{7.27}$$

where α is the linear temperature coefficient of resistivity. We postulate that thermal energy causes lattice atoms to oscillate about their equilibrium positions, thus increasing the incoherent scattering of the electron waves (or equivalently, increasing the number of electron-atom collisions). The **residual resistivity**, ρ_{res}, is interpreted to be due to imperfections in the crystal, such as impurities, vacancies, grain boundaries, or dislocations. The residual resistivity is essentially not temperature-dependent. According to **Matthiessen's rule** the resistivity arises from independent scattering processes which are additive, i.e.,

$$\rho = \rho_{th} + \rho_{imp} + \rho_{def} = \rho_{th} + \rho_{res}. \tag{7.28}$$

The thermally induced part of the resistivity, ρ_{th}, is called the *ideal* resistivity, whereas the resistivity that has its origin in impurities (ρ_{imp}) and defects (ρ_{def}) is summed up in the residual resistivity. The number of impurity atoms is generally constant in a given metal or alloy. The number of vacancies or grain boundaries, however, can be changed by various heat treatments. For example, if a metal is annealed at temperatures close to its melting point and then rapidly quenched into water at room temperature, its room-temperature resistivity increases noticeably due to quenched-in vacancies. Frequently, this resistance increase diminishes during room-temperature aging or annealing at slightly elevated temperatures due to the annihilation of some vacancies. Likewise, recrystallization, grain growth, and many other metallurgical processes change the resistivity of metals. As a consequence of this, and due to its simple measurement, the resistivity is one of the most widely studied properties in materials research.

It is interesting to compare the thermally induced change in conductivity in light of the quantum mechanical and classical models. The number of free

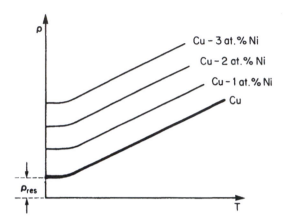

Figure 7.7. Schematic representation of the temperature dependence of the resistivity of copper and various copper–nickel alloys. ρ_{res} is the residual resistivity.

electrons, N_f, essentially does not change with temperature. Likewise, $N(E)$ changes very little with T. However, the mean free path, and thus the relaxation time, decreases with increasing temperature (due to a large rate of collisions between the drifting electrons and the vibrating lattice atoms). This, in turn, decreases σ according to (7.15) and (7.26), in agreement with the observations in Fig. 7.7. Thus, both models accurately describe the temperature dependence of the resistivity.

7.5.2. Alloys

The resistivity of alloys increases with increasing amount of solute content (Fig. 7.7). The slopes of the individual ρ versus T lines remain, however, essentially constant. Small additions of solute cause a linear shift of the ρ versus T curves to higher resistivity values in accordance with Matthiessen's rule. This resistivity increase has its origin in several mechanisms. First, atoms of different size cause a variation in the lattice parameter and, thus, in electron scattering. Second, atoms having different valences introduce a local charge difference that also increases the scattering probability. Third, solutes which have a different electron concentration compared to the host element alter the position of the Fermi energy. This, in turn, changes the population density $N(E)$ according to (6.8) and thus the conductivity, see (7.26).

Various solute elements might alter the resistivity of the host material to different degrees. This is demonstrated in Fig. 7.8. Experiments have shown that the resistivity of *dilute single-phase alloys* increases with the square of the valence difference between solute and solvent constituents (**Linde's rule**, Fig. 7.8(b)). Thus, the electron concentration of the solute element, i.e., the number of additional electrons the solute contributes, clearly plays a vital role in the resistance increase, as already mentioned above.

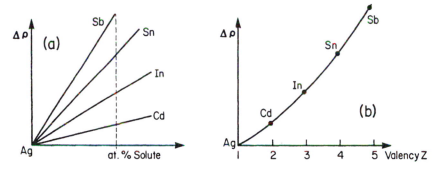

Figure 7.8. Resistivity change of various dilute silver alloys (schematic). Solvent and solute are all from the fifth period. (a) Resistivity change versus atomic % solute and (b) resistivity change due to 1 atomic % of solute.

The isothermal resistivity of *concentrated* single-phase alloys often has a maximum near 50% solute content, as shown in Fig. 7.9 (solid line). Specifically, the residual resistivity of these alloys depends, according to **Nordheim's rule**, on the fractional atomic compositions (X_A and X_B) of the constituents

$$\rho = X_A\rho_A + X_B\rho_B + CX_AX_B, \tag{7.29}$$

where C is a materials constant. Nordheim's rule holds strictly only for a few selected binary systems, because it does not take into consideration the changes in the density of states with composition. This is particularly true for alloys containing a transition metal.

The resistivity of *two-phase alloys* is, in many instances, the sum of the resistivities of each of the components, taking the volume fractions of each phase into consideration. However, additional factors, such as the crystal

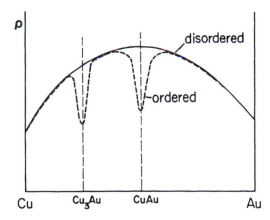

Figure 7.9. Schematic representation of the resistivity of ordered and disordered copper–gold alloys.

structure and the kind of distribution of the phases in each other, must also be considered. The concentration dependence of the resistivity of two-phase alloys does not exhibit a maximum, as in Fig. 7.9, but resembles instead a linear interpolation between the resistivities of the individual phases.

Some alloys (copper with small amounts of iron, for example) show a minimum in the resistivity at low temperatures. This anomaly is due to additional scattering of electrons by the magnetic moments of the solutes and is a deviation from the Matthiessen rule (**Kondo effect**).

The property of certain materials to conduct electricity, albeit with some resistance, is utilized for **resistors** in electrical circuits (to limit the current flow), or for generating heat (strip heaters, portable radiators, furnaces, etc.). The "**Joule heating**", or power, P, thus produced is proportional to the resistance of the wire and the square of the current:

$$P = I^2 R. \tag{7.30}$$

One common type of resistor is made from carbon-composites. Others are wire-wound, for example, around a ceramic body. They employ alloys of high resistivity (about 10^{-4} Ωcm), such as *nichrome* (nickel-chromium), and need to withstand corrosion and be suitable for high temperatures. Other resistors may consist of metal films on glass or ceramic substrates. Integrated circuits use silicon technology for the same purpose. Resistors having a fixed value are color-coded to indicate their nominal resistance, the tolerance of this value, and the rated wattage (see table in Appendix 4). Variable resistors, having a sliding contact, are either wire-wound or of the carbon-composite type.

7.5.3. Ordering

Solute atoms are generally randomly distributed in the solvent. Thus, the number of centers where incoherent scattering occurs increases proportionally with the number of substitutional atoms. If, however, the solute atoms are periodically arranged in the matrix, i.e., if, for example, in a 50/50 alloy the A and B atoms alternately occupy successive lattice sites, then the electron waves are coherently scattered. This causes a decrease in resistivity (and an increase in the mean free path) (Fig. 7.9). Only selected alloys, such as Cu_3Au, $CuAu$, Au_3Mn, etc., show a tendency towards **long-range ordering**.

The ordered state can be achieved by annealing an alloy of appropriate composition slightly below the order–disorder transition temperature (about 395°C in Cu_3Au) followed by a moderate cooling rate, or by slowly cooling from above the transition temperature. Long-range ordering causes superlattice lines in X-ray patterns.

The disordered state can be obtained at room temperature by quenching the alloy rapidly in ice brine from slightly above the transition temperature. Annealing above this transition temperature destroys the ordering effect. In some alloys, however, such as in CuAu, the tendency towards ordering is so strong that even near the melting point some ordering remains.

Some alloys, such as α-copper–aluminum, exhibit a much smaller resistance decrease by annealing below a certain ordering temperature. This effect is called **short-range ordering** and has been found to be due to small domains in which the atoms are arranged in an ordered fashion. In the short-range ordered state the A–B interactions are slightly stronger than the A–A or B–B interactions. (Short-range ordering can be identified by using small-angle X-ray scattering. It causes small and broad intensity increases between the regular diffraction lines.[4])

7.6. Superconductivity

Superconductors are materials whose resistivities become immeasurably small or actually become zero below a critical temperature, T_c. The most sensitive measurements have shown that the resistance of these materials in the superconducting state is at least 10^{16} times smaller than their room temperature values. (See, in this context, Fig. 7.1.) So far, 27 elements, numerous alloys, ceramic materials (containing copper oxide), and organic compounds (based, e.g., on selenium or sulfur) have been discovered to possess superconductivity (see Table 7.1). Their T_c values range between 0.01 K and 134 K.

---

[4] H. Warlimont, ed., *Order–Disorder Transformations in Alloys*, Springer-Verlag, Berlin (1974).

Table 7.1. Critical Temperatures of Some Superconducting Materials.

Materials	T_c [K]	Remarks
Tungsten	0.01	—
Mercury	4.15	H.K. Onnes (1911)
Sulfur-based organic superconductor	8	S.S.P. Parkin et al. (1983)
Nb_3Sn and Nb–Ti	9	Bell Labs (1961), Type II
V_3Si	17.1	J.K. Hulm (1953)
Nb_3Ge	23.2	(1973)
La–Ba–Cu–O	40	Bednorz and Müller (1986)
$YBa_2Cu_3O_{7-x}$[a]	92	Wu, Chu, and others (1987)
$RBa_2Cu_3O_{7-x}$[a]	~92	R = Gd, Dy, Ho, Er, Tm, Yb, Lu
$Bi_2Sr_2Ca_2Cu_3O_{10+\delta}$	113	Maeda et al. (1988)
$Tl_2CaBa_2Cu_2O_{10+\delta}$	125	Hermann et al. (1988)
$HgBa_2Ca_2Cu_3O_{8+\delta}$	134	R. Ott et al. (1995)

[a] The designation "1-2-3 compound" refers to the molar ratios of rare earth to alkaline earth to copper. (See chemical formula.)

Some metals such as cesium become superconducting only if a large pressure is applied to them. The superconducting transition is reversible. The superconducting state has to be considered as a separate state, distinct from the liquid, solid, or gaseous states. It has a higher degree of order—the entropy is zero.

75 years after the first discovery of superconductivity in **mercury** (H.K. Onnes, Leiden/Holland, 1911) a new class of superconductors was found by Bednorz and Müller (Zürich/Switzerland, 1986) which involved **copper oxide–based ceramics**. These materials displayed a transition temperature almost twice that of what has been known so far. This observation triggered an immense research effort virtually everywhere in the world involving billions of dollars in research money and thousands of scientists who competed for finding the most advantageous superconducting compound. As a result of this endeavor, within a few years, new copper oxide–based compounds were found that were named 1-2-3 superconductors because of the characteristic molar ratios between rare earth to alkaline earth to copper (see Table 7.1). Eventually, ceramic materials having critical temperatures above 77 K were synthesized, which were euphorically called "**high-T_c superconductors**." Superconductors having a T_c above 77 K (boiling point of liquid nitrogen) are technologically interesting because they do not require liquid helium (boiling point 4 K) or liquid hydrogen (boiling point 20 K) for cooling.

A zero resistance combined with high current densities makes superconductors useful for strong electromagnets, as needed, e.g., in magnetic resonance imaging devices (used in medicine), high-energy particle accelerators, or electric power storage devices. (The latter can be appreciated by knowing that once an electrical current has been induced in a loop consisting of a superconducting wire, it continues to flow without significant decay for several weeks.) Further potential applications are lossless power transmission lines, high-speed levitated trains, more compact and faster computers, or switching devices called cryotrons. (The latter device is based on the destruction of the superconducting state in a strong magnetic field, see below).

Despite the above-mentioned discoveries and achievements, superconducting electromagnets for high magnetic fields are, as of this writing, still manufactured from "old-fashioned" Nb–Ti or Nb_3Sn alloys (and not from ceramic superconductors) for reasons which will be discussed in the next section. The wires for the electromagnets are composed of fine filaments of a Nb–Ti alloy, each of which is only micrometers in diameter. They are imbedded in a matrix of nearly pure copper (for flexibility). We shall cover the basic concepts for these applications in the following sections.

7.6.1. Experimental Results

When the temperature of a superconducting material is lowered, the transition into the superconducting state is generally quite sharp for pure and

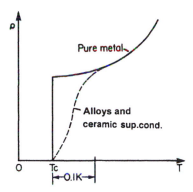

Figure 7.10. Schematic representation of the resistivity of pure and impure superconducting elements. T_c is the transition or critical temperature.

structurally perfect elements (Fig. 7.10). A temperature range of less than 10^{-5} K has been observed in pure gallium. In alloys, however, the transition may be spread over a range of about 0.1 K. Ceramic superconductors generally display an even wider spread in transition temperatures.

The **transition temperature**, T_c, often varies with the atomic mass, m_a, according to

$$m_a^\alpha \cdot T_c = \text{const.,} \qquad (7.31)$$

where α is a materials constant (**Isotope effect**). As an example, T_c for mercury varies from 4.185 K to 4.146 K when m_a changes from 199.5 to 203.4 atomic mass units.

Elimination of the superconducting state does not only occur by raising the temperature, but also by subjecting the material to a strong magnetic field. The **critical magnetic field strength**, H_c, above which superconductivity is destroyed, depends upon the temperature at which the material is held. In general, the lower the sample temperature, the higher the critical field H_c (Fig. 7.11(a)). One finds

$$H_c = H_0 \left(1 - \frac{T^2}{T_c^2} \right), \qquad (7.32)$$

where H_0 is the critical magnetic field strength at 0 K. Ceramic superconductors usually have a smaller H_c than metallic superconductors, i.e., they are more vulnerable to lose superconductivity by a moderate magnetic field.

As already mentioned above, one of the main applications of superconductors is in wires for the windings of high-strength electromagnets. We will learn in Chapter 14 that considerable currents are needed for these large field strengths. Now, conventional wires, when passed by large currents, generate substantial amounts of resistive heating, see (7.30), which needs to

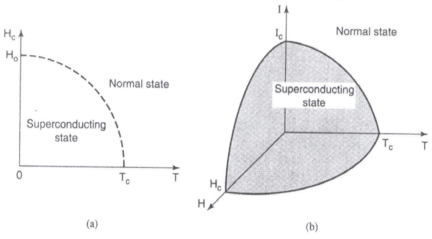

(a)   (b)

Figure 7.11. (a) Dependence of critical field strength, H_c, at which superconductivity is destroyed, in relation to the temperature of the specimen. (b) The limits of superconductivity are defined in a critical T-H-I-diagram.

be removed somehow, for example, by water cooling. On the other hand, superconducting wires that have a zero resistance below T_c are free of the resistive power loss. In this case, however, a cooling below T_c is still needed. In practice, it is a weighting between acquisition price and operation cost which commands the decision whether a superconducting or a normal electromagnet is used.

One limiting factor for ultrahigh field strengths is that the magnetic field thus produced can reach H_c, so that the superconducting state is eventually destroyed by its own magnetic field. Moreover, another limiting parameter exists, namely, the **critical current**, I_c, above which superconductivity disappears. All taken, an interrelationship between temperature, current, and magnetic field strength is observed: an increase in one of these parameters decreases the critical value of the remaining two. In other words, superconductivity is only present when temperature, magnetic field strength, and current remain within a "critical space" in a T-H-I-diagram, as depicted in Fig. 7.11(b).

Two classes of superconducting materials are distinguished. In **type I superconductors** the destruction of the superconducting state by a magnetic field, i.e., the transition between the superconducting and normal state, occurs sharply (Fig. 7.12). The critical field strength H_c is relatively low. Thus, type I superconductors are generally not used for coils for superconducting magnets. In **type II superconductors** the elimination of the superconducting state by a magnetic field is gradual. The superconducting properties are extended to a field H_{c2}, which might be 100 times higher than H_{c1} (Fig. 7.13(a)). Because of this stronger resistance against the magnetically induced destruction of the superconducting state, type II superconductors are mainly utilized for superconducting solenoids. Magnetic fields of several

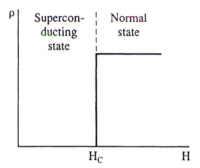

Figure 7.12. Schematic representation of the resistivity of a *type I* (or soft) super-conductor when a magnetic field of field strength H is applied. These solids behave like normal conductors above H_c.

tens of tesla (hundreds of kilogauss) have been achieved with these materials. Among the type II superconductors are transition metals and alloys consisting of niobium, aluminum, silicon, vanadium, lead, tin, titanium, and, in particular, Nb_3Sn or Nb–Ti. Ceramic superconductors also belong to this group. (The terms "type I or type II superconductors" are often used likewise when the abrupt or gradual transition with respect to *temperature* is described, see Fig. 7.10).

The interval between H_{c1} and H_{c2} represents a state in which superconducting and normal conducting areas are mixed in the solid. Specifically, one observes small circular regions, called **vortices** or **fluxoids**, which are in the normal state and which carry the smallest possible unit of a magnetic flux, called a **flux quantum**,

$$\phi_0 = \frac{h}{2e} = 2.07 \times 10^{-15} \; (T \cdot m^2). \tag{7.33}$$

The vortices are surrounded by large, superconducting regions.

The fluxoids are parallel to the magnetic field lines and are regularly arranged in space, thus forming essentially a two-dimensional superlattice (Fig. 7.13(b)). (The regular arrangement of the fluxoids stems mainly from the fact that they repel each other.) One would therefore expect that a current which flows perpendicular to these fluxoids (as is the case for electromagnets) would always find an unobstructed path through the super-conducting matrix and thus would exhibit unlimited superconductivity. However, since the current in an electromagnet flows at a right angle to the magnetic field, a so-called **Lorentz force** is created, which pushes the fluxoids perpendicular to the current and the magnetic field directions see Fig. 8.11. Thus, the moving fluxoids may become obstacles for the drifting electrons. As a result, the current is reduced, or equivalently, the electrical resistance is increased. The obstruction does not occur, however, when the fluxoids are pinned to their positions, for example, by microstructural inhomogeneities in

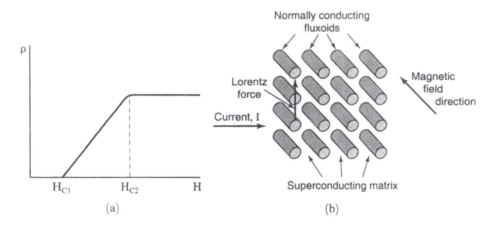

Figure 7.13. (a) Schematic representation of the resistivity of a *type II* (or hard) superconductor. The region between H_{c1} and H_{c2} is called the *vortex state*. Above H_{c2}, the solid behaves like a normal conductor. (b) Schematic representation of fluxoids in a superconducting matrix.

the matrix, such as grain boundaries, dislocations, or fine particles of the alloying components. This **fluxoid pinning** has been achieved by heat treatment and by plastic deformation, for example, by wire drawing. It is the basis for the presently used Nb_3Sn superconducting magnets.

Fluxoid pinning and resultant large critical currents have not yet been achieved in ceramic superconductors. The reason for this lies in the fact that thermally induced lattice vibrations make fluxoid pinning at higher temperatures (100 K) considerably more difficult than at much lower temperatures.

It is noted in passing that superconducting materials have **exceptional magnetic properties**. For example, a permanent magnet levitates in mid-air above a piece of a superconducting material that is cooled below T_c. We shall return to the magnetic properties of superconductors in Section 15.1.1.

Ceramic superconductors seem to be characterized by two-dimensional sheets of atoms, a Cu–O nonstoichiometry (i.e., a limited amount of an oxygen deficiency, see Fig. 7.14), a reduced lattice parameter between the copper atoms, and a tetragonal (high temperature) to orthorhombic (below room temperature) transition. Only the **orthorhombic** modification is superconducting. Further, ceramic superconductors appear to be antiferromagnetic (see Section 15.1.4). Thus, the superconductivity is most likely connected to the entire lattice structure.

Despite their considerably higher transition temperatures, ceramic superconductors have not yet revolutionized new technologies, mainly because of their inherent brittleness, their incapability of carrying high current densities, and their environmental instability. These obstacles may be overcome eventually, e.g., by using bismuth-based materials that are capable of carrying high currents when cooled to about 20 K or by utilizing composite materials,

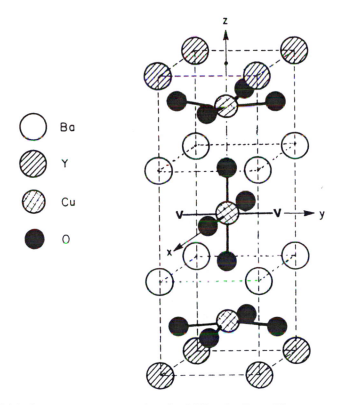

Figure 7.14. Room-temperature unit cell of $YBa_2Cu_3O_{7-x}$. The structure is an orthorhombic layered perovskite ($BaTiO_3$) containing periodic oxygen vacancies. Two examples for oxygen vacancies are indicated by a "V." Adapted from M. Stavola, *Phys. Rev. B*, **36**, 850 (1987).

i.e., by inserting the ingredient oxide powders into silver tubes and sintering them *after* plastic deformation (e.g., wire pulling). Other techniques employ depositions of ceramic superconducting films on ductile substrates. Additions of silver into some ceramic superconductors improve their environmental stability (by reducing the porosity of the material) without lowering T_c. In any event, the further development of superconducting materials should be followed with great anticipation.

*7.6.2. Theory

Attempts to explain superconductivity have been made since its discovery in 1911. One of these theories makes use of the two-fluid model, which postulates superelectrons that experience no scattering, have zero entropy (perfect order), and have long coherence lengths, i.e., an area 1000 nm wide over which the superelectrons are spread. The London theory is semiphenome-

Figure 7.15. Schematic of a Cooper pair.

nological and dwells basically on the electrodynamic properties. The **BCS theory** (which was developed in 1957 by Bardeen, Cooper, and Schrieffer) is capable of explaining the properties of conventional superconductors reasonably well. However, it does not seem to satisfactorily interpret high-temperature (ceramic) superconductors. The BCS theory is quite involved. Phenomenological descriptions of the concepts leading to this theory are probably simplifications of the actual mechanisms which govern superconduction and may thus provide temptations for misleading conclusions. (As is so often the case in quantum mechanics, the mathematics is right—it is only our lack of imagination that holds us back from correctly interpreting the equations.) Nevertheless, a conceptual description of the BCS theory and its results is attempted.

One key to the understanding of the BCS theory is accepting the existence of a pair of electrons (**Cooper pair**) that has a lower energy than two individual electrons. Imagine an electron in a metal at $T = 0$ K (no lattice vibrations). This electron perturbs the lattice slightly in its neighborhood. When such an electron drifts through a crystal the perturbation is only momentary, and, after passing, a displaced ion reverts back into its original position. One can consider this ion to be held by springs in its lattice position, so that after the electron has passed by, the ion does not simply return to its original site, but overshoots and eventually oscillates around its rest position. A *phonon* is created.[5] This phonon in turn interacts quickly with a second electron, which takes advantage of the deformation and lowers its

---

[5] A *phonon* is a lattice vibration quantum. We will describe the properties of phonons in Chapter 20.

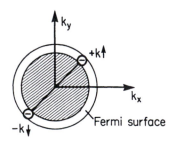

Figure 7.16. Fermi sphere, Fermi surface, and Cooper pair in a metal.

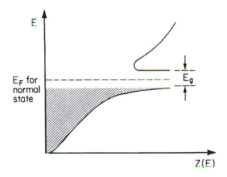

Figure 7.17. Density of states, $Z(E)$, versus electron energy in the superconducting state.

energy. Electron 2 finally emits a phonon by itself, which interacts with the first electron and so on. It is this passing back and forth of phonons which couples the two electrons together and brings them into a lower energy state (Fig. 7.15). One can visualize that all electrons on the Fermi surface having opposite momentum and opposite spin (i.e., $k\uparrow$ and $-k\downarrow$) form those Cooper pairs (Fig. 7.16), so that these electrons form a cloud of Cooper pairs which drift cooperatively through the crystal. Thus, the superconducting state is an *ordered* state of the *conduction electrons*. The scattering on the lattice atoms is eliminated, thus causing a zero resistance, as described similarly in Section 7.5.3 where we observed that ordering of the *atoms* in a crystal lattice reduces the resistivity.

One further aspect has to be considered. We just mentioned that the electrons of a Cooper pair have a lower energy than two unpaired electrons. Thus, the Fermi energy in the superconducting state may be considered to be lower than that for the nonsuperconducting state. This lower state is separated from the normal state by an energy gap, E_g (Fig. 7.17). The energy gap stabilizes the Cooper pairs against small changes of net momentum, i.e., prevents them from breaking apart. Such an energy gap of about 10^{-4} eV has indeed been observed by impinging IR radiation on a superconductor at temperatures below T_c and observing an onset of absorption of the IR radiation.

An alternate method for measuring this gap energy is by utilizing the **Josephson effect**. The experiment involves two pieces of metal, one in the superconducting state and the other in the normal state. They are separated by a thin insulating film of about 1 nm thickness (Fig. 7.18(a)). A small voltage of proper polarity in the millivolt range applied to this device raises the energy bands in the superconductor. Increasing this voltage eventually leads to a configuration where some filled electron states in the superconductor are opposite to empty states in the normal conductor (Fig. 7.18(b)). Then the Cooper pairs are capable of tunneling across the junction similarly as described in Section 4.3. The gap energy is calculated from the threshold voltage at which the tunneling current starts to flow.

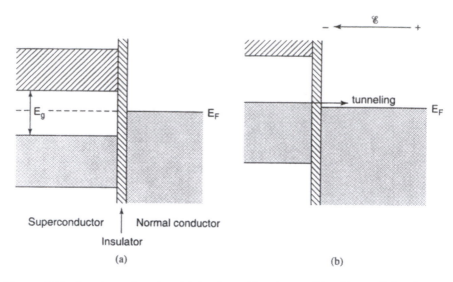

Figure 7.18. Josephson junction (a) in the unbiased state (b) with applied voltage across the junction which facilitates tunneling in the indicated direction.

In closing, we would like to revisit the electron–phonon coupling mechanism, which is believed to be the essential concept for the interpretation of superconduction, at least for metals and alloys. It has been explained above that in the normal state of conduction (above T_c) strong interactions between electrons and phonons would lead to collisions (or scattering of the electron waves), and thus to electrical resistance, whereas at low temperatures the same interactions would cause Cooper pairs to form and thus promote superconduction. This would explain why the noble metals (which have small electron–phonon interactions) are not superconducting. In other words, poor conductors in the normal state of conduction are potential candidates for high-T_c superconductors (and vice versa). Ceramic and organic superconductors fit into this scheme. Still, some scientists believe that phonons are involved in the coupling process only at very low temperatures (e.g., below 40 K). At somewhat higher temperatures, when phonons cause substantial scattering of the electrons, excitons (i.e., electron-hole pairs) may link electrons to form Cooper pairs, as suggested by A. Little for organic superconductors. Still other scientists propose resonating valence bonds as a coupling mechanism for high-T_c superconductors.

7.7 Thermoelectric Phenomena

Assume that two different types of materials (e.g., a copper and an iron wire) are connected at their ends to form a loop, as shown in Figure 7.19. One of

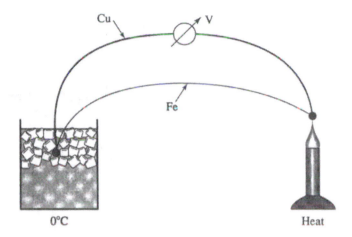

Figure 7.19. Schematic representation of two thermocouples made of copper and iron which are brought in contact with each other (Seebeck effect).

the junctions is brought to a higher temperature than the other. Then a potential difference, ΔV, between these two **thermocouples** is observed which is essentially proportional to the temperature difference, ΔT, where

$$\frac{\Delta V}{\Delta T} = S \tag{7.34}$$

is called the **thermoelectric power**, or the **Seebeck coefficient** (after its inventor, T.J. Seebeck, a German physicist who discovered, in 1821, that a thermoelectric circuit like the one just described deflected a close-by compass needle). A thermoelectric power of several microvolts per degree is commonly observed. As an example, the frequently used copper/constantan (Cu–45% Ni) combination yields about 43 μV/K. It has a useful range between -180 and $+400°$C. For higher temperatures, thermocouples of chromel (90%Ni–10%Cr) and alumel (95%Ni–2%Mn–2%Al) or platinum/Pt–13%Rh (up to 1700°C) are available. Some semiconductors have Seebeck coefficients that reach into the millivolt per degree range, that is, they are one or two orders of magnitude higher than for metals and alloys. Among them are bismuth telluride (Bi_2Te_3), lead telluride ($PbTe$), and silicon–30% germanium alloys.

Thermocouples made of metal wires are utilized as rigid, inexpensive, and fast probes for measuring temperatures even at otherwise not easily accessible places. **Thermoelectric power generators** (utilizing the above-mentioned semiconductors) are used particularly in remote locations of the earth (Siberia, Alaska, etc.). They contain, for example, a ring of thermocouples, arranged over the glass chimney of a kerosene lamp which is concomitantly used for lighting. The temperature difference of 300°C thus achieved yields electric power of a few watts or sometimes more, which can be used for radios or communication purposes. Heat produced by the decay of radioisotopes or by

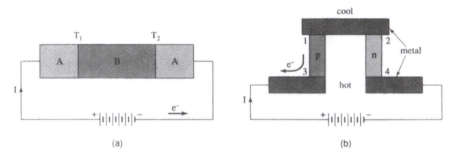

Figure 7.20. Thermoelectric refrigeration devices which make use of the Peltier effect. (a) Principle arrangement. (b) Efficient device utilizing p- and n-type semi-conductors (see Section 8.3) in conjunction with metals.

small nuclear reactors yields thermoelectric power for scientific instruments on the moon (e.g., to record moon quakes) and for relaying the information back to earth. In solar thermoelectric generators sunlight is concentrated by concave mirrors on thermocouples. Most of the above-described devices have an efficiency between 5 and 10%.

A reversion of the Seebeck effect is the **Peltier effect**: A direct electric current that flows through junctions made of different materials causes one junction to be cooled and the other to heat up (depending on the direction of the current); see Figure 7.20(a). Lead telluride or bismuth telluride in combination with metals are frequently used. One particularly effective device for which temperature differences up to 70°C have been achieved is shown in Figure 7.20(b). It utilizes n- and p-type semiconductors (see Section 8.3) in conjunction with metals. Cooling occurs on those junctions that are connected to the upper metal plate (1 and 2), whereas heat develops on the lower junctions 3 and 4. The heat on the lower plate is removed by water or air cooling. The above-quoted temperature drop can even be enhanced by cascading several devices, that is, by joining multiple **thermoelectric refrigerators** for which each stage acts as the heat sink for the next.

The thermoelectric effects can be explained by applying elements of electron theory as described in the previous sections: When two different types of conducting materials are brought into contact, electrons are transferred from the material with higher Fermi energy (E_F) "down" into the material having a lower E_F until both Fermi energies are equal. As a consequence, the material that had the smaller E_F assumes a negative charge with respect to the other. This results in the above-mentioned **contact potential** between the materials. The contact potential is temperature-dependent. Specifically, when a material is heated, a substantial number of electrons are excited across the Fermi energy to higher energy levels. These extra electrons drift to the cold junction, which becomes negatively charged compared to the hot junction. The equivalent is true for the Peltier effect: The electrons having a larger energy (that is, those having a higher E_F) are caused by the current to transfer their extra energy into the material having a lower E_F, which in turn heats up. Concomitantly, the material having a higher E_F is caused to lose energy and thus becomes colder.

Problems

1. Calculate the number of free electrons for gold using its density and its atomic mass.

2. Does the conductivity of an alloy change when long-range ordering takes place? Explain.

3. Calculate the time between two collisions and the mean free path for pure copper at room temperature. Discuss whether or not this result makes sense. *Hint*: Take the velocity to be the Fermi velocity, v_F, which can be calculated from the Fermi energy of copper $E_F = 7$ eV. Use otherwise classical considerations and $N_f = N_a$.

4. Electron waves are "coherently scattered" in ideal crystals at $T = 0$. What does this mean? Explain why in an ideal crystal at $T = 0$ the resistivity is small.

5. Calculate the number of free electrons per cubic centimeter (and per atom) for sodium from resistance data (relaxation time 3.1×10^{-14} s).

6. Give examples for coherent and incoherent scattering.

7. When calculating the population density of electrons for a metal by using (7.26), a value much larger than immediately expected results. Why does the result, after all, make sense? (Take $\sigma = 5 \times 10^5$ $1/\Omega$ cm; $v_F = 10^8$ cm/s and $\tau = 3 \times 10^{-14}$ s.)

8. Solve the differential equation

$$m\frac{dv}{dt} + \frac{e\mathscr{E}}{v_F}v = e\mathscr{E} \tag{7.10}$$

and compare your result with (7.11).

9. Consider the conductivity equation obtained from the classical electron theory. According to this equation, a bivalent metal, such as zinc, should have a larger conductivity than a monovalent metal, such as copper, because zinc has about twice as many free electrons as copper. Resolve this discrepancy by considering the quantum mechanical equation for conductivity.

CHAPTER 8

Semiconductors

8.1. Band Structure

We have seen in Chapter 7 that *metals* are characterized by *partially* filled **valence bands** and that the electrons in these bands give rise to electrical conduction. On the other hand, the valence bands of *insulators* are *completely* filled with electrons. Semiconductors, finally, represent in some respect a position between metals and insulators. We mentioned in Chapter 6 that semiconductors have, at low temperatures, a completely filled valence band and a narrow gap between this and the next higher, unfilled band. The latter one is called the **conduction band**. We discuss this now in more detail.

Because of band overlapping, the valence as well as the conduction bands of semiconductors consist of mixed (hybrid) s- and p-states. The eight highest $s + p$ states (two s- and six p-states)[6] split into two separate $(s + p)$ bands,[6] each of which consists of one s- and three p-states (see Fig. 8.1). The lower s-state can accommodate one electron per atom, whereas the three lower p-states can accommodate three electrons per atom. The valence band can, therefore, accommodate $4N_a$ electrons. (The same is true for the conduction band.) Because germanium and silicon possess four valence electrons per atom (group IV of the Periodic Table), the valence band is completely filled with electrons and the conduction band remains empty.

A deeper understanding of this can be gained from Fig. 8.2, which depicts part of a calculated band structure for silicon. Consider at first that electrons are "filled" into these bands like water being poured into a vessel. Then, of course, the lowest s-state will be occupied first. Since no energy gap exists

---

[6] See Appendix 3.

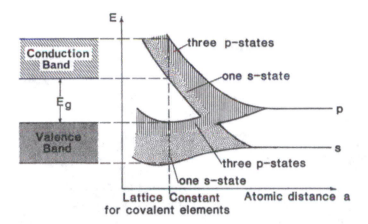

Figure 8.1. Sharp energy levels, widening into bands, and band overlapping with decreasing atomic distance for covalent elements. (Compare with Fig. 4.14.)

between the top of the s-state and the next higher p-state, additional electrons will immediately start to occupy the p-states. This process proceeds until all three lower p-states are filled. All of the $4N_a$ electrons of the semiconductor are accommodated now. Note that no higher energy band touches the p-states of the valence band. Thus, an energy gap exists between the filled valence and the empty conduction band. (As was shown in Fig. 5.23, the bands in different directions in k-space usually have different shapes so that a complete assessment can only be made by inspecting the entire band structure.)

All materials which have bonds characterized by electron sharing (co-

Figure 8.2. Schematic band structure of silicon in the k_x (or X) direction (plotted in the reduced zone scheme). The separation of the two highest p-states in the valence band is strongly exaggerated. Compare with the complete band structure of Fig. 5.23.

Table 8.1. Gap Energies for
Some Group IV Elements at
0 K (see also Appendix 4).

Element	E_g [eV]
C (diamond)	5.48
Si	1.17
Ge	0.74
Sn (gray)	0.08

valent bonds) have in common the above-mentioned hybrid bands (Fig. 8.1). An important difference is the magnitude of the gap energy, E_g, between the conduction band and the valence band. As can be seen from Table 8.1, the gap energies for group IV elements decrease with increasing atomic number. Diamond, for example, has a gap energy of 5.48 eV and is, therefore, an insulator (at least at and below room temperature) whereas the E_g for silicon and germanium is around 1 eV. Gray tin, finally, has an energy gap of only 0.08 eV. (It should be noted in passing that the utilization of diamond as an extrinsic semiconductor has been recently contemplated.)

The gap energy is slightly temperature dependent according to the empirical equation

$$E_{gT} = E_{g0} - \frac{\xi T^2}{T + \theta_D},$$

(8.1)

where E_{g0} is the band gap energy at $T = 0$ K, $\xi \approx 5 \times 10^{-4}$ eV/K, and θ_D is the Debye temperature (see Table 19.2). It is noted that E_g becomes smaller with increasing temperature. For example, the temperature dependence of E_g for Si is -2.4×10^{-4} eV/K (see Appendix 4).

8.2. Intrinsic Semiconductors

Semiconductors become conducting at elevated temperatures. In an intrinsic semiconductor, the conduction mechanism is predominated by the properties of the *pure* crystal. In order for a semiconductor to become conducting, electrons have to be excited from the valence band into the conduction band where they can be accelerated by an external electric field. Likewise, the electron holes which are left behind in the valence band contribute to the conduction. They migrate in the opposite direction to the electrons. The energy for the excitation of the electrons from the valence band into the conduction band stems usually from thermal energy. The electrons are transferred from one band into the next by **interband transitions**.

We turn now to a discussion of the **Fermi energy** in semiconductors. We learned in Section 6.2 that the Fermi energy is that energy for which the

Figure 8.3. Schematic Fermi distribution function and Fermi energy for an intrinsic semiconductor for $T > 0$ K. The "smearing out" of the Fermi distribution function at E_0 and E_V is exaggerated. For reasons of convenience, the zero point of the energy scale is placed at the bottom of the conduction band.

Fermi distribution function equals $\frac{1}{2}$. (It is advisable to keep only this "definition" of the Fermi energy in mind. Any other definition which might give a correct understanding for metals could cause confusion for semiconductors!) The probability that any state in the valence band of an intrinsic semiconductor at $T = 0$ K is occupied by electrons is 100%, i.e., $F(E) = 1$ for $E < E_v$ (Fig. 8.3). At higher temperatures, however, some of the electrons close to the top of the valence band have been excited into the conduction band. As a consequence, the probability function $F(E)$ is slightly reduced at the top of the valence band for $T > 0$ K.

On the other hand, no electrons are found at $T = 0$ K in the conduction band. Thus, the Fermi distribution function for $E > E_0$ must be zero. Again, for higher temperatures, a small deviation from $F(E) = 0$ near the bottom of the conduction band is expected (Fig. 8.3). The connection between the two branches of the $F(E)$ curve just discussed is marked with a dashed line in Fig. 8.3. This connecting line does *not* imply that electrons can be found in the forbidden band since $F(E)$ is merely the probability of occupancy of an *available* energy state. (A detailed calculation provides a slightly modified $F(E)$ curve whose vertical branches extend further into the forbidden band.)

Our discussion leads to the conclusion that the Fermi energy, E_F (i.e., that energy where $F(E) = \frac{1}{2}$), is located in the center of the forbidden band. In other words, for intrinsic semiconductors we find $E_F = -E_g/2$ when the zero point of the Energy scale is placed at the bottom of the conduction band.

We may also argue somewhat differently: For $T > 0$ K the same amount of current carriers can be found in the valence as well as in the conduction band. Thus, the *average* Fermi energy has to be halfway between these bands. A simple calculation confirms this statement. (Problem 3 in this chapter should be worked at this point to deepen the understanding.) We implied in our consideration that the effective masses of electrons and holes are alike (which is not the case; see Appendix 4).

Of special interest to us is the number of electrons in the conduction band. From the discussion carried out above, we immediately suspect that a large number of electrons can be found in the conduction band if E_g is small and, in addition, if the temperature is high. In other words, we suspect that the number of electrons in the conduction band is a function of E_g and T. A detailed calculation, which we will carry out now, verifies this suspicion.

In Section 6.4 we defined N^* to be the number of electrons that have an energy equal to or smaller than a given energy, E_n. For an energy interval between E and $E + dE$, we obtained (6.9),

$$dN^* = N(E)\, dE, \tag{8.2}$$

where

$$N(E) = 2 \cdot Z(E) \cdot F(E) \tag{8.3}$$

was called the population density (6.7) and

$$Z(E) = \frac{V}{4\pi^2} \left(\frac{2m}{\hbar^2}\right)^{3/2} E^{1/2} \tag{8.4}$$

is the density of states (6.5). In our particular case, the Fermi distribution function, $F(E)$, can be approximated by

$$F(E) = \frac{1}{\exp\left(\dfrac{E - E_F}{k_B T}\right) + 1} \simeq \exp\left[-\left(\frac{E - E_F}{k_B T}\right)\right] \tag{8.5}$$

because $E - E_F$ is about 0.5 eV and $k_B T$ at room temperature is of the order of 10^{-2} eV. Therefore, the exponential factor is large compared to 1 (Boltzmann tail). We integrate over all available electrons that have energies larger than the energy at the bottom of the conduction band ($E = 0$), and obtain, with (8.2), (8.4), and (8.5),[7]

$$N^* = \frac{V}{2\pi^2} \cdot \left(\frac{2m}{\hbar^2}\right)^{3/2} \int_0^\infty E^{1/2} \cdot \exp\left[-\left(\frac{E - E_F}{k_B T}\right)\right] dE$$

or

$$N^* = \frac{V}{2\pi^2} \cdot \left(\frac{2m}{\hbar^2}\right)^{3/2} \exp\left(\frac{E_F}{k_B T}\right) \int_0^\infty E^{1/2} \cdot \exp\left[-\left(\frac{E}{k_B T}\right)\right] dE. \tag{8.6}$$

---

[7] The integration should actually be done over the states in the conduction band only. However, since the probability factor $F(E)$ is rapidly approaching zero for energies $E > E_F$, the substitution of infinity for the upper limit does not change the result appreciably. This substitution brings the integral into a standard form, namely:

$$\int_0^\infty x^{1/2} e^{-nx}\, dx = (1/2n)\sqrt{\pi/n}.$$

Integration[7] yields

$$N^* = \frac{V}{2\pi^2}\left(\frac{2m}{\hbar^2}\right)^{3/2}\exp\left(\frac{E_F}{k_BT}\right)\frac{k_BT}{2}(\pi k_BT)^{1/2}$$

$$= \frac{V}{4}\left(\frac{2mk_BT}{\pi\hbar^2}\right)^{3/2}\exp\left(\frac{E_F}{k_BT}\right). \tag{8.7}$$

Introducing $E_F = -E_g/2$ (see above) and the effective mass ratio[8] m_e^*/m_0 we then obtain, for the number of conduction-band electrons per unit volume, $N_e = N^*/V$,

$$N_e = \frac{1}{4}\left(\frac{2m_0k_B}{\pi\hbar^2}\right)^{3/2}\left(\frac{m_e^*}{m_0}\right)^{3/2}T^{3/2}\exp\left[-\left(\frac{E_g}{2k_BT}\right)\right]. \tag{8.8}$$

The constant factor $\frac{1}{4}\left(\frac{2m_0k_B}{\pi\hbar^2}\right)^{3/2}$ has the value 4.84×10^{15} (cm^{-3} K$^{-3/2}$). Thus, we can write for (8.8)

$$\boxed{N_e = 4.84 \times 10^{15}\left(\frac{m_e^*}{m_0}\right)^{3/2}T^{3/2}\exp\left[-\left(\frac{E_g}{2k_BT}\right)\right].} \tag{8.9}$$

We see from (8.9) that the **number of electrons in the conduction band per cm^3** is a function of the energy gap and the temperature, as expected. We further notice that the contribution of a temperature increase to N_e resides mostly in the exponential term and only to a lesser extent in the term $T^{3/2}$. A numerical evaluation of (8.9) tells us that the number of electrons per cubic centimeter in silicon at room temperature is about 10^9 (see Problem 1). In other words, at room temperature, only one in every 10^{13} atoms contributes an electron to the conduction. This explains the poor conduction of Si, see Fig. 7.1. We shall see in the next section that in extrinsic semiconductors many more electrons can be found in the conduction band.

The electron and hole density is shown in Fig. 8.4 for an intrinsic semiconductor. The number of electrons is given by the area enclosed by the $Z(E)$ curve and $F(E) = \exp[-(E - E_F)/k_BT]$ (8.6).

As implied before, the number of electrons in the conduction band must equal the number of holes in the valence band. This means that an identical equation to (8.8) can be written for the holes if we assume $m_e^* \equiv m_h^*$, which is not strictly true.[9] (An additional term, which is usually neglected, modifies E_F slightly.)

The conductivity[9] of an intrinsic semiconductor is not determined by the number of electrons and holes alone. The **mobility**[9], μ, of the current

---

[8] Note that $m = m_0$, see Section 6.7 and Footnote 17 in Section 6.

[9] For numerical values, see the tables in Appendix 4.

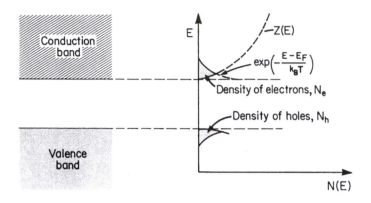

Figure 8.4. Density of electrons (N_e) and holes (N_h) for an intrinsic semiconductor.

carriers,

$$\mu = \frac{v}{\mathscr{E}}, \qquad (8.10)$$

i.e., their (drift) velocity per unit electric field, also contributes its share to the **conductivity**, σ. An expression for the conductivity is found by combining (7.2),

$$j = \sigma\mathscr{E}, \qquad (8.11)$$

and (7.4),

$$j = Nve, \qquad (8.12)$$

with (8.10), which yields

$$\sigma = N\frac{v}{\mathscr{E}}e = N\mu e. \qquad (8.13)$$

Taking both electrons and holes into consideration we can write

$$\sigma = N_e e\mu_e + N_h e\mu_h,$$

$$\sigma = 4.84 \times 10^{15}\left(\frac{m^*}{m_0}\right)^{3/2} T^{3/2}e(\mu_e + \mu_h)\exp\left[-\left(\frac{E_g}{2k_B T}\right)\right], \qquad (8.14)$$

where the subscripts e and h stand for electrons and holes, respectively. With increasing temperatures, the mobility of the current carriers is reduced by lattice vibrations (Fig. 8.5(a)). On the other hand, around room temperature, an increasing number of electrons are excited from the valence band into the conduction band, thus strongly increasing the number of current carriers, N_e and N_h (Fig. 8.5(b)). The conductivity is, according to (8.14), a function of these two factors whereby N is dominating (Fig. 8.5(c)).

At low temperatures the electrons are incoherently scattered by impurity atoms and lattice defects. It is therefore imperative that semiconductor

Figure 8.5. Schematic representation of the temperature dependence of (a) electron and hole mobilities, (b) number of carriers in an intrinsic semiconductor, and (c) conductivity for an intrinsic semiconductor. (T is given in Kelvin.)

materials are of extreme purity. Methods to achieve this high purity will be discussed in Section 8.7.11.

8.3. Extrinsic Semiconductors

8.3.1. Donors and Acceptors

We learned in the previous section that in intrinsic semiconductors only a very small number of electrons (about 10^9 electrons per cubic centimeter) contribute to the conduction of the electric current. In most semiconductor devices, a considerably higher number of charge carriers are, however, present. They are introduced by **doping**, i.e., by adding small amounts of impurities to the semiconductor material. In most cases, elements of group III or V of the periodic table are used as dopants. They replace some regular lattice atoms in a substitutional manner. Let us start our discussion by considering the case where a small amount of phosphorus (e.g., 0.0001%) is added to silicon. Phosphorus has five valence electrons, i.e., one valence electron more than silicon. Four of these valence electrons form regular electron-pair bonds with their neighboring silicon atoms (Fig. 8.6). The fifth electron, however, is only loosely bound to silicon, i.e., the binding energy is about 0.045 eV (see Appendix 4 and Problem 10.) At slightly elevated temperatures this extra electron becomes disassociated from its atom and drifts through the crystal as a conduction electron when a voltage is applied to the crystal. Extra electrons of this type are called "**donor electrons.**" They populate the conduction band of a semiconductor, thus providing a contribution to the conduction process.

It has to be noted that at sufficiently high temperatures, in addition to these donor electrons, some electrons from the valence band are also excited into the conduction band in an intrinsic manner. The conduction band contains, therefore, electrons from two sources, the amount of which depends on

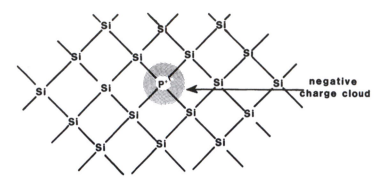

Figure 8.6. Two-dimensional representation of the silicon lattice. An impurity atom of group V of the periodic table (*P*) is shown to replace a silicon atom. The charge cloud around the phosphorus atom stems from the extra phosphorus electron. Each electron pair between two silicon atoms constitutes a covalent bond (electron sharing). The two electrons of such a pair are indistinguishable, but must have opposite spin to satisfy the Pauli principle.

the device temperature (see Section 8.3.3). Since the conduction mechanism in semiconductors with donor impurities (P, As, Sb) is predominated by negative charge carriers (electrons) these materials are called **n-type semiconductors**. The electrons are the **majority carriers**.

A similar consideration may be done with impurities from the third group of the Periodic Chart (B, Al, Ga, In). They possess one electron less than silicon and, therefore, introduce a *positive* charge cloud into the crystal around the impurity atom. The conduction mechanism in these semiconductors with **acceptor** impurities is predominated by positive carriers (holes) which are introduced into the valence band. They are therefore called **p-type semiconductors**.

8.3.2. Band Structure

The band structure of impurity or *extrinsic* semiconductors is essentially the same as for intrinsic semiconductors. It is desirable, however, to represent in some way the presence of the impurity atoms by *impurity states*. It is common to introduce into the forbidden band so-called **donor or acceptor levels** (Fig. 8.7). The distance between the donor level and the conduction band represents the energy that is needed to transfer the extra electrons into the conduction band. (The same is true for the acceptor level and valence band.) It has to be emphasized, however, that the introduction of these impurity levels does not mean that mobile electrons or holes are found in the forbidden band of, say, silicon. The impurity states are only used as a convenient means to remind the reader of the presence of extra electrons or holes in the crystal.

Figure 8.7. (a) Donor and (b) acceptor levels in extrinsic semiconductors.

8.3.3. Temperature Dependence of the Number of Carriers

At 0 K the excess electrons of the donor impurities remain in close proximity to the impurity atom and do not contribute to the electric conduction. We express this fact by stating that all donor levels are filled. With increasing temperature, the donor electrons overcome the small potential barrier (Fig. 8.7(a)) and are excited into the conduction band. Thus, the donor levels are increasingly emptied and the number of negative charge carriers in the conduction band increases exponentially, obeying an equation similar to (8.9). Once all electrons have been excited from the donor levels into the conduction band, any further temperature increase does not create additional electrons and the N_e versus T curve levels off (Fig. 8.8). As mentioned before, at still higher temperatures intrinsic effects create additional electrons which, depending on the amount of doping, can outnumber the electrons supplied by the impurity atoms.

Similarly, the acceptor levels do not contain any electrons at 0 K. At in-

Figure 8.8. Schematic representation of the number of electrons per cubic centimeter in the conduction band versus temperature for an extrinsic semiconductor with low doping.

creasing temperatures, electrons are excited from the valence band into the acceptor levels, leaving behind positive charge carriers. Once all acceptor levels are filled, the number of holes in the valence band is not increased further until intrinsic effects set in.

8.3.4. Conductivity

The conductivity of extrinsic semiconductors can be calculated, similarly as in the previous section (8.13), by multiplying the number of carriers by the mobility, μ, and electron charge, e. Around room temperature, however, only the majority carriers need to be considered. For electron conduction, for example, one obtains

$$\sigma = N_{de} e \mu_e, \qquad (8.15)$$

where N_{de} is the number of donor electrons and μ_e is the mobility of the donor electrons in the conduction band. As mentioned above, it is reasonable to assume that, at room temperature, essentially all donor electrons have been excited from the donor levels into the conduction band (Fig. 8.8). Thus, for pure n-type semiconductors, N_{de} is essentially identical to the number of impurities (i.e., donor atoms), N_d. At substantially lower temperatures, i.e., at around 100 K, the number of conduction electrons needs to be calculated using an equation similar to (8.8).

Figure 8.9 shows the temperature dependence of the conductivity. We notice that the magnitude of the conductivity, as well as the temperature dependence of σ, is different for various doping levels. For low doping rates and low temperatures, for example, the conductivity decreases with increasing temperature (Fig. 8.9(b)). This is similar to the case of metals, where the lattice vibrations present an obstacle to the drifting electrons (or, expressed differently, where the mobility of the carriers is decreased by incoherent

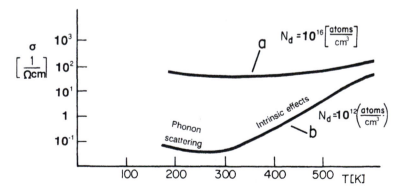

Figure 8.9. Conductivity of two extrinsic semiconductors, (a) relatively high doping and (b) low doping. N_d = number of donor atoms per cubic centimeter.

scattering of the electrons). However, at room temperature intrinsic effects set in, which increase the number of carriers and therefore enhance the conductivity. As a consequence, two competing effects determine the conductivity above room temperature: an increase of σ due to an increase in the number of electrons, and a decrease of σ due to a decrease in mobility. (It should be mentioned that the mobility of electrons or holes also decreases slightly when impurity atoms are added to a semiconductor.) For high doping levels, the temperature dependence of σ is less pronounced due to the already higher number of carriers (Fig. 8.9(a)).

8.3.5. Fermi Energy

In an n-type semiconductor, more electrons can be found in the conduction band than holes in the valence band. This is particularly true at low temperatures. The Fermi energy must therefore be between the donor level and the conduction band (Fig. 8.10). With increasing temperatures, an extrinsic semiconductor becomes progressively intrinsic and the Fermi energy approaches the value for an intrinsic semiconductor, i.e., $-(E_g/2)$. [Similarly, the Fermi energy for a p-type semiconductor rises with increasing temperature from below the acceptor level to $-(E_g/2)$.]

Figure 8.10. Fermi level of an n-type semiconductor as a function of temperature. $N_d \approx 10^{16}$ (atoms per cubic centimeter).

*8.4. Effective Mass

Some semiconductor properties can be better understood and calculated by evaluating the effective mass of the charge carriers. We mentioned in Section 6.7 that m^* is inversely proportional to the curvature of an electron band. We now make use of this finding.

Let us first inspect the upper portion of the valence bands for silicon near Γ (Fig. 8.2). We notice that the curvatures of these bands are convex downward. It is known from Fig. 6.8 that in this case the charge carriers have a negative effective mass, i.e., these bands can be considered to be populated by electron holes. Further, we observe that the curvatures of the individual bands are slightly different. Thus, the effective masses of the holes in these bands must likewise be different. One distinguishes appropriately between **light holes** and **heavy holes**. Since two of the bands, namely, those having the smaller curvature, are almost identical, we conclude that two out of the three types of holes are heavy holes.

We turn now to the *conduction* band of silicon and focus our attention on the lowest band (Fig. 8.2). We notice a minimum (or valley) at about 85% between the Γ and X points. Since the curvature at that location is convex upward, we expect this band to be populated by electrons. (The energy surface near the minimum is actually a spheroid. This leads to **longitudinal** and **transverse masses** m_l^* and m_t^*.) Values for the effective masses are given in Appendix 4. Occasionally, *average effective masses* are listed in the literature. They may be utilized for estimates.

8.5. Hall Effect

The number and type of charge carriers (electrons or holes) that were calculated in the preceding sections can be elegantly measured by making use of the Hall effect. Actually, it is quite possible to measure concentrations of less than 10^{12} electrons per cubic centimeter in doped silicon, i.e., one can measure one donor electron (and therefore one donor atom) per 10^{10} silicon atoms. This sensitivity is several orders of magnitude better than in any chemical analysis.

We assume for our discussion an *n*-type semiconductor in which the conduction is predominated by electrons. Suppose an electric current having a current density j flows in the positive x-direction and a magnetic field (of magnetic induction B) is applied normal to this electric field in the z-direction (Fig. 8.11). Each electron is then subjected to a force, called the **Lorentz force**, which causes the electron paths to bend, as shown in Fig. 8.11. As a consequence, the electrons accumulate on one side of the slab (in Fig. 8.11 on the right side) and are deficient on the other side. Thus, an electric field is created in the (negative) y-direction which is called the Hall field. In equilibrium, the **Hall force**

$$F_\mathrm{H} = -e\mathscr{E}_y \tag{8.16}$$

balances the above-mentioned Lorentz force

$$F_\mathrm{L} = v_x B_z e, \tag{8.17}$$

Figure 8.11. Schematic representation of the Hall effect in an n-type semiconductor (or a metal in which electrons are the predominant current carriers).

where v_x is the velocity of the electrons, and e is the electron charge. $F_H + F_L = 0$ yields, for the Hall field,

$$\mathscr{E}_y = v_x B_z. \qquad (8.18)$$

Combining (8.18) with (7.4) (and knowing that the current is defined to be directed in the opposite direction to the electron flow)

$$j_x = -N v_x e \qquad (8.19)$$

yields for the number of conduction electrons (per unit volume)

$$N = \frac{j_x B_z}{e \mathscr{E}_y} = \frac{I_x B_z L_y}{A_x e V_y}, \qquad (8.20)$$

where A_x is the area perpendicular to the electron flow and V_y is the **Hall voltage** measured in the y-direction.

The variables on the right side of (8.20) can all be easily measured and the number of conduction electrons can then be calculated. Quite often, a **Hall constant**

$$R_H = -\frac{1}{Ne} \qquad (8.21)$$

is defined which is inversely proportional to the density of charge carriers, N. The sign of the Hall constant indicates whether electrons or holes predominate in the conduction process. R_H is negative when electrons are the predominant charge carriers. (The electron holes are deflected in the same direction as the electrons but travel in the opposite direction.)

8.6. Compound Semiconductors

Gallium arsenide (a compound of group III and group V elements of the Periodic Table) is of great technical interest, partially because of its large band gap,[10] which essentially prevents intrinsic contributions in impurity semiconductors even at elevated temperatures, partially because of its larger electron mobility,[10] which aids in high-speed applications, and particularly because of its optical properties, which result from the fact that GaAs is a "direct-band gap" material (see Chapter 12). The large electron mobility in GaAs is caused by a small value for the electron effective mass, which in turn results from a comparatively large convex upward curvature of the conduction electron band near Γ. (See in this context the band structure of GaAs in Fig. 5.24.) The electrons which have been excited into the conduction band (mostly from donor levels) most likely populate this high curvature region near Γ.

The atomic bonding in III–V and II–VI semiconductors resembles that of the group IV elements (covalent) with the additional feature that the bonding is partially ionic because of the different valences of the participating elements. The ionization energies[10] of donor and acceptor impurities in GaAs are as a rule one order of magnitude smaller than in germanium or silicon, which ensures complete ionization even at relatively low temperatures. The crystal structure of GaAs is similar to that of silicon. The gallium atoms substitute for the corner and face atoms, whereas arsenic takes the places of the four interior sites (zinc-blende structure).

The high expectations that have been set for GaAs as the semiconductor material of the future have not yet materialized to date. It is true that GaAs devices are two and a half times faster than silicon-based devices, and that the "noise" and the vulnerability to cosmic radiation is considerably reduced in GaAs because of its larger band gap. On the other hand, its ten-times higher price and its much greater weight ($\delta_{Si} = 2.3$ g/cm^3 compared to $\delta_{GaAs} = 5.3$ g/cm^3) are serious obstacles to broad computer-chip usage or for solar panels. Thus, GaAs is predominantly utilized for special applications, such as high-frequency devices (e.g., 10 GHz), certain military projects, or satellite preamplifiers. One of the few places, however, where GaAs seems to be, so far, without serious competition is in optoelectronics (though even this domain appears to be challenged according to the most recent research results).

We will learn in Part III that only direct band-gap materials such as GaAs are useful for lasers and light-emitting diodes (LED). Indirect–band gap materials, such as silicon, possess instead the property that part of the energy of an excited electron is removed by lattice vibrations (phonons). Thus, this energy is not available for light emission. We shall return to GaAs devices in Section 8.7.9.

---

[10] See the tables in Appendix 4.

GaAs is, of course, not the only compound semiconductor material which has been heavily researched or is being used. Indeed, most compounds consisting of elements of groups III and V of the periodic table are of some interest. Among them are GaP, GaN, InP, InAs, InSb, and AlSb, to mention a few.[10] But also, group II–VI compounds, such as ZnO, ZnS, ZnSe, CdS, CdTe, or HgS are considered for applications. These compounds have in common that the combination of the individual elements possesses an average of four valence electrons per atom because they are located at equal distances from either side of the fourth column. Another class of compound semiconductors is the group IV–VI materials,[10] which include PbS, PbSe, and PbTe. Finally, ternary alloys, such as $Al_xGa_{1-x}As$, or quaternary alloys, such as $Al_xGa_{1-x}As_ySb_{1-y}$, are used. Most of the compounds and alloys are utilized in optoelectronic devices, e.g., $GaAs_{1-x}P_x$ for LEDs, which emit light in the visible spectrum (see Part III). $Al_xGa_{1-x}As$ is also used in modulation-doped field-effect transistors (MODFET).

Finally, silicon carbide is the most important representative of the group IV–IV compounds. Since its band gap is around 3 eV, α-SiC can be used for very-high-temperature (700°C) device applications and for LEDs that emit light in the blue end of the visible spectrum. SiC is, however, expensive and cannot yet be manufactured with reproducible properties. Ga–N–In have now replaced SiC as blue-emitting LEDs, see p. 280.

Doping of GaAs is accomplished, for example, by an excess of Ga atoms (p-type) or an excess of As (n-type). Si acts as a donor if it replaces Ga atoms and as an acceptor by substituting for As atoms. The recently refined technique of molecular beam epitaxy (MBE) allows the production of the wanted compounds and dopings.

8.7. Semiconductor Devices

8.7.1. Metal–Semiconductor Contacts

If a semiconductor is coated on one side with a metal, a **rectifying contact** or an *ohmic* contact is formed, depending on the type of metal used. Both cases are equally important. Rectifiers are widely utilized in electronic devices, e.g., to convert alternating current into direct current. However, the type discussed here has been mostly replaced by $p - n$ rectifiers. On the other hand, all semiconductor devices need contacts in which the electrons can easily flow in both directions. They are called **ohmic contacts** because their current–voltage characteristic obeys Ohm's law (7.1).

At the beginning of our discussion let us assume that the surface of an n-type semiconductor has somehow been negatively charged. The negative charge repels the free electrons that had been near the surface and leaves positively charged donor ions behind (e.g., As^+). Any electron that drifts toward the surface (negative x-direction in Fig. 8.12(a)) "feels" this repelling force. As a consequence, the region near the surface has fewer free electrons

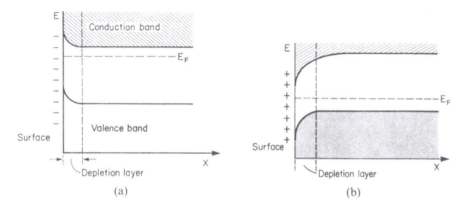

Figure 8.12. (a) Band diagram for an *n*-type semiconductor whose surface has been negatively charged. (b) Band diagram for a *p*-type semiconductor, the surface of which is positively charged. X is the distance from the surface.

than the interior of the solid. This region is called the **depletion layer** (or sometimes *space-charge region*).

In order to illustrate the repelling force of an external negative charge, it is customary to curve the electron bands upward near the surface. The depletion can then be understood by stating that the electrons assume the lowest possible energy state (or colloquially expressed: "The electrons like to roll downhill"). The depletion layer is a **potential barrier** for electrons.

Similarly, if a *p*-type semiconductor is positively charged at the surface, the positive carriers (holes) are repelled toward the inner part of the crystal and the band edges are bent downward (Fig. 8.12(b)). This represents a potential barrier for holes (because holes "want to drift upward" like a hydrogen-filled balloon).

8.7.2. Rectifying Contacts (Schottky Barrier Contacts)

It is essential for further discussion to introduce the **work function**, ϕ, which is the energy difference between the Fermi energy and the ionization energy. In other words, ϕ is the energy which is necessary to transport an electron from E_F to infinity. (Values for ϕ are given in Appendix 4.)

Let us consider a metal and an *n*-type semiconductor before they are brought into contact. In Fig. 8.13(a) the Fermi energy of a metal is shown to be lower than the Fermi energy of the semiconductor, i.e., $\phi_M > \phi_S$. Immediately after the metal and semiconductor have been brought into contact, electrons start to flow from the semiconductor "down" into the metal until the Fermi energies of both solids are equal (Fig. 8.13(b)). As a consequence, the metal will be charged negatively and a potential barrier is formed just as shown in Fig. 8.12. This means that the energy bands in the bulk semiconductor are lowered by the amount $\phi_M - \phi_S$ with respect to a point A.

Figure 8.13. Energy bands for a metal and an *n*-type semiconductor (a) before and (b) after contact. $\phi_M > \phi_S$. The potential barrier is marked with heavy lines. χ is the *electron affinity*.

In the equilibrium state, electrons from both sides cross the potential barrier. This electron flow constitutes the so-called **diffusion current**. The number of electrons diffusing in both directions must be identical for the following reason: the metal contains more free electrons, but these electrons have to climb a higher potential barrier than the electrons in the semiconductor, whose conduction band contains fewer free electrons.

Similarly, if a *p*-type semiconductor is brought into contact with a metal and $\phi_M < \phi_S$, then electrons diffuse from the metal into the semiconductor, thus charging the metal and, therefore, the surface of the semiconductor positively. Consequently, a "downward" potential barrier (for the holes) is formed (Fig. 8.14).

In addition to the diffusion current just mentioned, a "**drift current**" needs to be taken into consideration. Let us assume that an electron–hole pair was thermally created in or near the depletion layer. Then, the thermally created electron in the conduction band is immediately swept down the barrier, and the hole in the valence band is swept up the barrier. This drift current is usually very small (particularly if the band gap is large, such as in GaAs) and is relatively insensitive to the height of the potential barrier. The total current across a junction is the sum of drift and diffusion components.

The potential barrier height for an electron diffusing from the semiconductor into the metal is $\phi_M - \phi_S$ (see Fig. 8.13(b)). This potential difference is called the **contact potential**. The height of the potential barrier from the metal side is $\phi_M - \chi$, where χ is the **electron affinity**, measured from the bottom of the conduction band to the ionization energy (vacuum level) (Fig. 8.13(a)).

We shall now estimate the net current that flows across the potential barrier when a metal and an *n*-type semiconductor are connected to a d.c. source (*biasing*). At first, the metal is assumed to be connected to the negative

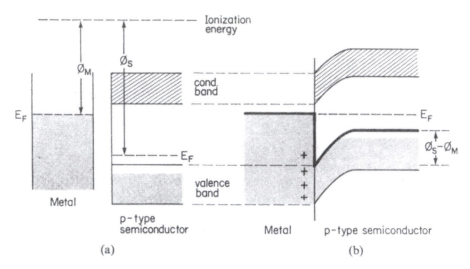

Figure 8.14. Energy bands for a metal and a p-type semiconductor (a) before and (b) after contact. $\phi_M < \phi_S$.

terminal of a battery. As a result, the metal is charged even more negatively than without bias. Thus, the electrons in the semiconductor are repelled even more, and the potential barrier is increased (Fig. 8.15(a)). Further, the depletion layer becomes wider. Because both barriers are now relatively high, the diffusion currents in both directions are negligible. However, the small and essentially voltage-independent drift current still exists, which results in a very small and constant net electron current from the metal into the semiconductor (**reverse bias**, Fig. 8.15(a)).

Figure 8.15. Metal–semiconductor contact with two polarities: (a) reverse bias and (b) forward bias. The number of electrons that flow in both directions and the net current is indicated by the length of the arrows. The potential barriers are marked by heavy lines.

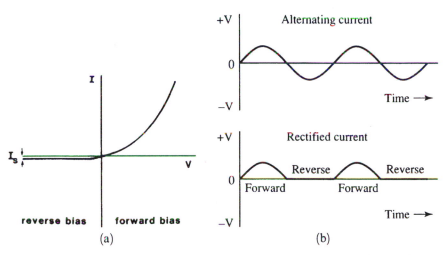

Figure 8.16. (a) Characteristic of a rectifier. The reverse current is grossly exaggerated! (b) Voltage versus time curves to demonstrate the behavior of an alternating current and a current for which the negative voltage has been eliminated.

If the polarity of the battery is reversed, the potential barrier in the semiconductor is reduced, i.e., the electrons are "driven" across the barrier so that a large net current from the semiconductor into the metal results (**forward bias**). The depletion layer is narrow (Fig. 8.15(b)). The voltage–current characteristic of a rectifier is shown in Fig. 8.16(a). Rectifiers of this type are used to convert alternating current into direct current, Fig. 8.16(b).

The current that flows from the metal into the semiconductor is

$$I_{\text{MS}} = ACT^2 \exp\left[-\left(\frac{\phi_{\text{M}} - \chi}{k_{\text{B}}T}\right)\right], \tag{8.22}$$

(see Fig. 8.13b) where A is the area of the contact and C is a constant. The current flowing from the semiconductor into the metal is

$$I_{\text{SM}} = ACT^2 \exp\left[-\left(\frac{\phi_{\text{M}} - \phi_{\text{S}} - eV}{k_{\text{B}}T}\right)\right], \tag{8.23}$$

where V is the bias voltage (which has the sign of the polarity of the metal) and e is the electronic charge. The net current $I_{\text{net}} = I_{\text{SM}} - I_{\text{MS}}$ consists of two parts, namely, the **saturation current** (occasionally called the generation current)[11]

$$I_{\text{S}} = ACT^2 \exp\left[-\left(\frac{\phi_{\text{M}} - \phi_{\text{S}}}{k_{\text{B}}T}\right)\right] \tag{8.24}$$

and a voltage-dependent term. The net current is then obtained by combin-

---

[11] For low enough temperatures, one can assume $\phi_{\text{S}} \approx \chi$; see Figs. 8.10 and 8.13.

ing (8.23), and (8.24),

$$I_{\text{net}} = I_{\text{S}} \left[\exp\left(\frac{eV}{k_{\text{B}}T} \right) - 1 \right]. \tag{8.25}$$

We see from (8.25) that for forward bias (positive V) the net current increases exponentially with voltage. Figure 8.16 reflects this behavior. On the other hand, for reverse bias (negative V) the current is essentially constant and equal to $-I_{\text{S}}$. The saturation current is about three orders of magnitude smaller than the forward current. (It is shown exaggerated in Fig. 8.16.)

We shall learn in Section 8.7.4 that the same rectifying effect as discussed above can also be achieved by using a p–n diode. There are, however, a few advantages in using the metal/semiconductor rectifier. First, the conduction in a metal/semiconductor device involves, naturally, one type of conduction carrier (e.g., electrons) only. Thus, no mutual annihilation of electrons and holes can occur. As a consequence of this lack of "carrier recombination," the device may be switched more quickly from forward to reverse bias and is therefore better suited for microwave-frequency detectors. Second, the metal base provides better heat removal than a mere semiconductor chip, which is helpful in high-power devices.

8.7.3. Ohmic Contacts (Metallizations)

In Fig. 8.17(a) and (b), band diagrams are shown for the case where a metal is brought into contact with an n-type semiconductor. It is assumed that $\phi_{\text{M}} < \phi_{\text{S}}$. Thus, electrons flow from the metal into the semiconductor, charging the metal positively. The bands of the semiconductor bend "downward" and no barrier exists for the flow of electrons in either direction. In other words, this configuration allows the injection of a current into and out of the semiconductor without suffering a sizable power loss. The

Figure 8.17. Ohmic contact between metal and n-type semiconductor ($\phi_{\text{M}} < \phi_{\text{S}}$). (a) Metal and semiconductor are separate. (b) Metal and semiconductor are in contact. (c) Current–voltage characteristic.

current increases, in essence, linearly with increasing voltage and is symmetric about the origin as Ohm's law requires (Fig. 8.17(c)). Accordingly, this junction is called an **ohmic contact**. A similar situation exists for a p-type semiconductor and $\phi_M > \phi_S$.

Aluminum is frequently used for making the contact between a device (e.g., the p-region of a rectifier) and the external leads. Aluminum bonds readily to Si or SiO_2 if the device is briefly heated to about 550°C after Al deposition. Since aluminum has a larger work function than silicon (see Appendix 4) the contact to a p-region is ohmic. Additionally, the diffusion of aluminum into silicon yields a shallow and highly conductive p^+-region.[12]

Now, aluminum is likewise used as a contact material for n-type silicon. To prevent a rectifying contact in this case, one usually lays down a heavily doped and shallow n^+-layer[12] on top of the n-region. Since this n^+-layer is highly conductive and is made to be very thin, tunneling through the barrier accomplishes the unhindered electron flow (see Sections 4.3 and 8.7.8).

8.7.4. $p–n$ Rectifier (Diode)

We learned in Section 8.7.2 that when a metal is brought into contact with an extrinsic semiconductor, a potential barrier may be formed which gives rise to the rectifier action. A similar potential barrier is created when a p-type and an n-type semiconductor are joined.

As before, electrons flow from the higher level (n-type) "down" into the p-type semiconductor so that the p-side is negatively charged. This proceeds until equilibrium is reached and both Fermi energies are at the same level. The resulting band diagram is shown in Fig. 8.18.

---

[12] The superscript plus means *heavily doped region*.

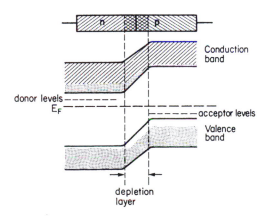

Figure 8.18. Schematic band diagram for a $p–n$ junction (diode) in equilibrium.

Consider first the conduction band only. The electrons that want to diffuse from the *n*-region into the *p*-region encounter a potential barrier near the junction. For statistical reasons, only a few of them have enough energy to climb the barrier and diffuse into the *p*-region. The electrons in the *p*-region, on the other hand, can easily diffuse "down" the potential barrier into the *n*-region. Note that only a few electrons exist in the conduction band of the *p*-region. (They have been thermally excited into this band by intrinsic effects.) In the equilibrium state the number of electrons crossing the junction in both directions is therefore identical. (The same is true for the holes in the valence band.)

When an external potential is applied to this device, effects similar to the ones described in Section 8.7.2 occur: connecting the positive terminal of a d.c. source to the *n*-side withdraws electrons and holes from the depletion area which becomes wider and the potential barrier grows higher (Fig. 8.19(a and b)). As a consequence, only a small drift current (from intrinsic effects) exists (**reverse bias**). On the other hand, if the *n*-side is charged negatively, the barrier decreases in height and the space charge region narrows. A large net electron flow occurs from the *n*-type region to the *p*-type region (**forward bias**, Fig. 8.19(c) and (d)).

In Fig. 8.19(a) and (c) "**quasi-Fermi levels**" for electrons and holes are shown. They are caused by the fact that the electron density varies in the junction from the *n*-side to the *p*-side by many orders of magnitude, while the electron current is almost constant. Consequently, the Fermi level must also be almost constant over the depletion layer.

It has to be emphasized that the current in a *p–n* rectifier is the sum of both electron and hole currents. The net current may be calculated by using an equation similar to (8.25) whereby the saturation current, I_S, in the present case is a function of the equilibrium concentration of the holes in the *n*-region (C_{hn}), the concentration of electrons in the *p*-region (C_{ep}), and other device parameters. The saturation current in the case of reverse bias is given by the **Shockley equation**, which is also called the **ideal diode law**:

$$I_S = Ae\left(\frac{C_{ep}D_{ep}}{L_{ep}} + \frac{C_{hn}D_{hn}}{L_{hn}}\right), \tag{8.26}$$

where the D's and L's are diffusion constants and diffusion lengths, respectively (e.g., D_{ep} = diffusion constant for electrons in the *p*-region, etc.). The diffusion constant is connected with the mobility, μ, through the **Einstein relation**:

$$D_{ep} = \frac{\mu_{ep}k_BT}{e} \tag{8.27}$$

(see textbooks on thermodynamics). The minority carrier diffusion length is given by a reinterpretation of a well-known equation of thermodynamics,

$$L_{ep} = \sqrt{D_{ep} \cdot \tau_{ep}}, \tag{8.28}$$

Figure 8.19. (a) + (b) Reverse and (c) + (d) forward biasing of a p–n junction (diode). (e) Symbol of a p–n rectifier in a circuit and designation of polarity in an actual rectifier. See also Figure 8.16(a).

where τ_{ep} is the lifetime of the electrons in the p-type region before these electrons are annihilated by recombination with holes. In order to keep the reverse current small, both C_{hn} and C_{ep} (minority carriers) have to be kept at low levels (compared to electrons and holes introduced by doping). This can be accomplished by selecting semiconductors having a large energy gap (see tables in Appendix 4) and by high doping.

8.7.5. Zener Diode

When the reverse voltage of a p–n diode is increased above a critical value, the high electric field strength causes some electrons to become accelerated to

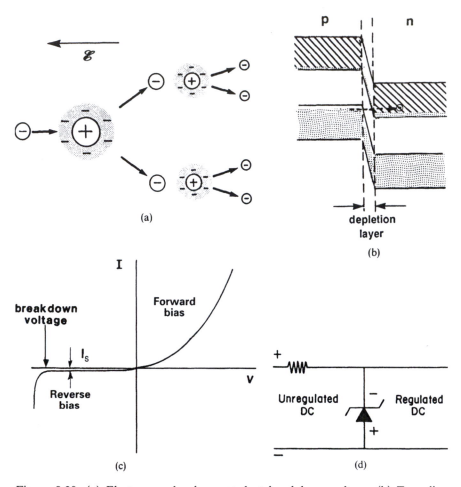

Figure 8.20. (a) Electron avalanche created at breakdown voltage. (b) Tunneling (Zener breakdown). (c) Voltage–current characteristic of a *p–n* diode exhibiting a breakdown voltage at a large reverse voltage. As in Fig. 8.16(a), I_s is shown grossly exaggerated. (d) Zener diode in a circuit for voltage regulation.

a velocity at which **impact ionization** occurs [Fig. 8.20(a)]. In other words, some electrons are excited by the electric field from the valence band into the conduction band, leaving behind an equal number of holes. The free electrons (and holes) thus created are likewise accelerated and create new electron–hole pairs, etc., until eventually a **breakdown** occurs, i.e., the reverse current increases quite rapidly (Fig. 8.20(c)). The breakdown voltage, which is the result of this **avalanching** process, depends on the degree of doping: the higher the doping, the lower the breakdown voltage. Alternatively to this avalanche mechanism, a different breakdown process may take place under certain conditions. It occurs when the doping is heavy and thus the barrier width becomes very thin (i.e., <10 nm). Applying a high enough reverse voltage causes the bands to shift to the degree that some electrons in the valence band of the *p*-side are opposite to empty states in the conduction band of the *n*-material. These electrons can then tunnel through the depletion layer, as described in Sections 4.3 and 8.7.8 and depicted in Fig. 8.20(b). **Tunneling** (or **Zener breakdown**) takes place usually at low reverse voltages (e.g., below about 4 volts for silicon-based diodes), whereas avalanching is the mechanism that occurs when the reverse voltage is large.

The breakdown effect just described is used in a circuit to hold a given voltage constant at a desired level (Fig. 8.20(d)). The Zener diode is therefore utilized as a circuit protection device. The Zener diode is generally not destroyed by the breakdown, unless excessive heat generation causes it to melt.

8.7.6. Solar Cell (Photodiode)

A photodiode consists of a *p–n* junction (Fig. 8.21). If light of sufficiently high energy falls on or near the depleted area, electrons are lifted from the valence band into the conduction band, leaving holes in the valence band. The electrons in the depleted area immediately "roll down" into the *n*-region, whereas the holes are swept into the *p*-region. These additional carriers can be measured in an external circuit (photographic exposure meter) or used to generate electrical energy. In order to increase the effective area of the junction, the *p*-type region is made extremely thin (1 μm) and light is radiated through the *p*-layer (Fig. 8.21(a)). Since the *p*-layer is thin, the electric energy must be collected on the front surface, utilizing narrow metal electrodes (e.g.,

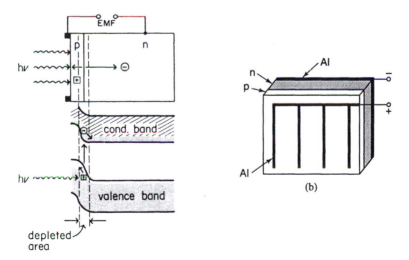

Figure 8.21. Solar cell; the *p*-region is only about 1 μm thick. (a) side view; (b) Front view.

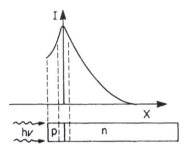

Figure 8.22. Schematic representation of the contribution of electrons and holes to the photocurrent (I) with respect to the distance x from the p–n junction.

Al) which are arranged in the form of stripes, see Fig. 8.21(b). A silicon photovoltaic device yields an inherent voltage of 0.6 V.

The electron–hole pairs that are created some distance away from the depleted region are generally not separated by the junction field and eventually recombine; they do not contribute to the electric current. However, some electrons or holes which are within a diffusion length from the depleted region drift into this area and thus contribute to the current. In semiconducting materials that contain only a few defects (such as grain boundaries, dislocations, and impurities) the electrons or holes may diffuse up to 200 μm before they get trapped, whereas in semiconducting materials containing a large number of defects the diffusion length decreases to 10 μm. The closer a carrier was created to the p–n boundary, the larger is its chance of contributing to the current (Fig. 8.22).

The thin p-type layer introduces an internal resistance to the collection current, which reduces the **efficiency** of the energy conversion. At present, the maximal efficiency of a photovoltaic device, made of crystalline silicon and involving a three-layer technology (see Part III), is about 20–36%. The energy needed to produce such a device (including mounting and installation) is recovered in about 6 years when the collector is located in North Africa or Central America. (Installation in central Europe or the northern states of the USA and Canada may double the energy recovery time.) The **cost** of photovoltaic devices (presently $8–$10 per installed watt) can be reduced by utilizing polycrystalline, less purified, or amorphous silicon, but at the expense of efficiency. As an example, photovoltaics made of commerical, hydrogen-doped amorphous silicon (see Section 9.4) have an efficiency of only 6–8%, but its invested energy for production and mounting is recovered in just 1 year. The efficiency of this device has been enhanced to 12% in laboratory experiments. The goal is to produce for terrestrial applications inexpensive solar cells having 20% efficiency or better and a lifetime of about 20 years. The lifetime is reduced when the metal contacts (grids) to the semiconductor corrode. Despite the fact that photovoltaics are still relatively inefficient, their worldwide sale has grown for the past 10 years by more

Figure 8.23. Schematic of a transverse-type photodiode that is connected to a light-carrying medium such as an optical fiber or a waveguide ($L \approx 100\,\text{nm}$).

than 15% per year and has reached now the \$2 billion mark, while the cost has steadily decreased. The most recent development employs dye-coated titanium dioxide and an electrochemical cell which mimics the role of chlorophyll in photosynthesis.

The photovoltaic cell depicted in Fig. 8.21 has one inherent disadvantage: the impinging light has to travel first through the p-type layer (however thin it may be) before it eventually reaches the depleted (active) area. This attenuates its intensity to a certain degree. In addition, the incoming light is somewhat blocked by the metal electrodes, which cover part of the face of the cell. The resulting loss in efficiency is a trade-off for a large surface area (which is often desirable to increase power). For telecommunication applications however, for which high *efficiency* is more important, a rather ingenious alternative design can be used. Imagine that the light impinges *transversely* on (or better, *along*) the depletion layer. For this the beam is channeled-in from the *side* by a light-conducting device such as an optical fiber or a wave guide (Fig. 8.23). In order to increase the effective area, i.e., the width, W, of the depletion region, the photodiode is strongly reverse-biased and the doping of one of the semiconductors is comparatively light. (For details refer to Fig. 8.19(a).) The efficiency is further maximized by increasing the length of the depletion layer, L. This device yields almost 100% quantum efficiency.

The **quantum efficiency** can be calculated by the equation

$$\eta = 1 - \frac{\exp(-\alpha W)}{1 + \alpha L}, \tag{8.29}$$

where α is a parameter that determines the degree of photon absorption by the electrons (α is defined in (10.21a)). As an example, for a GaAs photodiode the n-region is lightly doped because the *electron* mobility in GaAs is much larger than the hole mobility, see Appendix 4. This shifts the depleted region towards the n-side. On the other hand, the p-region is *heavily* doped (and thin) in order to minimize its resistance.

The incoming light that is modulated by information (such as the spoken word in telecommunications) modulates, in turn, the electrical current in the photodiode. This transforms a signal which is transmitted by light into an electrical signal. We shall return to this topic and to other *optoelectronic devices* in Part III.

*8.7.7. Avalanche Photodiode

This device is a *p–n* photodiode that is operated in a high reverse bias mode, i.e., at near-breakdown voltage. The electrons and holes that were created by transitions from the valence band into the conduction band by the incident light are accelerated through the depleted area with a high velocity. As a consequence, they ionize the lattice atoms and generate secondary hole–electron pairs, which, in turn, are accelerated, thus generating even more hole–electron pairs. The result is a photocurrent gain, which may be between 10 and 1000. The avalanche photodiode is ideally suited for low-light-level applications, because of its high signal-to-noise ratio, and for very high frequencies (GHz). It is particularly used for detectors in long-distance, fiber-optics telecommunication systems. See in this context Fig. 8.23.

*8.7.8. Tunnel Diode

So far, we have restricted our discussion mostly to the case for which the electrons drift from the *n*-type to the *p*-type semiconductor by way of "climbing" a potential barrier. Another electron transfer mechanism is possible, however. If the depleted area is very narrow (approximately 10 nm) and if certain other requirements (see below) are fulfilled, electrons may tunnel *through* the potential barrier. (See in this context Fig. 4.7, Fig. 8.20(b), and equation (4.39).) Heavy doping (e.g., 10^{20} impurity atoms per cubic centimeter) yields this condition.

The situation can best be understood by inspecting Fig. 8.24(a), in which a schematic band diagram of a tunnel diode is shown. Because of the high doping level, the Fermi energy extends into the valence band of the *p*-type semiconductor and into the conduction band of the *n*-type semiconductor. In the equilibrium state, the same amount of electrons is tunneling through the potential barrier in both directions, i.e., no net current flows.

If a small *reverse* bias is applied to this device (Fig. 8.24(b)), the potential barrier is increased as usual and the Fermi energy, along with the top and bottom of the bands in the *p*-area, is raised. This creates empty electron states in the conduction band of the *n*-type semiconductor opposite from filled states in the valence band of the *p*-type semiconductor. As a consequence, some electrons tunnel from the *p*-type to the *n*-type semiconductor, as indicated by an arrow. An increase in the *reverse* voltage yields an increase in the electron current through the device (see Fig. 8.24(f)).

Let us now consider several forward voltages. A small forward bias (Fig.

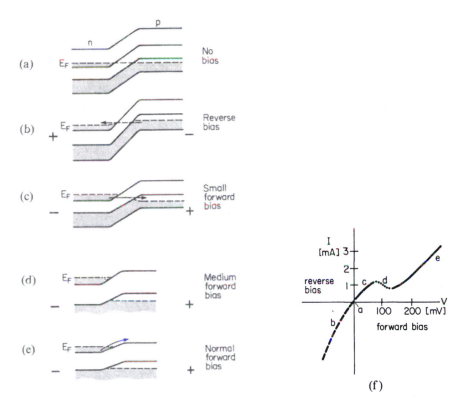

Figure 8.24. (a)–(e) Schematic energy band diagrams for highly doped *n*- and *p*-type semiconductors (tunnel diode). (a) No bias. (b) Reverse bias. (c) Small forward bias. (d) Medium forward bias. (e) "Normal" forward bias. (f) Voltage–current characteristic for a tunnel diode.

8.24(c)) creates just the opposite of that seen in Fig. 8.24(b). Electrons are tunneling through the potential barrier from the conduction band of the *n*-type semiconductor into empty states of the valence band of the *p*-type semiconductor. The applied voltage needs to be only several millivolts and it produces a forward current of about one milliamp.

If, however, the voltage is increased to, say, 100 mV, the potential barrier might be decreased so much that, opposite to the filled *n*-conduction states, no allowed empty states in the *p*-area are present [Fig. 8.24(d)]. (The area opposite to the filled *n*-conduction states may be the forbidden band.) In this case, no tunneling takes place. As a consequence of this, the current decreases with increasing forward voltage, as shown in Fig. 8.24(f). We experience a *negative current–voltage characteristic*.

Finally, if the forward voltage is increased even more, the electrons in the conduction band of the *n*-type semiconductor obtain enough energy to climb the potential barrier to the *p*-side just as in a regular *p–n* junction. As a consequence, the current increases with voltage, just as in Fig. 8.16(a).

Of particular interest is the range in which a negative voltage–current characteristic is experienced. One has to bear in mind that all other electrical devices have a positive voltage–current characteristic, i.e., they dissipate energy. Therefore, if a tunnel diode is connected to properly dimensioned resistors and capacitors, a simple oscillator can be built which does not lose energy because the net resistance is zero. Those devices can oscillate at frequencies up to 10^{11} cycles per second.

8.7.9. Transistors

Bipolar Junction Transistor. An *n–p–n* transistor may be considered to be an *n–p* diode back-to-back with a *p–n* diode. A schematic band diagram for an unbiased *n–p–n* transistor is shown in Fig. 8.25. The three connections of the transistor are called emitter (E), base (B), and collector (C).

If the transistor is used for the amplification of a signal, the "diode" consisting of emitter and base is forward biased, whereas the base–collector "diode" is strongly reverse biased (Fig. 8.26(a)). The electrons injected into the emitter, therefore, need to have enough energy to be able to "climb" the potential barrier into the base region. Once there, the electrons diffuse through the base area until they have reached the depletion region between base and collector. Here, the electrons are accelerated in the strong electric field produced by the collector voltage (Fig. 8.26(b)). This acceleration causes amplification of the input a.c. signal.

One may consider this amplification from a more quantitative point of view. The forward biased emitter–base diode is made to have a small resistivity (approximately 10^{-3} Ω cm), whereas the reverse biased base–collector diode has a much larger resistivity (about 10 Ω cm). Since the current flowing through the device is practically identical in both parts, the power ($P = I^2R$) is larger in the collector circuit. This results in a power gain.

Figure 8.25. Schematic band diagram of an unbiased *n–p–n* bipolar junction transistor.

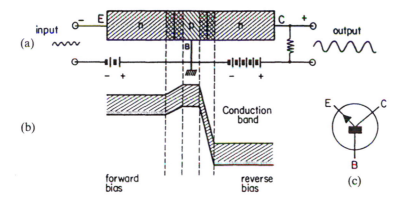

Figure 8.26. (a) Biasing of an *n–p–n* bipolar transistor. (b) Schematic band diagram (partial) of a biased *n–p–n* bipolar transistor. (c) Symbol used for a bipolar *n–p–n* transistor.

The electron flow from emitter to collector can be controlled by the bias voltage on the base: a *large* positive (forward) bias decreases the potential barrier and the width of the depleted region between emitter and base (Fig. 8.19). As a consequence, the electron injection into the *p*-area is relatively high. In contrast, a *small*, but still positive base voltage results in a comparatively larger barrier height and in a wider depletion area, which causes a smaller electron injection from the emitter into the base area. In short, the voltage applied between emitter and base modulates the transfer of the electrons from the emitter into the base area. As a consequence, the strong collector signal mimics the waveform of the input signal. This feature is utilized for the amplification of music or voice, etc.

In another application, a transistor may be used as an electronic switch. The electron flow from emitter to collector can be stopped completely (or turned on) by an appropriate base voltage. This virtue is used for logic and memory functions in computers (see Section 8.7.12).

The device shown in Fig. 8.26 is called a "bipolar transistor"; the current passes in series through *n*-type as well as through *p*-type semiconductor materials.

Some details need to be added about technical features of the bipolar transistor. In order to obtain a large electron density in the emitter, this area is heavily doped. In the *p*-doped base area, the drifting electrons are subject to possible recombination with holes. Therefore, the number of holes there has to be kept to a minimum, which is accomplished by light doping. (Light doping also reduces the unwanted injection of hole current into the base.) Recombination is further decreased by making the base region extremely

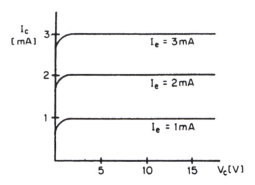

Figure 8.27. Schematic collector voltage–current characteristics of a transistor for various emitter currents. I_c = collector current, I_e = emitter current, and V_c = collector voltage.

thin, i.e., 10^{-5}–10^{-7} m. A narrow base region has a beneficial side effect: it increases the frequency response. (The reciprocal of the electron transit time equals the highest possible frequency at which amplification can be achieved.) The doping rate of the collector area is in general not critical. Usually, the doping is light for high gain and low capacitance of the device. The voltage–current characteristics for a transistor are shown in Fig. 8.27.

In p–n–p transistors, the majority carriers are holes. The function and features of a p–n–p transistor are similar to an n–p–n transistor.

Metal–Oxide–Semiconductor Field-Effect Transistor (MOSFET). A field-effect transistor consists of a *channel* through which the charge carriers (e.g., electrons in Fig. 8.28) need to pass on their way from a *source* (S) to the *drain* (D). The conducting path (source, channel, and drain) is made of the same kind of semiconducting material only, e.g., n-type. (This is in contrast to the *bipolar* transistor shown in Fig. 8.26, in which the current passes in series through n-type as well as through p-type semiconductor materials.) Field-effect transistors are therefore designated as *unipolar*. The electrons that flow from the source to the drain can be controlled by an electric field which is established by applying a voltage to the so-called *gate* (G).

A periodic variation of the gate voltage varies the source to drain current in the same manner (quite similar to the way the electron flow between emitter and collector in a bipolar transistor is modulated by the base voltage). The gate electrode is electrically insulated from the channel by a thin oxide layer which prevents a d.c. current to flow from gate to channel.

Two types of MOSFETs are common: The depletion-type MOSFET depicted in Fig. 8.28(a) consists of high-doped source and drain regions and a low-doped channel, all of the same polarity (e.g. n-type). (The high doping facilitates low-resistance connections.) The n-channel MOSFET is laid down on a p-type substrate called the body.

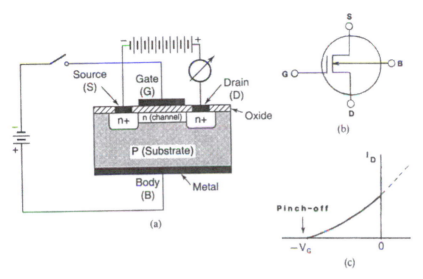

Figure 8.28. (a) Schematic representation of an *n*-channel depletion- (normally on) type MOSFET. The dark areas symbolize the (aluminum) metallizations. The "oxide" layer may consist of SiO_2, nitrides (Si_3N_4), oxinitrides ($Si_3N_4-SiO_2$), or multilayers of these substances. This layer is about 10 nm thick. The gate voltage is applied between terminals G and B. Quite often the B and S terminals are interconnected. (b) Circuit symbol for *n*-channel depletion-type MOSFET. (c) Gate voltage/Drain current characteristic ("Transfer" characteristic). For positive gate voltages (dashed portion of the curve) the device can operate in the "enhancement mode" (see Fig. 8.29(c)).

The channel width is controlled by the voltage between gate and body. Specifically, a negative charge on the gate drives the channel electrons away from the gate and towards the substrate, similarly as is illustrated in Fig. 8.12. In short, the channel can be made to be partially depleted of electrons, i.e., the conductive region of the channel becomes narrowed by a negative gate voltage. The more negative the gate voltage (V_G), the smaller the current through the channel from source to drain until eventually the current is *pinched off* (see Fig. 8.28(c).) For the above reasons, this device is called a *depletion-type metal-oxide semiconductor field-effect transistor* or "**normally on**" MOSFET.

An alternative to the depletion-type MOSFET that we just discussed is the **enhancement-type MOSFET**. Figure 8.29 shows that this device does not possess a built-in channel for electron conduction, i.e., at least as long as no gate voltage is applied. In essence, there is no electron flow from source to drain for a zero gate voltage. The device is therefore called a "**normally-off**" MOSFET. If, however, a large enough positive voltage is applied to the gate, most of the holes immediately below the gate oxide are repelled, i.e., they are driven into the substrate, thus removing possible recombination sites. Concomitantly, negative charge carriers are attracted into this channel (called

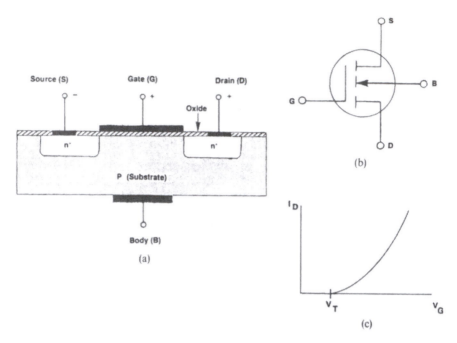

Figure 8.29. (a) Enhancement (normally-off)-type n-channel MOSFET. For details, see the caption of Fig. 8.28. (b) Circuit symbol. (The broken line indicates that the path between S and D is normally interrupted.) (c) Gate voltage (V_G)/drain current (I_D) characteristic. V_T is the threshold gate voltage above which a drain current sets in.

the *inversion layer*). In short, a path (or a bridge) for the electrons between source and drain can be created by a positive gate voltage. The metal-oxide semiconductor technology, particularly, the enhancement-type MOSFETs, dominate the integrated circuit industry at present. They are utilized in memories, microcomputers, logic circuits, amplifiers, analog switches, and operational amplifiers. They possess very high input impedances,[13] thus minimizing Joule heating.

Depletion-type and enhancement-type MOSFET technologies that utilize n-channels (as depicted in Figs. 8.28 and 8.29 are summarized by the name "NMOSFET" (in contrast to "PMOSFET", which employs devices with p-channels). If both an n-channel and a p-channel device are integrated on one chip and wired in series, the technology is labeled "**CMOSFET**" which stands for *complementary MOSFET*. This tandem device has become the

---

[13] The term *impedance* is used to describe the a.c. resistance, which may consist of ohmic, capacitive, and inductive parts.

dominant technology for information processing, because of its low operating voltage (0.1 V), low power consumption (heat!), and short channel length with accompanying high speed. Alternative names for MOSFET are MOST (metal–oxide–semiconductor transistor) or MISFET (metal–insulator–semiconductor field-effect transistor).

A few words on device geometry, etc., of a MOSFET, as shown in Fig. 8.28, may be useful. In order to obtain a short switching time and a high-frequency response, the channel length has to be short. The highest possible frequency at which amplification can be achieved equals the inverse of the electron source-to-drain transit time. The width of the device has to be kept small in order to reduce the cross-sectional area and, thus, the power density. (This reduces the heat which needs to be removed.) As an example, the channel length may be about 1 μm, the device width may be a few micro-meters, and the field oxide thickness may be near 0.05 μm. The doping of the p-area needs to be small to sustain a high resistance and thus, a high electric field ($\sim 10^6$ V/cm) across the junction without current breakdown. The metal layer is generally made of aluminum. Alternate materials are highly doped silicon, refractory metals such as tungsten, or silicides of refractory metals such as TiSi or MoSi.

*Junction Field-Effect Transistor (JFET). The JFET consists again of a channel through which the carriers (electrons in Fig. 8.30) pass from source to drain. This electron flow is controlled by an electric field which is established by applying a negative voltage to the p-doped gate, to stay within the example of Fig. 8.30. In other words, the p–n gate-to-channel diode is

Figure 8.30. (a) Schematic representation of an n-channel junction field-effect transistor. The dark areas symbolize the metal contacts (e.g., aluminum). (b) Circuit symbol for an n-channel JFET. *Note:* In a p-channel JFET the arrow points away from the channel.

reverse biased. This reverse biasing increases the width of the depletion layer (see Fig. 8.19) thus causing the conducting channel to become narrower. (Close to the drain terminal, the *p–n* junction is more reverse biased which results in a wider depletion layer near the drain.) A zero bias voltage on the gate results in a maximal source-to-drain current. A reverse voltage on the gate *depletes* the source-to-drain electron flow. A very large reverse current eventually *pinches* the current *off*. Junction field-effect transistors are therefore said to be of the *depletion* or "*normally-on*" type.

Junction field-effect transistors can be used as amplifiers, exploiting the effect that a small change in the gate voltage causes a large change in the channel current. Since the gate-to-channel *p–n* junction is reverse biased, only a minute current flows in the gate/source circuit (Fig. 8.16). The input impedance[13] is therefore high (but not as high as in a MOSFET).

JFETs which use *n*-type semiconductors for the channel material, as depicted in Fig. 8.30, are appropriately called *n-channel field-effect transistors*. The reader may correctly suspect that a *p-channel field-effect transistor* uses holes as charge carriers, *n*-type semiconductors as gate materials, and a reversal of the polarities of all voltages for its operation. The arrow in the circuit symbol (Fig. 8.30(b)) for *p*-channel transistors points away from the gate.

Bipolar transistors in combination with JFETs are called "BIFETs." They are used in high-performance linear circuits. If a JFET structure employs a metal–semiconductor junction, often in combination with *n*-type GaAs, a "MESFET" device is created, which is used for amplifiers and logic circuits in the gigahertz range (see next section).

A MODFET (modulation-doped field-effect transistor) consists of a thin layer of aluminum–gallium–arsenide deposited on an undoped GaAs substrate. This device is even faster than a MESFET, because the absence of impurity atoms increases the distance that an electron or a hole can travel before a collision with a foreign atom occurs.

*Gallium Arsenide Metal–Semiconductor Field-Effect Transistor (MESFET). Users of computers demand still higher switching speeds than the present 10^{-9} s cut-off or cut-on times achieved with silicon technology. Gallium arsenide, with its almost sixfold larger electron mobility compared to silicon (see Appendix 4), seems to be the answer. A quick inspection of the relevant band diagrams (Figs. 5.23 and 5.24) indeed confirms that the curvature of the conduction band near Γ is larger for GaAs than the comparable band for silicon (close to the X symmetry point) which translates into a smaller effective mass and, thus, into the just-mentioned larger electron mobility for GaAs. However, the *upper valence* bands for both materials are almost identical and fairly flat. Thus, the effective masses of the *holes* for GaAs and silicon are rather large and their hole mobilities are consequently small (see also Appendix 4). A transistor that aims to exploit the higher electron mobility in GaAs should therefore utilize *n*-type GaAs only.

Figure 8.31. Schematic representation of a GaAs MESFET (Metal–semiconductor field-effect transistor). Source and drain metallizations (dark areas) are selected to form ohmic contacts with the *n*-doped GaAs. The gate metal forms, with the *n*-doped GaAs, a Schottky-barrier contact.

Figure 8.31 depicts a metal–semiconductor field-effect transistor (MESFET), which consists of an *n*-doped, thin GaAs *active layer* situated over a semi-insulating (Cr-doped) GaAs slab. Three metal contacts provide the source, the gate, and the drain areas. The gate metal forms, together with the underlying semiconductor, a Schottky barrier (see Section 8.7.2). If ϕ_M is larger than ϕ_S and the gate metal is negatively charged, a reverse bias results (Fig. 8.15(a)). The larger the reverse bias, the wider the depletion region. If the depletion region is caused to fill essentially the entire active layer, any attempted electron flow from source to drain is stopped (or *pinched off*). A small negative gate voltage (or no gate voltage at all) allows an almost unhindered source-to-drain electron flow. The device shown in Fig. 8.31 is therefore a *depletion-* (or *normally-on*) *type* FET (see also Fig. 8.28(c)).

For high-speed, low-power applications, however, the *normally-off* GaAs MESFET is even better suited. For this device, the active layer is made so thin that the depletion area between the metal and the GaAs (Fig. 8.15) fills the entire active layer.[14] As a consequence, the active layer below the gate metal electrode is depleted of electrons without necessitating an applied voltage. A *positive* gate voltage is then required to attract electrons into the depletion area, thus making it conductive. Given the above-described GaAs device, the speed, i.e., the response time of the source-to-drain current to a change in the gate voltage, can be further increased by decreasing the length of the gate, which is presently about 1 μm.

Several effects may, however, offset the superior electron mobility in GaAs. First, the time required to reach the breakdown voltage under the influence of a reverse voltage (see Fig. 8.20(c)) is only two and a half times faster than in silicon. As we know from Fig. 8.20(a), this breakdown electric field triggers a helpful self-ionizing avalanche that multiplies the number of

---

[14] The depletion layer width in GaAs varies with impurity concentration between 3 μm for 10^{14} cm^{-3} and 0.05 μm for 10^{18} impurity atoms per cubic centimeter.

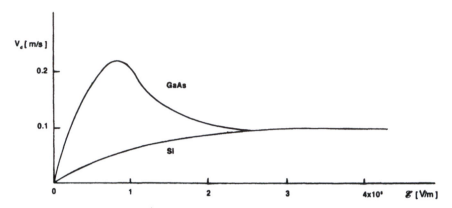

Figure 8.32. Average electron drift velocity as a function of electric field strength for GaAs and silicon.

electrons. Second, a transistor of any type can be made to switch faster by applying more power to it. This, in turn, increases the heat which needs to be dissipated. Now, silicon has a three-times larger thermal conductivity than GaAs (see Appendix 4). Thus, silicon switches can be made much smaller than those made of GaAs. Since the speed of a device also depends on the length the electrons have to travel, a very small silicon device may well switch as fast as a large device made of GaAs. Third, the electron drift velocity depends upon the electric field strength. At low field strengths, the GaAs drift velocity is indeed substantially larger than for silicon (Fig. 8.32). However, as the field strength increases, the drift velocity for silicon and GaAs becomes nearly identical. This has its reason in the extra and slightly higher energy states that silicon possesses near the X-symmetry point (Fig. 5.23), in which electrons can be scattered after they have collided with structural imperfections of the crystal lattice.

Knowing the facts presented above, it seems understandable why some leading semiconductor manufacturers have left the GaAs field. However, the pendulum may soon swing in the other direction, as suggested in the next section.

*8.7.10. Quantum Semiconductor Devices

It is the ultimate goal of industry to make semiconductor switches for computer applications as small, as fast, as inexpensive, and as efficient as possible. Conventional field-effect transistors pose, ultimately, certain limitations towards progressive miniaturization: the smaller they become, the less effective they switch, owing to current leakage, and particularly because of impurities or lattice defects that scatter the moving electrons in ultrasmall devices to an intolerable degree. There are also processing limitations caused by the pres-

(a)

(b)   (c)

Figure 8.33. (a) Schematic representation of a quantum dot structure. (b) Energy levels for GaAs for the quantum dot structure depicted in (a). (*Note*: The gap energy difference between GaAs ($E_g = 1.42$ eV) and AlGaAs is greatly exaggerated. This difference may be as small as 0.2 eV.) (c) Discontinuous density of energy states for a quantum dot structure. The dashed parabola indicates the density of states for a bulk crystal, as is known from Fig. 6.4.

ently used photolithography techniques. Quantum structures are said to be the devices of the future that may overcome these shortcomings.

In order to explain the nature of a quantum device, we need first to recall that the electron states for bulk crystalline solids consist of continuous energy bands, such as the valence band or the conduction band (Fig. 8.2). We also recall that the density-of-states curve has a parabolic shape in this case (Fig. 6.4). If, however, the dimensions of a crystalline solid are reduced to the size of the wavelength of electrons (e.g., 20 nm for GaAs), the formerly continuous energy bands split into discrete energy levels, similarly as is known from Section 4.2, where we treated the behavior of *one* electron in a potential

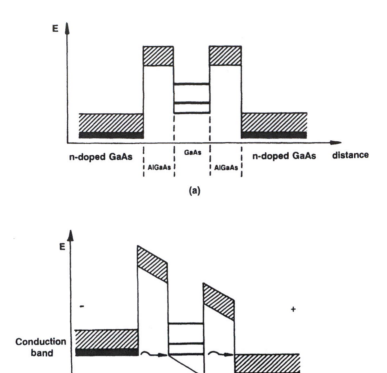

Figure 8.34. Parts of two energy band structures for the quantum device shown in Fig. 8.33. For simplicity, only the conduction bands are shown. (a) No applied voltage. (b) With applied voltage, which facilitates electron tunneling from the conduction band of the n-doped GaAs into an empty energy level of the center GaAs region.

well. In essence, the same type of calculation presented in Section 4.2 is carried out for quantum devices. Thus, results equivalent to (4.18) are obtained. Further, when the dimensions are reduced to the degree as outlined above, and under certain other conditions (see below), the density of states becomes discontinuous, i.e., $Z(E)$ also becomes quantized (see Fig. 8.33(c)). The mechanism associated with these effects is, therefore, quite appropriately called *size quantization*.

Let us demonstrate size quantization for a particular case in which a small–band gap material is sandwiched between two layers of a "wide"– band gap material. Specifically, a cube-shaped piece of GaAs whose lateral dimensions are made to be about 20 nm is layered between two similarly shaped cubes made of aluminum–gallium–arsenide, which in turn are sandwiched between two longer slabs consisting of n-doped GaAs (Fig. 8.33(a)). This configuration, for which all three dimensions of the center materials

have values near the electron wavelength, is called a **quantum dot** (in contrast
to a two-dimensional "confinement," which is termed **quantum wire**, or a
one-dimensional confinement, named **quantum well**).

Figure 8.33(b) depicts simplified electron bands for the quantum dot
structure shown in Fig. 8.33(a). AlGaAs is a "wide"–band gap material
whose electron affinity (Fig. 8.13) is smaller than that of GaAs. Thus, its
conduction band is at a higher energy compared to the conduction band of
GaAs. This results in a *potential barrier* between the two GaAs regions. In
general, an electron in the n-doped GaAs area does not possess enough energy for climbing this potential barrier or otherwise diffusing into the adjacent regions (Fig. 8.34(a)). If, however, a sufficiently large voltage is applied
to this device, the conduction band of the n-doped GaAs is raised to a level
at which its conduction electrons are at the same height as an empty energy
state of the center GaAs region (Fig. 8.34(b)). At this point the electrons are
capable of tunneling through the potential barrier formed by the AlGaAs
region and thus reach one of these discrete energy levels. The tunneling is
quite effective because of the large density of states that is associated with
these quantum states (Fig. 8.33(c)).

If a slightly higher (or somewhat smaller) voltage is applied, the electrons
of the n-doped GaAs are no longer at par with an empty energy level and the
tunneling comes to a near standstill. This causes a current–voltage characteristic with negative differential resistance, i.e., a region in which the current
decreases as the applied voltage increases (see Fig. 8.35).

An interrelated effect to size quantization is *resonance*, which enhances the
tunneling current. Once a specific voltage, the resonating voltage, has been
reached, the electron waves inside the center region are reflected back and
forth between the walls. In essence, constructive interference occurs between
the waves traveling in opposite directions.

A further advancement of the quantum device introduced so far consists
of an *array* of a multitude of quantum wells stacked on top of each other.
This periodic arrangement of wide–band gap and narrow–band gap materials is called a *superlattice*. It introduces an artificial periodicity into the

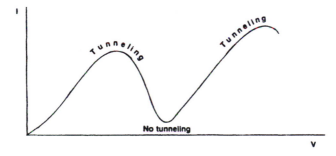

Figure 8.35. Current–voltage characteristic of a quantum dot device as depicted in
Figs. 8.33 and 8.34.

solid, caused by the multiple atomic layers of one type of material in sequence with multiple atomic layers of another type. By this mode of varying the structural parameters of a solid, new electronic properties can be engineered.

Quantum devices are about one-hundredth of the size of presently known FETs. Thus, major problems have still to be overcome concerning interconnections, device architecture, and fabrication of three-terminal devices. It has been speculated, however, that once these problems have been solved the reduction in cost per function might be as large as ten-thousand-fold.

8.7.11. Semiconductor Device Fabrication

The evolution of solid-state microelectronic technology started in 1947 with the invention of the germanium point contact transistor by Bardeen, Brattain, and Shockley at Bell Laboratories. Until then, electronic devices used vacuum tubes invented in 1906 by Lee deForest, as well as silicon, copper oxide, or germanium rectifiers. (The latter was discovered in 1915 by M. Benedicks). The development went via the *germanium junction transistor* (Shockley, 1950), the *silicon transistor* (Shockley, 1954), the first *integrated circuit* (Kilby, Texas Instruments, 1959), the *planar transistor* (Noyce and Fairchild, 1962), and the *planar epitaxial transistor* (Texas Instruments, 1963) to the *ultra-large-scale integration* (ULSI) of today with several millions of transistors on one chip. Attempts are now made to reach one billion transistors per chip, called *gigascale integration* (GSI). We have discussed in the previous sections some obstacles to this goal, which are imposed to a large degree by the *"materials barrier."* (However, device limits, circuit limits, and system limits likewise play a role.) Silicon has been the principal semiconductor material used in the past 50 years even though solid-state electronics technology actually started with *germanium*, which could be manufactured in these early days in comparatively ultrapure form. No other electronic material has a combination of so many favorable properties. Most of all, silicon is abundant; 28% of the earth's crust consists of silicon in one way or another. (Silicon is behind oxygen, the second most abundant chemical element.) The raw material (sand, i.e., quartzite) is inexpensive. The native oxide, silicon dioxide (SiO_2), is an excellent insulator. The band gap is large enough to guarantee stable electrical properties at moderate temperatures. The heat conductivity is relatively large. Further, silicon forms almost perfect (dislocation-free) single crystals. And finally, silicon is nontoxic, i.e., environmentally safe. Still, for special applications and possibly gigascale integration, compound semiconductors need to be considered, as discussed in the previous chapters.

The starting material for silicon wafer fabrication is sand (SiO_2), which is electromet reduced (in an arc furnace) with coal, etc., to 98% silicon. This powdered raw silicon is reacted with hydrogen chloride to form trichlorosilane gas ($Si + 3HCl \rightarrow SiHCl_3 + H_2$), which is fractionally distilled for purification and subsequently reduced with hydrogen to polycrystalline silicon

(a)

(b)

Seed crystal

Furnace Travel

As

Ingot

618°C

1250°C

(c)

Figure 8.36. Techniques for single-crystal growth. (a) Czochralski method. Heating is performed by radio frequency coils or (for big crucibles) by resistance heating. (b) Float zone method. (c) Bridgman method (demonstrated for GaAs). (d) A 300 mm (12 inch) silicon single crystal is removed from the crucible. (Courtesy Wacker Siltronic AG)

($SiHCl_3 + H_2 \rightarrow Si + 3HCl$). From here on, several methods for single crystal growth are used. In the predominantly utilized crystal pulling process, invented in 1918 by J. **Czochralski**, the high-purity silicon is melted in a fused-silica (SiO_2) crucible, which is, in turn, supported by a carbon crucible (Fig. 8.36(a)). A seed crystal (mainly (100) or (111) orientation), held on a rod, initially touches the melt and is then slowly lifted, employing a with-

(d)

Figure 8.36. (*Continued*)

drawal speed of about 1 mm per minute. Concomitantly, the crucible as well as the pulling rod are rotated in opposite directions at about 50 revolutions per minute. The entire system is enclosed in a chamber that is either slightly evacuated (a few Torrs) or backfilled with argon or helium. The starting crystal must initially have a thin neck to produce a dislocation-free crystal (invented in 1959 by W. Dash). Proper cooling and pulling speeds allow one to control the diameter of the evolving single crystal rod. Specifically, the initial pulling speed needs to be large so that the dislocations are frozen-in and thus cannot propagate further into the single-crystal rod.

Since the crucible consists of SiO_2 and of carbon, some oxygen and carbon are introduced into the silicon during melting (about 5×10^{17} oxygen atoms and about 2×10^{16} carbon atoms per cubic centimeter). Other foreign elements of high-purity silicon are generally in the 10^{10}–10^{13} per cubic centimeter range. Oxygen and carbon impurities are electrically inactive because they form inert compounds with silicon (e.g., SiO_2 or SiC). However, their presence in high concentrations leads to the premature breakdown of *p–n* junctions. Harmful impurities and tiny defects can be trapped (*gettered*) either at a specially prepared back side of the wafer (e.g., by mechanically introduced dislocations) or inside the crystal on very small SiO_2 precipitates. All taken, the surface layer (several μm thick) on which the transistor is manufactured (Fig. 8.29) needs to be free of oxygen atoms, whereas inside the wafer a high defect density is beneficial for gettering. This configuration is achieved by heating a wafer near 1000°C, which causes the migration of the mobile oxygen atoms to the surface where a large number of them are removed through evaporation.

A lower oxygen concentration (10^{16} atoms/cm^3) can be achieved involving the crucibleless **float-zone technique** (Fig. 8.36(b)). At first, a pure, poly-crystalline silicon rod is manufactured by resistive-heating a silicon filament (8 mm wide, 2 m long) in a trichlorosilane and hydrogen atmosphere. As mentioned above, this reduces the $SiHCl_3$ to silicon, which is slowly de-posited on the 1000°C hot silicon filament. The polysilicon rod thus grown is vertically inserted into a vacuum chamber with a single crystalline seed crystal at its bottom, and then rotated. An induction-heated ring-shaped furnace is slowly moved along the rod, which melts, at first, a part of the seed crystal, and then consecutive small zones (a few cm long) from the bottom up, thus eventually forming a large single crystal as an extension of the seed crystal. The float-zone technique is also used for purification pur-poses (**zone refining**). Wafers produced by this method are substantially more expensive than Czochralski wafers with the added disadvantage that less oxygen gettering can take place. However, float-zone single crystals are a necessity when the whole wafer thickness is required for the electrical func-tion of discrete transistors, such as for high-power applications.

The **Bridgman technique** is rarely used for silicon production. It is, how-ever, frequently applied to grow single-crystalline GaAs. The Bridgman method involves the melting of polycrystalline material in a long (silicon–nitride coated) carbon crucible or fused quartz crucible which, in turn, is placed into a horizontally arranged, sealed quartz tube. A traveling furnace with two different heating zones melts the ingot as well as part of a single-crystal seed which is placed next to it. In the case of GaAs (Fig. 8.36(c)), some extra arsenic, located in the low-temperature (618°C) part of the tube, provides an overpressure of arsenic to maintain stoichiometry. Moving the hot zone of the furnace slowly away from the melt causes gradual solidifi-cation and eventually an extension of the single crystal into the entire rod.

The two-zone furnace is also used to melt separately arsenic and gallium in individual boats, which facilitates the synthesis of GaAs over a period of many hours.

Once the rods have been obtained, they are sliced, lapped, etched, and polished to obtain the 0.3–0.4 mm thick wafers. At present, up to 200 mm (8 in.) diameter silicon wafers are commercially available; the trend goes, however, towards the 300 mm (12 in.) disc, see Fig. 8.36(d).

Next, the devices are fabricated on (or in) these wafers in extremely clean rooms, applying surface oxidation, photolithography, etching, and (most of all) by introducing various dopants involving successive and often quite

Figure 8.37. Photoresist (PR) masking sequence to obtain a p–n–p bipolar transistor.

elaborate manufacturing steps. The most important of these production steps are illustrated in Fig. 8.37 and described in detail below. A simplified example of the final product is depicted in Fig. 8.38, which contains some basic electronic components on one common substrate. Millions of these elements are squeezed on a, say, 6×6 mm^2 area, and hundreds of these complete circuits (properly interconnected) are fabricated together on one wafer. After electrical testing, the individual chips are cut apart with a

Figure 8.38. Basic components of integrated circuits (bipolar). The dark areas are the contact pads.

diamond-tipped saw. The functional chips are finally packaged in hermetically sealed containers and sold.

The individual manufacturing steps shown in Fig. 8.37 are as follows:

Oxidation. Silicon dioxide forms readily on silicon by placing the wafer into a tube furnace, heated between 900° and 1200°C, and exposing it to water vapor and possibly oxygen. This *wet* or *steam oxidation* is much faster than dry oxidation for which oxygen without steam is reacted with the silicon slice. Occasionally, silicon nitride replaces SiO_2.

Photolithography. In order to be able to etch small openings through an SiO_2 layer at a desired place, the silicon dioxide needs first to be coated with a protective layer called the *photoresist*, which, after exposure to UV light and subsequent developing, remains on the substrate. Thus, a mask (comparable to a photographic negative) has to be produced that contains a pattern of nontransparent areas. The mask/photoresist/wafer sandwich is then exposed to UV light. During the subsequent developing process, the unexposed photoresist is dissolved at the places where the SiO_2 needs to be removed. The remaining photoresist protects the SiO_2 from the etching solution.

Oxide Etch. Wet chemical removal of the SiO_2 layer is accomplished by applying hydrofluoric acid (HF) at room temperature. The underlying silicon is not attacked by this etchant (this would require a HNO_3/HF solution). *Wet chemical etching* poses, however, some problems if submicron geometries need to be produced, since the etchant attacks not only vertically but also laterally, causing line broadening by undercutting. Thus, *dry etching* techniques, such as ion etching or reactive plasma etching, are increasingly utilized. Ion etching involves the removal of the exposed material by bombardment with energetic noble gas ions (such as argon). Since ion etching removes SiO_2 as well as the underlying silicon, the bombarding time has to be carefully controlled. Plasma etching, on the other hand, uses a chemical reaction that converts the substance to be etched into a volatile compound by utilizing a chemically active gas such as halocarbons (CF_4) in a plasma chamber.

Photoresist Strip. This process is accomplished by a simple chemical dissolution reaction.

Doping. In the pioneering times of semiconductor fabrication, the infusion of donor or acceptor elements into silicon was mainly done out of the gas phase with subsequent drive-in diffusion at temperatures near or above 1000°C. Though this process worked quite well initially, increasing miniaturization demanded a more precise doping technique. Thus, ion implantation has mainly been utilized since the early 1970s. This technique involves the ionization of the species to be implanted and their subsequent acceleration towards the substrate in an electric field. (The silicon needs to be shielded where implantation is unwanted.) The range where the heavy dopants come to rest in the silicon substrate obeys a Gaussian distribution. The collision-induced lattice damage, needs to be removed in a subsequent processing step. Annealing between 700° and 1000°C for a short time restores the original lattice symmetry and also causes the dopants to become electrically active.

Another method used for special purposes (low-level, extremely homogeneous doping) utilizes a neutron-irradiation-induced process that transforms Si into P. Specifically, the silicon isotope $^{30}_{14}Si$, which accounts for 3.1% of the atoms in common Si, is bombarded with a neutron, thus forming $^{31}_{14}P$, which transforms with a half-time of 2.6 hours (under emission of an electron) into the stable $^{31}_{15}P$.

Metallization. The internal connections between the individual transistors, etc., are accomplished by narrow and thin (about $1\mu m$ thick) metal films. Until recently, these metallizations consisted of aluminum with 2 to 4% Cu and possibly 1% Si. Due to the high current densities involved (several 10^6 A/cm^2) and the sharp bends of theses strips, some holes and extrusions (called hillocks) may be formed after a certain time of operation, causing interruption of the current paths and thus failure of the device. This process is dubbed *electromigration*. In essence, a momentum exchange between the accelerated electrons and the metal ions pushes some metal ions from the negative side to the positive end of the thin film stripe, causing voids to form near the cathode. Copper metallizations laid down by electroplating (*damascene* process) rather than by physical vapor deposition or sputtering provide some enhancement of the lifetime and a better conductivity which translates into less Joule-heating. However, the copper needs to be prevented from diffusing into the silicon device where it would cause trapping of charge carriers. Thus, a barrier layer between Cu and Si consisting of, for example, Ti–N or Ta–N or pure Ta needs to be inserted. For high-temperature applications and GaAs devices metal silicides are occasionally utilized as connecting stripes.

Passivation. The last layer on a chip is designed to protect the device from the environment and mechanical damage during packaging. Silicon dioxide has been used in the past for this purpose. However, SiO_2 is extremely brittle and cracks readily. Thus, more ductile insulators, such as silicon nitride, phosphosilicate glass, or polyimide, are now used. Still, phosphorus from the phosphosilicate glass could leach out in time and when combined with residual absorbed water may form phosphoric acid, which causes corrosion. On the other hand, polyimide is somewhat permeable and, thus, may allow water penetration.

Packaging. The individual chips (after having been cut from the wafer using a diamond blade saw) are bonded to *headers* and then sealed. 85% of the chips are currently encapsulated in plastics even though this packaging technique does not provide a complete hermetic seal from possible hostile environments. Ceramic packaging consisting of aluminum nitride, silicon carbide, or glass, having an internal cavity for chip mounting, provides a better seal, but this is more expensive. For plastic packaging, the chip is commonly bonded to a metal frame by a eutectic alloy (low melting point), or by polyimide adhesives, or a heat-conducting epoxy. These adhesives must provide for adequate removal of the heat that is generated in the chip.

Next, electrical connections between chip pads and the external leads are performed, utilizing extremely thin (30 μm) gold wires (thermal-sonic bonding). Subsequently, the device is encapsulated in a dense and rigid plastic, using for example a transfer molding process that requires pressure and heat (about 175°C). Among other techniques are reaction-injection, or radial spread molding. Materials for the envelopes include epoxy, silicone, thermoplastics, and certain polymer blends called interpenetrating polymer networks. It is of utmost importance that chip and envelope do not possess a thermal expansion mismatch in order to prevent stress and cracking of the chip.

Other packaging methods involve hermetically sealed metal cans, all ceramics, or the *glob top* process, in which a drop of epoxy resin is placed on top of a chip on a ceramic header which, when cured, forms a shiny black hemisphere.

The pins extruding from the packaged device are either arranged on the two opposing long sides, called the *dual-in-line* package (DIP), or on all four sides, dubbed the *quadpack*. Up to 200 pins per device, having a separation distance of only 0.64 mm, are possible when a second row of pins inside the outer row is used. The pins are eventually plugged into a socket or soldered to copper pads located on the surface of a circuit board (surface mount technology). In another design, the pins are inserted through holes on a circuit board and then connected by soldering.

The bottleneck for accomplishing ever-increasing speeds in computers is the transfer time of the electrons from one chip to the next. Thus, multichip modules (MCM), i.e., the mounting of many chips into one package, will be increasingly used in the future to eliminate interchip delays. The drawback of this technique is that a single defective chip could make all the remaining chips worthless. In addition, it requires intercompany cooperation since one manufacturer does not usually produce all types of chips.

Recent Developments. The trend towards increasing the number of devices per chip continues. For example, in 1985 one million storage elements (1Mbit) went into mass production. The 1Mbit chip eventually was succeeded by 4, 16, 64, and 256 Mbit devices. The 1 gigabit chip, hosting one billion elements, is already on the horizon and is expected to be in production by 2001. In general, the number of transistors on a chip doubles every 12 to 18 months (**Moore's rule**). Each further advancement step, however, creates new problems which need to be solved. Specifically, when the number of devices per chip is increased, the width of the structures is generally reduced in order to keep the total size of the chips at reasonable dimensions. Along with the rise in the number of devices per chip, the size of the structures has continuously fallen from 2 μm for the 64Kbit chip to 0.25 μm for the 256 Mbit chip. This poses increasingly higher demands on the wafer quality. For example, the allowable deviation from a completely flat wafer may be only fractions of micrometers (e.g., 0.13 μm). An improved flatness is accomplished by double-sided lapping, grinding, and particularly polishing. An even better surface quality can be achieved by epitaxially growing a silicon layer onto a silicon wafer. For this, silicon is slowly deposited at high temperatures out of the gas phase on a Si wafer at a rate of micrometers per minute so that each silicon atom has enough time to find its proper place in the lattice of the growing interface. (Pulling a crystal from the crucible at this rate would require about one year.) An epitaxial Si layer on a Si substrate has an added advantage: The low electrical resistance between epi-layer and substrate prevents a "latchup," that is, a coupling effect between $p–n–p$ and neighboring $n–p–n$ transistors through recombination of the different types of charge carriers.

In order to achieve the just-mentioned ultra-small structures, optical photolithography will be eventually replaced by electron beam or X-ray lithography, which allow a much finer definition of the device features because of the smaller wavelengths involved.

A further increase in the number of elements per chip area is achieved by stacking the elements in several levels. This task is certainly not trivial and causes in addition heat removal problems. Nevertheless, up to three levels have been accomplished so far. A typical chip requires about 600 processing steps and about 2 months of production time.

At the end of this chapter, a few remarks on **economics** may be of interest. The fabrication of solid-state microelectronic devices (chips) was in 1996 an $850 billion per year industry (worldwide). It is expected to rise to $1.3 trillion by the year 2000. This figure does *not* include the factory sales of complete electronic systems in which these chips are incorporated (which is a factor of 10 higher). The increase in value of the packaged circuit compared to the raw material is roughly 10 millionfold if the starting material (sand) is valued at a transportation cost of 3 cents per kilogram. The raw silicon can be purchased for a few dollars per kg. The 98% pure polysilicon already represents a value of $60 per kilogram and the silicon wafer sells for $2,000 per kilogram (about $100 for an 8 in. wafer or $1,000 per 12 inch wafer). The next steps are big jumps: the processed wafer is valued at $25,000 per kilogram, the raw chip costs $110,000, and the packaged chip is finally sold to the computer manufacturers for $300,000 per kilogram.

Let us look at the economic picture from another point of view. As one might expect, it takes a substantial amount of energy to produce a wafer. The largest part (about 400 kWh/kg) is already consumed to produce the polysilicon. Or, put differently: the energy consumption for melting and purification alone is 1000 kWh for 1 m^2 of wafer surface. Production of single crystals by the Czochralski method requires another 150 kWh for 1 m^2 of silicon surface. Doping, etc., consumes 25–50 kWh/m^2 (depending on the complexity of the device). All taken, including packaging, etc., roughly 1400 kWh are expended by the time microelectronic devices have been fabricated on a 1 m^2 silicon surface. (See, in this context, also Sections 8.7.6 and 9.4.)

*8.7.12. Digital Circuits and Memory Devices

The reader might legitimately wonder at this point how transistors are used in computers and similar devices. Even though this topic sidetracks the flow of our presentation somewhat, a few introductory remarks on switching devices, information processing, and information storage may nevertheless be of interest. We need to start with the recognition that electronic data-processing systems use binary digits, i.e., *zeros* and *ones* as carriers for information. As an example, the numeral sequence "0010" means in the binary system the decimal number "two," whereas 0101 represents the decimal number 5. The first digit at the right of a binary number represents 2^0, the next digits represent, consecutively, $2^1, 2^2, 2^3$, etc. A **bi**nary *digi*t, or a **bit**, is the smallest possible piece of information. (A group of related bits, e.g., 8 bits for word processing, is called a **byte**.) A "zero" in the present context means that the electric current is off, whereas a "one" means that the current is on. So much about preliminaries.

Let us begin with an "AND" device. We inspect the normally-off MOSFET in Fig. 8.29 and see that a voltage on the drain terminal is *only* obtained if we apply voltages *simultaneously* to the source *and* the gate terminals. In other words, a source voltage *and* a gate voltage cause a voltage on the drain terminal, see Fig. 8.39(a). The circuit resembles a gate in a fence and is therefore called an **AND gate**. The circuit symbol for an AND gate is depicted in Fig. 8.39(b).

Next, we discuss the **inverter** circuit. It consists of two *normally-off* MOS transistors which are wired in series (Fig. 8.40). The upper or *load transistor* (whose channel is made long and narrow to restrict the current flow) is always kept "on" by connecting the driving voltage to its gate. If a high

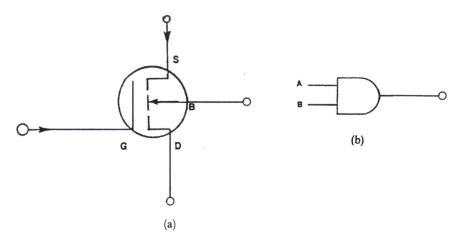

(a)

(b)

Figure 8.39. (a) AND gate and (b) circuit symbol for an AND gate. (Compare to Fig. 8.29)

enough voltage is simultaneously applied to the gate of the *lower* or *input* transistor, then this lower MOSFET likewise becomes conducting and the driving voltage drains through both transistors into the ground. As a consequence, the output voltage at terminal Q is nearly zero. Thus, the inverter circuit inverts a "one" signal on the input terminal into a "zero" signal on the output terminal (and vice versa). The circuit symbol for an inverter (or "**NOT gate**") is shown in Fig. 8.40(b).

(a)

(b)

Figure 8.40. Inverter made of two "normally-off" (*n*-channel, enhancement-type) MOSFETs (NOT gate). (a) circuit; (b) symbol in wiring diagram. (V_{DD} means "Drain power supply voltage".) The load transistor may be replaced by a (poly-silicon) resistor or an enhancement-type *p*-channel MOSFET.

We have just mentioned that the current through the load transistor is relatively small due to its special design. Still, an inverter which essentially does not consume *any* power (except during switching) would be even more desirable. This is accomplished by CMOS technology, i.e., by using an enhancement-type *p*-channel MOSFET as a load transistor, and an enhancement-type *n*-channel MOSFET as an input transistor. Unless the circuit is switching, one MOSFET is always off (not conducting current) whereas the other is on. Since the two MOSFETs are connected in series similarly as in Fig. 8.40, little power (except due to leakage current) is consumed. It is left to the reader to draw up and discuss the appropriate circuit diagram. (See Problem 18.)

The next logic device that we discuss is a "NAND" circuit. It consists of a load MOSFET, wired in series with two (or more) input transistors. All MOSFETs shown in Fig. 8.41 are of the "normally-off" type. Assume that high enough voltages are applied to the gates of *both* input transistors to make them conducting. Thus, the output terminal Q is connected to ground, i.e., the output voltage is almost zero. Since input voltages on the gates of the A *and* the B transistors inverts the input signal from "one" to "zero," we call the present logic building block an "AND" gate combined with a "NOT circuit" and term the entire digital function "NOT–AND" or a "NAND" gate for short. The reader may convince himself that the output is always "one" when at least one of the input voltages is "zero." On the other hand, if both inputs are "one," the output is "zero."

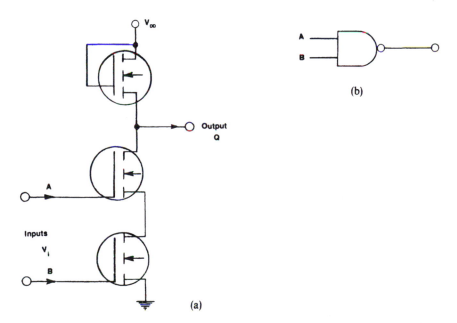

Figure 8.41. (a) NAND gate and (b) circuit symbol for a NAND digital function.

Figure 8.42. OR logic circuit with circuit symbol.

In an "**OR**" **gate** (Fig. 8.42) the output voltage Q is "one" when either A or B (or both) possess a voltage. Otherwise Q is "zero."

Finally, in a "**NOR**" circuit, the input transistors are again wired in parallel (Fig. 8.43). Applying high enough gate voltages to one *or* all of them causes the output voltage to be "zero." The circuit is appropriately called "NOT–OR" or "NOR." Evidently, the output voltage is only "one" if all input voltages are "zero."

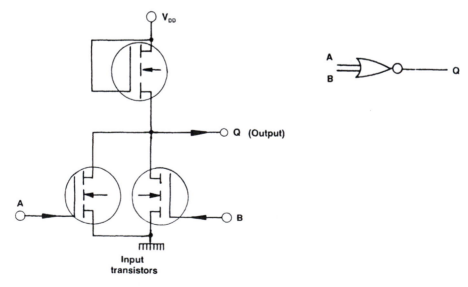

Figure 8.43. NOR logic circuit with circuit symbol.

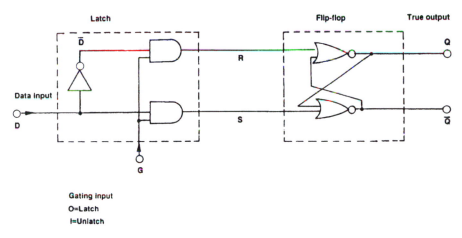

Figure 8.44. SRAM memory device called *R–S flip-flop with latch*. (The bar on a letter signifies the complement information.)

In short, the five basic *building blocks* that are obtained by properly circuiting one or more transistors are AND, NAND, OR, NOR, and NOT. (See also Problem 21.)

We are ultimately interested in knowing how a memory device works, i.e., we are interested in a unit made of transistors which can store information. For comparison, a toggle switch that turns a light on or off can be considered to be a digital information storage device. It can be flipped on and then flipped off to change the content of its information. Appropriately, the device that we are going to discuss is called a flip-flop. Let us assume that a flip-flop has a built-in latch to prevent the accidental change of information. This is done electronically by combining a NOT gate with two AND gates, as shown in the left part of Fig. 8.44. (The output of an AND gate is zero as long as *one* of the inputs is zero!) It is left to the reader to figure out the various combinations. As an example, if the gate is unlatched (1) and the data input is 1, we obtain "zero" on the R terminal and "one" on the S terminal.

The flip-flop on the right part of Fig. 8.44 consists of two NOR gates which are cross-coupled. (Remember that the *output* of a NOR is always *zero* when *at least one input is one*, and it is *one* when *both inputs are zero*.) The above example with R = 0 and S = 1 yields a "one" at the output terminal Q (and a "zero" at the complement output $\bar{Q}$), i.e., D and Q are identical. The information that is momentarily fed to the D terminal and into the system is *permanently stored* in the flip-flop even when the wires R and S are cut off. The cross-coupling keeps the two NOR gates mutually in the same state at least as long as a driving voltage remains on the devices. In short, one bit of information has been stored.

Let us now latch the gating network, i.e., let G be "zero." Whatever option the data input D will assume in this case, the output Q will always be "one," as the reader should verify. In other words, the latching prevents an

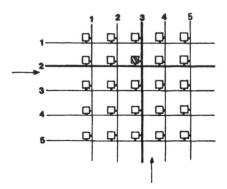

Figure 8.45. Schematic representation of a two-dimensional memory addressing system. By activating the #2 row wire and the #3 column wire, the content of the cross-hatched memory element (situated at their intersection) can be changed.

accidental change of the stored information. On the other hand, latching and unlatching by itself does not change the information content of the flip-flop either!

The device described above is called a **static random-access memory** or, in short, an "**S-RAM**", because the information remains permanently in the storage unit. The memory cell shown in Fig. 8.44 can evidently store only *one* bit of information. Let us now imagine a two-dimensional array of these storage elements, connected in a number of horizontal and vertical lines (Fig. 8.45). A designated memory element, being located at the crosspoint of a specific row and a specific column wire, can then be exclusively addressed by sending an electrical impulse through *both* wires simultaneously. One of these wires operates on the gating input (Fig. 8.44), the other one activates the data input. As we have discussed above, only a simultaneous activation of both input wires can change the information content of a flip-flop. An array of 32 columns and 32 rows of memory elements constitutes 1024 bits of information storage, or one *kilobit*. (Yes, a Kbit is *not* 1,000 bits.)

In order to reduce the area on a chip and the power consumption of a storage device, a memory cell different from the above-introduced flip-flop is frequently used. It is called the *one-transistor* **dynamic random-access memory** (**DRAM**, pronounced D-RAM). The information is stored in a capacitor, which can be accessed through an enhancement-type transistor (Fig. 8.46). Only concomitant voltages on gate and source allow access to the capacitor. Since the stored charge in a capacitor leaks out in a few milliseconds, the information has to be "refreshed" every 2 milliseconds by means of refresh circuits. No voltage on the capacitor is used as a "zero," whereas a certain voltage on the capacitor represents the "one" logic.

The 256 *Megabit chip* combines a multitude of these or similar building blocks through ultra-large-scale integration (ULSI) on one piece of silicon the size of a finger nail.

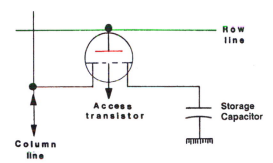

Figure 8.46. One-transistor dynamic random-access memory (DRAM). The information flows in and out through the column line.

The memory devices discussed so far are of the *"volatile"* type, i.e., they lose their stored information once the electric power of the computer is interrupted. In *nonvolatile memories*, such as the *read-only memory* (ROM), information is permanently stored in the device. Let us consider, for example, the MOSFET depicted in Fig. 8.29(b). Assume that the connection to the gate has been permanently interrupted during fabrication. Then the transistor will never transmit current from source to drain. Thus, a "zero" is permanently stored without the necessity to maintain a driving voltage. If, on the other hand, the path to the gate is left intact, the MOSFET can be addressed and current between source and drain may flow, which constitutes a "one." The stored information can be read, but it cannot be altered.

In the **programmable read-only memory (PROM)** the information may be written by the user, for example, by blowing selected fuse links to the gate. As above, the alteration is permanent and the information can be read only.

The **erasable-programmable read-only memory (EPROM)** allows the user to program the device as well as erase the stored information. An EPROM contains a *"floating gate,"* i.e., a gate (consisting of heavily doped polysilicon) which is completely imbedded in SiO_2, see Fig. 8.47. For programming, the drain-substrate junction is strongly reverse biased until avalanche-breakdown sets in (see Fig. 8.20), and electrons are injected from the drain region into the SiO_2 layer. A large voltage (~ 25 V) between a second gate, the control gate, and the substrate allows some electrons to cross the insulator, thus negatively and permanently charging the floating gate. The oxide thickness is on the order of 100 nm, which assures a charge retention time of about 100 years. A permanent charge on the floating gate constitutes a "1"; no charge represents the zero state. Exposure of the EPROM to ultraviolet light or X-rays through a window (not shown in Fig. 8.47) increases the conductivity of the insulator and allows the charge to leak out of the floating gate, thus erasing any stored information.

For electrical erasure, a large positive voltage can be applied to the control gate which removes the stored charge from the floating gate. This returns the **"electrically erasable-programmable ROM"** (**EEPROM**) to the zero state.

Figure 8.47. Electrically erasable-programmable read-only memory device (EEPR-OM), also called stacked-gate avalanche-injected MOS (SAMOS), or, with some modifications, flash memory device.

Alternately, electrical erasure can be performed by applying a positive voltage to the source, which also pulls charge from the floating gate. The *"flash memory"* device uses this erasure method while employing a thinner (10 nm) and higher-quality oxide below the floating gate. This improves efficiency and reliability. **Flash memory cards**, which are the size of a credit card, are predicted to cut substantially into the floppy (magnetic) disk and perhaps the hard disk memory market because of their small weight, their lower power consumption, and their fast access time (twenty times faster than a floppy disk). These virtues are important features for laptop computers. Sales for 2000 are projected to be about $2.5 billion.

It should be mentioned in closing that magnetic storage devices are discussed in Section 17.4. Optical storage devices are explained in Section 13.10.

Problems

Intrinsic Semiconductors

1. Calculate the number of electrons in the conduction band for silicon at $T = 300$ K. (Assume $m_e^*/m_0 = 1$.)

2. Would germanium still be a semiconductor if the band gap was 4 eV wide? Explain! (*Hint*: Calculate N_e at various temperatures. Also discuss extrinsic effects.)

3. Calculate the Fermi energy of an intrinsic semiconductor at $T \neq 0$ K. (*Hint*: Give a mathematical expression for the fact that the probability of finding an electron at the top of the valence band plus the probability of finding an electron at the bottom of the conduction band must be 1.) Let $N_e \equiv N_p$ and $m_e^* \equiv m_h^*$.

4. At what (hypothetical) temperature would all 10^{22} (cm^{-3}) valence electrons be excited to the conduction band in a semiconductor with $E_g = 1$ eV? *Hint*: Use a programmable calculator.

5. The outer electron configuration of neutral germanium in its ground state is listed in a textbook as $4\,s^2 4p^2$. Is this information correct? Someone argues against this configuration stating that the p-states hold six electrons. Thus, the p-states in germanium and therefore the valence band are only partially filled. Who is right?

6. In the figure below, σ is plotted as a function of the reciprocal temperature for an intrinsic semiconductor. Calculate the gap energy. (*Hint*: Use (8.14) and take the ln from the resulting equation.)

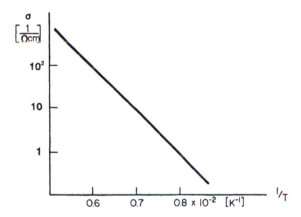

Extrinsic Semiconductors

7. Calculate the Fermi energy and the conductivity at room temperature for germanium containing 5×10^{16} arsenic atoms per cubic centimeter. (*Hint*: Use the mobility of the electrons in the host material.)

8. Consider a silicon crystal containing 10^{12} phosphorous atoms per cubic centimeter. Is the conductivity increasing or decreasing when the temperature is raised from 300° to 350°C? Explain by giving numerical values for the mechanisms involved.

9. Consider a semiconductor with 10^{13} donors/cm^3 which have a binding energy of 10 meV.
   (a) What is the concentration of extrinsic conduction electrons at 300 K?

(b) Assuming a gap energy of 1 eV (and $m^* \equiv m_0$), what is the concentration of intrinsic conduction electrons?

(c) Which contribution is larger?

10. The binding energy of a donor electron can be calculated by assuming that the extra electron moves in a hydrogen-like orbit. Estimate the donor binding energy of an n-type impurity in a semiconductor by applying the modified equation (4.18a)

$$E = \frac{m^* e^4}{2(4\pi\varepsilon_0 \hbar)^2 \varepsilon^2},$$

where $\varepsilon = 16$ is the dielectric constant of the semiconductor. Assume $m^* = 0.8$ m_0. Compare your result with experimental values listed in Appendix 4.

11. What happens when a semiconductor contains both donor and acceptor impurities? What happens with the acceptor level in the case of a predominance of donor impurities?

Semiconductor Devices

12. You are given a p-type doped silicon crystal and are asked to make an ohmic contact. What material would you use?

13. Describe the band diagram and function of a p–n–p transistor.

14. Can you make a solar cell from metals only? Explain!

*15. A cadmium sulfide photodetector is irradiated over a receiving area of 4×10^{-2} cm^2 by light of wavelength 0.4×10^{-6} m and intensity of 20 W m^{-2}.
(a) If the energy gap of cadmium sulfide is 2.4 eV, confirm that electron–hole pairs will be generated.
(b) Assuming each quantum generates an electron–hole pair, calculate the number of pairs generated per second.

16. Calculate the room-temperature saturation current and the forward current at 0.3 V for a silver/n-doped silicon Schottky-type diode. Take for the active area 10^{-8} m^2 and $C = 10^{19}$ A/m^2 K^2.

17. Draw up a circuit diagram and discuss the function of an inverter made with CMOS technology. (*Hint*: An enhancement-type p–n–p MOSFET needs a negative gate voltage to become conducting; an enhancement-type n–p–n MOSFET needs for this a positive gate voltage.)

18. Draw up a circuit diagram for an inverter which contains a *normally-on* and a *normally-off* MOSFET. Discuss its function.

19. Convince yourself that the unit in (8.26) is indeed the ampere.

20. Calculate the thermal energy provided to the electrons at room temperature. You will find that this energy is much smaller than the band gap of silicon. Thus, no intrinsic electrons should be in the conduction band of silicon at room

temperature. Still, according to your calculations in Problem 1, there is a sizable amount of intrinsic electrons in the conduction band at $T = 300$ K. Why?

21. Explain the "OR" logic circuit.

22. Calculate the lateral dimensions of a quantum well structure made of GaAs. (*Hint*: Keep in mind that the lateral dimension has to equal the wavelength of the electrons in this material.) Refer to Section 4.2 and Fig. 4.4(a). Use the data contained in the tables of Appendix 4.

23. Calculate the number of electrons and holes per incident photon, i.e., the quantum efficiency, in a transverse photodiode. Take $W = 8$ μm, $L = 8$ mm, and $\alpha = 40$ cm^{-1}.

CHAPTER 9

Electrical Properties of Polymers, Ceramics, Dielectrics, and Amorphous Materials

9.1. Conducting Polymers and Organic Metals

Materials which are electrical (and thermal) insulators are of great technical importance and are, therefore, used in large quantities in the electronics industry, e.g., as handles for a variety of tools, as coatings for wires, or as casings for electrical equipment. Most polymeric materials have the required insulating properties and have been used for decades for this purpose. It came, therefore, as a surprise when it was discovered in the late 1970's that some polymers and organic substances may have electrical properties which resemble those of conventional semiconductors, metals, or even super-conductors. We shall focus our attention mainly on these materials. This does not imply that the predominance of applications of polymers is in the conductor field. Quite the contrary is true. Nevertheless, conducting poly-mers (also called synthetic metals) steadily gain ground compared to in-sulating polymers.

Initially, conducting polymers were unstable in air or above room tem-perature. In addition, some dopants, used to impart a greater conductivity, were toxic, and the doping made the material brittle. However, more re-cently, stable conducting polymers were synthesized which have as an added benefit an optical transparency across the entire visible spectrum. Among them, poly(3,4-ethylenedioxythiophene) (PEDT) and its derivatives enjoy now multi-ton productions, in particular for antistatic layers in photographic films. These conducting layers are beneficial for the prevention of friction-induced static electricity, which causes, on discharge, flashes of light, thus, pre-exposing the light-sensitive emulsion. Other uses of PEDT are transpar-ent electrodes for inorganic electroluminescent devices, anti-static treatments of plastics and cathode ray tubes, electrodes for capacitors, sensors, rechargeable batteries, photovoltaic devices, and packaging of electronic components. PEDT films can be heated in air at 100°C for over 1000 hours with hardly any change in conductivity.

We now attempt to discuss conducting polymers in the light of solid-state physics. Conventional solid-state physics deals preferably with the properties of well-defined regular arrays of atoms. We have learned in Chapter 7 that a periodic array of lattice atoms is imperative for coherent scattering of electron waves and thus for a high conductivity. Further, the periodic arrangement of atoms in a crystal and the strong interactions between these atoms causes, as explained in Section 4.4, a widening of energy levels into energy bands.

We know that highly conducting materials such as metals are characterized by partially filled bands, which allow a free motion of the conduction electrons in an electric field. Insulators and semiconductors, on the other hand, possess (at least at 0 K) completely filled valence bands and empty conduction bands. The difference in band structure between crystalline insulators and semiconductors is a matter of degree rather than of kind: insulators have wide gaps between valence and conduction bands whereas the energy gaps for semiconductors are narrow. Thus, in the case of semiconductors, the thermal energy is large enough to excite some electrons across the gap into the conduction band. The conductivity in pure semiconductors is known to *increase* (exponentially) with increasing temperature and decreasing gap energy (8.14), whereas the conductivity in metals *decreases* with increasing temperature (Fig. 7.7). Interestingly enough, most conducting polymers have a temperature dependence of the conductivity similar to that of semiconductors. This suggests that certain aspects of semiconductor theory may be applied to conducting polymers. The situation regarding polymers cannot be described, however, without certain modifications to the band model brought forward in the previous chapters. This is due to the fact that polymeric materials may exist in amorphous as well as in crystalline form or, more commonly, as a mixture of both. This needs to be discussed in some detail.

Polymers consist of molecules which are long and chainlike. The atoms that partake in such a chain (or macromolecule) are regularly arranged along the chain. Several atoms combine and form a specific building block, called a monomer, and thousands of monomers combine to a polymer. As an example, we depict polyethylene, which consists of repeat units of one carbon atom and two hydrogen atoms, Fig. 9.1(a). If one out of four hydrogen atoms in polyethylene is replaced by a chlorine atom, polyvinylchloride (PVC) is formed upon polymerization (Fig. 9.1(b)). In polystyrene, one hydrogen atom is replaced by a benzene ring. More complicated macromolecules may contain side chains attached to the main link. They are appropriately named "branched polymers." Macromolecules whose backbones consist largely of carbon atoms, as in Fig. 9.1, are called "organic" polymers.

Figure 9.1. (a) Polyethylene. (b) Polyvinylchloride. (The dashed enclosures mark the repeat unit. Polyethylene is frequently depicted as two CH_2 repeat units for historical reasons.

The binding forces that hold the individual atoms in polymers together are usually covalent and sometimes ionic in nature. Covalent forces are much stronger than the binding forces in metals. They are based on the same interactions that are responsible for forming a hydrogen molecule from two hydrogen atoms. Quantum mechanics explains covalent bonds by showing that a lower energy state is achieved when two equal atomic systems are closely coupled and in this way exchange their energy (see Section 16.2). In organic polymers each carbon atom is often bound to four atoms (see Fig. 9.1) because carbon has four valencies.

In contrast to the strong binding forces between the atoms within a polymeric chain, the secondary interactions between the individual macro-molecules are usually weak. The latter are of the Van der Waals type, i.e., they are based on forces which induce dipole moments in the molecules. (Similar weak interactions exist for noble gases such as argon, neon, etc.)

In order to better understand the electronic properties of polymers by means of the electron theory and the band structure concept, one needs to know the degree of order or the degree of periodicity of the atoms, because only ordered and strongly interacting atoms or molecules lead, as we know, to distinct and wide electron bands. Now, it has been observed that the degree of order in polymers depends on the length of the molecules and on the regularity of the molecular structure. Certain heat treatments may influ-ence some structural parameters. For example, if a simple polymer is slowly cooled below its melting point, one might observe that some macromolecules align parallel to each other. The individual chains are separated by regions of supercooled liquid, i.e., of amorphous material (Fig. 9.2). Actually, slow cooling yields, for certain polymers, a highly crystalline structure.

In other polymers, the cooling procedure might cause the entire material to go into a supercooled-liquid state. In this state the molecules can be considered to be randomly arranged. After further cooling, below a glass transition temperature, the polymer might transform itself into a glassy amorphous solid which is strong, brittle, and insulating. However, as stated before, we shall concern ourselves mainly with polymers that have a high degree of crystallinity. Amorphous materials will be discussed in Section 9.4.

A high degree of crystallinity and a relatively high conductivity have been

Figure 9.2. Simplified representation of a semicrystalline polymer (folded-chain model).

found in polyacetylene, which is the simplest **conjugated organic polymer**. It is considered to be the prototype of a conducting polymer. A conjugated polymer has alternating single and double bonds between the carbons (see Fig. 9.3, which should be compared to Fig. 9.1(a)). Two principle isomers are important: in the *trans* form, the hydrogen atoms are alternately bound to opposite sides of the carbons (Fig. 9.3(b)), whereas in the *cis* form the hydrogen atoms are situated on the same side of the double-bond carbons (Fig. 9.3(a)). **Trans-polyacetylene** is obtained as a silvery, flexible film that has a conductivity comparable to that of silicon (Fig. 9.4).

Figure 9.5 shows three band structures for *trans*-$(CH)_x$ assuming different distances between the carbon atoms. In Fig. 9.5(a) all carbon bond lengths are taken to be equal. The resulting band structure is found to be characteristic for a metal, i.e., one obtains distinct bands, the highest of which is *partially* filled by electrons. Where are the free electrons in the conduction band coming from? We realize that the electrons in the double bond of a conjugated polymer (called the π-electrons) can be considered to be only loosely bound to the neighboring carbon atoms. Thus, one of these electrons is easily disassociated from its carbon atom by a relatively small energy, which may be provided by thermal energy. The delocalized electrons may be accelerated as usual in an electric field.

Figure 9.3. Theoretical isomers of polyacetylene (a) *cis*-transoidal isomer, (b) *trans*-transoidal isomer. Polyacetylene is synthesized as *cis*-$(CH)_x$ and is then isomerized into the *trans*-configuration by heating it at 150°C for a few minutes.

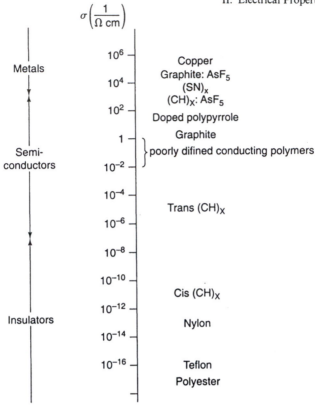

Figure 9.4. Conductivities of polymers in Ω^{-1} cm^{-1}. (Compare with Fig. 7.1.)

In reality, however, a uniform bond length between the carbon atoms does not exist in polyacetylene. Instead, the distances between the carbon atoms alternate because of the alternating single and double bonds. Band structure calculations for this case show, interestingly enough, some gaps between the individual energy bands. The resulting band structure is typical for a semiconductor (or an insulator)! The width of the band gap near the Fermi level depends mainly on the extent of alternating bond lengths (Fig. 9.5(b) and (c)).

It has been shown that the band structure in Fig. 9.5(b) best represents the experimental observations. Specifically, one finds a band gap of about 1.5 eV and a total width of the conduction band of 10–14 eV. The effective mass m^* is $0.6m_0$ at $k = 0$ and $0.1m_0$ at $k = \pi/a$. Assuming $\tau \to 10^{-14}$ s, the free carrier mobility, μ, along a chain is calculated to be about 200 cm^2/V s. The latter quantity is, however, hard to measure since the actual drift mobility in the entire solid is reduced by the trapping of the carriers which occurs during the "hopping" of the electrons between the individual macromolecules. In order to improve the conductivity of (CH)$_x$ one would attempt to decrease the disparity in the carbon–carbon bond lengths, thus eventually approaching the uniform bond length as shown in Fig. 9.5(a).

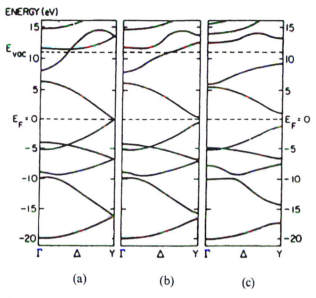

Figure 9.5. Calculated band structure of *trans*-(CH)$_x$ for different carbon–carbon bond lengths: (a) uniform (1.39 Å); (b) weakly alternating (C=C, 1.36 Å; C—C, 1.43 Å); and (c) strongly alternating (C=C, 1.34 Å; C—C, 1.54 Å). Note the band gaps at Y as bond alternation occurs. Reprinted with permission from P.M. Grant and I.P. Batra, *Solid State Comm.* **29**, 225 (1979).

Polyacetylene, as discussed so far, should be compared to conventional intrinsic semiconductors. Now, we know from Section 8.3 that the conductivity of semiconductors can be substantially increased by doping. The same is true for polymer-based semiconductors. Indeed, arsenic–pentafluoride-doped trans-polyacetylene has a conductivity which is about seven orders of magnitude larger than undoped trans-(CH)$_x$. Thus, σ approaches the conductivity of metals, as can be seen in Figs. 9.4 and 9.6. Many oxidants cause p-type semiconductors, whereas alkali metals are n-type dopants. The doping is achieved through the vapor phase or by electrochemical methods. The dopant molecules diffuse between the (CH)$_x$ chains and provide a charge transfer between the polymer and the dopant. The additional element ends up as an anion when it is an acceptor and as a cation when it is a donor. Among other (albeit nontoxic) dopants is n-dodecyl sulfonate (soap). A word of caution, when using the word "doping", should be added at this point. In semiconductor physics, doping means extremely small additions of impurity elements. In polymers, much larger quantities of additional substances are used, ranging from one tenth of a percent up to 20–40%.

A refinement in the description of the conduction mechanism in polyacetylene can be provided by introducing the concept of **solitons**. A soliton is a structural distortion in a conjugated polymer and is generated when a single bond meets another single bond, as shown in Fig. 9.7. At the distortion point

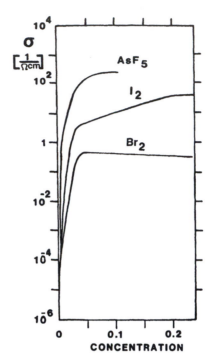

Figure 9.6. Conductivity change of polyacetylene as a result of doping.

a localized nonbonding electron state is generated, similar to an n-type impurity state in a silicon semiconductor. The result is a localized level in the center of the forbidden band. It is believed that when an electron is excited from the valence band into the conduction band (leaving a hole in the valence band) this electron–hole pair decays in about 10^{-12} s into a more stable soliton–antisoliton pair.

Near the center of a soliton, the bond lengths are equal. We recall that uniform bond lengths constitute a metal. Thus, when many solitons have been formed and their spheres of influence overlap, a metal-like conductor would result.

It is also conceivable that one of the double bonds next to a soliton switches

Figure 9.7. A broken symmetry in polyacetylene creates a *soliton*. (An *antisoliton* is the mirror image of a soliton.)

over to a single bond. If this switching occurs consecutively in one direction, a soliton wave results. This can be compared to a moving electron.

Up to now, we discussed mainly the properties of polyacetylene. Over the last 20 years additional conductive polymers have been discovered. They include polyanilines, polypyrroles, polythiophenes, polyphenylenes, poly(p-phenylene vinylene) and their derivatives. Of these, the **polyaniline** family can be easily processed at low cost but might yield toxic (carcinogenic) products upon degradation. Others are more "environmentally friendly" but are insoluble. On the other hand, the above-mentioned **PEDT** (developed by Bayer AG in Germany) can be made water-soluble by utilizing poly(styrenesulfonate) (PSS) as a dopant during polymerization. Its antistatic and other properties have been mentioned already above. The chains in inorganic **poly(sulfur nitride)** consist of alternating sulfur and nitrogen atoms. Because of the different valencies of the S^{2-} and N^{3-} ions, $(SN)_x$ is an electron-deficient material with an alternating bond structure. The bond length alternation is not severe, so that $(SN)_x$ has a room-temperature conductivity of about 10^3 ohm^{-1} cm^{-1} along the chain direction. The conductivity increases with a reduction in temperature. At temperatures close to 0 K, poly(sulfur nitride) becomes superconducting. In brominated $(SN)_x$ the Br_3^- and Br_2^- ions are aligned along the chain axis, giving rise to a one-dimensional superlattice.

In **graphite**, an individual "molecule" consists of a "sheet" of carbon atoms. The conductivity is found to be nearly metallic, at least parallel to the layers (Fig. 9.4). AsF$_5$-doped graphite has an even higher conductivity. The conduction is increased by producing a mixture of easily ionized electron donors and electron acceptors. The charge is then shared between the donors and acceptors. These materials are called charge-transfer complexes.

Another class of organic conductors is the **charge-transfer salts**, in which a donor molecule, such as tetrathiafulvalene (TTF), transfers electrons to an acceptor molecule, such as tetracyanoquinodimethane (TCNQ). The planar molecules stack on top of each other in sheets, thus allowing an overlap of wave functions and a formation of conduction bands that are partially filled with electrons due to the charge transfer. It is assumed that, because of the sheetlike structure, the charge-transfer compounds are quasi-one-dimensional. Along the stacks, conductivities as high as 2×10^3 Ω^{-1} cm^{-1} have been observed at room temperature. Below room temperature, the metallic conductors often transform into semiconductors or insulators. Even superconduction has been observed at very low temperatures (about 13 K). In the presence of a magnetic field and at low temperatures, these materials undergo, occasionally, a transition from a metallic, nonmagnetic state into a semimetallic, magnetic state. Organic metals are generally prepared by electrochemical growth in a solution. They are, as a rule, quite brittle, single crystalline, and relatively small. Other materials of this type include doped complexes of C_{60} (so-called Buckyballs) which exhibit superconductivity at low temperatures.

Replacing metals with lightweight conducting polymers (for wires) seems to be, in the present state of the art, nearly impossible, mainly because of their poor stability. However, this very drawback (i.e., the high reactivity of some conducting polymers) concomitant with a change in conductivity can be profitably utilized in devices such as remote gas sensors, biosensors, or other remotely readable indicators that detect changes in humidity, radiation dosage, mechanical abuse, or chemical release. As an example, polypyrrole noticeably changes its conductivity when exposed to only 0.1% NH_3, NO_2, or H_2S. Further, experiments have been undertaken to utilize $(CH)_x$ for measuring the concentration of glucose in solutions.

9.2. Ionic Conduction

In ionic crystals (such as the alkali halides), the individual lattice atoms transfer electrons between each other to form positively charged cations and negatively charged anions. The binding forces between the ions are electrostatic in nature and are thus very strong. The room-temperature conductivity of ionic crystals is about twenty-two orders of magnitude smaller than the conductivity of typical metallic conductors (Fig. 7.1). This large difference in σ can be understood by realizing that the wide band gap in insulators allows only extremely few electrons to become excited from the valence band into the conduction band.

The main contribution to the electrical conduction in ionic crystals (as little as it may be) is, however, due to a mechanism that we have not yet discussed, namely, ionic conduction. Ionic conduction is caused by the movement of some negatively (or positively) charged ions which *hop* from lattice site to lattice site under the influence of an electric field, see Figure 9.8 (b). (This type of conduction is similar to that which is known to occur in aqueous electrolytes.) This ionic conductivity

$$\sigma_{\text{ion}} = N_{\text{ion}} e \mu_{\text{ion}} \tag{9.1}$$

is, as outlined before (8.13), the product of three quantities. In the present case, N_{ion} is the number of ions per unit volume that can change their position under the influence of an electric field and μ_{ion} is the mobility of these ions.

In order for ions to move through a crystalline solid, they must have sufficient energy to pass over an *energy barrier* (Fig. 9.8). Further, an equivalent lattice site next to a given ion must be empty in order for an ion to be able to change its position. Thus, N_{ion} in (9.1) depends on the vacancy concentration in the crystal (i.e., on the number of *Schottky defects*[16]). In short, the theory of ionic conduction contains essential elements of diffusion theory, with which the reader might be familiar.

---

[16] A *Schottky defect* is formed when an anion as well as a cation of the same absolute valency are missing (to preserve charge neutrality).

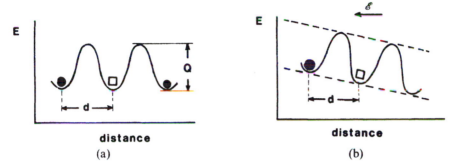

Figure 9.8. Schematic representation of a potential barrier, which an ion ($\bullet$) has to overcome to exchange its site with a vacancy ($\square$). (a) Without an external electric field; (b) with an external electric field. d = distance between two adjacent, equivalent lattice sites; Q = activation energy.

Diffusion theory links the mobility of the ions, which is contained in (9.1), with the diffusion coefficient, D, through the Einstein relation,

$$\mu_{ion} = \frac{De}{k_B T}. \tag{9.2}$$

(Note that (9.2) implies that *one* charge unit per atom is transported.)

The diffusion coefficient varies with temperature; this dependence is commonly expressed by an Arrhenius equation,

$$D = D_0 \exp\left[-\left(\frac{Q}{k_B T}\right)\right], \tag{9.3}$$

where Q is the activation energy for the process under consideration (Fig. 9.8), and D_0 is a pre-exponential factor that depends on the vibrational frequency of the atoms and some structural parameters. Combining (9.1) through (9.3) yields

$$\sigma_{ion} = \frac{N_{ion}e^2 D_0}{k_B T} \exp\left[-\left(\frac{Q}{k_B T}\right)\right]. \tag{9.4}$$

Equation (9.4) is shortened by combining the pre-exponential constants into σ_0:

$$\sigma_{ion} = \sigma_0 \exp\left[-\left(\frac{Q}{k_B T}\right)\right]. \tag{9.5}$$

Taking the natural logarithm yields

$$\ln \sigma_{ion} = \ln \sigma_0 - \left(\frac{Q}{k_B}\right)\frac{1}{T}. \tag{9.6}$$

Figure 9.9. Schematic representation of $\ln \sigma$ versus $1/T$ for Na^+ ions in sodium chloride. (Arrhenius plot.)

Equation (9.6) suggests that if $\ln \sigma_{ion}$ is plotted verus $1/T$, a straight line with a negative slope would result. Figure 9.9 depicts schematically a plot of $\ln \sigma$ versus $1/T$ as experimentally found for alkali halides. The linear $\ln \sigma$ versus $1/T$ relationship indicates that Fig. 9.9 is an actual representation of (9.6). The slopes of the straight lines in Arrhenius plots are utilized to calculate the activation energy of the processes under consideration. We notice in Fig. 9.9 two temperature regions representing two different activation energies: at low temperatures, the activation energy is small, the thermal energy is just sufficient to allow the hopping of ions into already existing vacancy sites. This temperature range is commonly called the **extrinsic region**. On the other hand, at high temperatures, the thermal energy is large enough to create additional vacancies. The related activation energy is thus the sum of the activation energies for vacancy creation and ion movement. This temperature range is called the **intrinsic region**.

So far, we have not been very specific in describing the circumstances of vacancy formation in an ionic crystal. Now, we have to realize that whenever vacant lattice sites are created, an overall charge neutrality needs to be maintained. The latter is the case when both a cation and an anion are removed from a lattice. Another permissible mechanism is the formation of a vacancy-interstitial pair (**Frenkel defect**). More often, however, vacancies are created as a consequence of introducing differently charged impurity atoms into an ionic lattice, i.e., by replacing, say, a monovalent metal atom with a divalent atom. In order to maintain charge neutrality in this case, a positively charged vacancy needs to be introduced. For example, if a divalent Mg^{2+} ion substitutes for a monovalent Na^+ ion, one extra Na^+ ion has to be removed to restore charge neutrality, see Fig. 9.10. Or, if zirconia (ZrO_2) is treated with CaO (to produce the technically important **calcia-stabilized zirconia**), the Ca^{2+} ions substitute for Zr^{4+} ions and an anion vacancy needs to be created to maintain charge neutrality. Nonstoichiometric compounds contain a high amount of vacancies even at relatively low temperatures, whereas in stoichiometric compounds vacancies need to be formed by elevating the temperature.

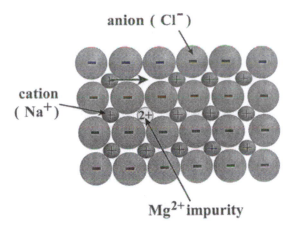

anion (Cl⁻)

cation
(Na⁺)

Mg²⁺ impurity

Figure 9.10. Schematic representation of a {100} plane of an ionic crystal having the NaCl structure. The diffusion of a cation into a cation vacancy is shown. Also depicted is the creation of a cation vacancy when replacing a Na⁺ ion with a Mg²⁺ ion.

In principle, both cations and anions are capable of moving simultaneously under the influence of an electric field. It turns out, however, that in most alkali halides the majority carriers are provided by the (smaller) metal ions, whereas in other materials, such as the lead halides, the conduction is predominantly performed by the halide ions.

So far, it was implied that the materials under consideration are single crystals. For polycrystalline materials, however, it appears reasonable to assume that the vacant lattice sites provided by the grain boundaries would be utilized by the ions as preferred paths for migration, thus enhancing the conductivity. This has indeed been experimentally observed for alkali ions.

One piercing question remains to be answered: If ionic conduction entails the transport of ions, i.e., of matter from one electrode to the other, would this not imply some segregation of the constituents? Indeed, a pile-up of mobile ions at the electrodes has been observed for long-lasting experiments with a concomitant induced electric field in the opposite direction to the externally applied field. As a consequence the conductivity decreases gradually over time. Of course, this does not happen when nonblocking electrodes are utilized which provide a source and a sink for the mobile species.

9.3. Conduction in Metal Oxides

Metal oxides do not actually represent a separate class of conducting materials on their own. Indeed, they can be insulating, such as TiO_2, have metallic conduction properties, such as TiO, or be semiconducting. For understanding the mechanisms involved in metal oxides, e.g., in the aforementioned titanium oxides, it is helpful to inspect the table in Appendix 3. Oxygen is seen there to have four $2p$-electrons in its outermost shell. Two more electrons will bring O^{2-} into the closed-shell configuration and four electrons are obviously needed to accomplish the same for two oxygen ions, such as in TiO_2. These four electrons are provided by the titanium from its $3d$- and $4s$-shells. Thus, in the case of TiO_2, all involved elements are in the noble gas configuration. Since ionic bonds are involved, any attempted removal of electrons would require a considerable amount of thermal energy. TiO_2 is, consequently, an insulator having a wide band gap. Not so for TiO. Since only two titanium valence electrons are needed to fill the $2p$-shell of *one* oxygen ion, two more titanium electrons are free to serve as conduction electrons. Thus, TiO has metallic properties with a σ in the $10^3 \; \Omega^{-1} \; cm^{-1}$ range.

A refinement of our understanding is obtained by considering the pertinent electron bands. TiO has, according to the aforementioned explanations, a filled oxygen $2p$-valence band and an essentially empty titanium $4s$-conduction band. Also involved is a narrow titanium $3d$-band which is partially filled by the above-mentioned two electrons. The conduction in TiO takes place, therefore, in the titanium $3d$-band, which can host, as we know, a total of 10 electrons.

We discuss **zinc oxide** as a next example. Zn in ZnO has two valence $4s$-electrons which transfer to the oxygen $2p$-band. ZnO, if strictly stoichiometric, has, thus, a filled valence $2p$-band and an empty zinc $4s$-band employing a gap energy of 3.3 eV. Stoichiometric ZnO is therefore an insulator or a wide-band-gap semiconductor. Now, if interstitial zinc atoms (or oxygen vacancies) are introduced into the lattice (by heating ZnO in a reducing atmosphere, which causes neutral oxygen to leave the crystal) then the valence electrons of these zinc interstitials are only loosely bound to their nuclei. One of these two electrons can easily be ionized (0.05 eV) and acts therefore as a donor. Nonstoichiometric ZnO is, consequently, an n-type semiconductor. The same is incidentally true for nonstoichiometric Cu_2O (see Appendix 3), an established semiconducting material from which Cu/Cu_2O Schottky-type rectifiers were manufactured long before silicon technology was invented.

Another interesting metal oxide is SnO_2 (sometimes doped with In_2O_3), which is transparent in the visible region and which is a reasonable conductor in the $1 \; \Omega^{-1} \; cm^{-1}$ range. It is used in optoelectronics to provide electrical contacts without blocking the light from reaching a device. It is known as **indium-tin-oxide or ITO.**

Finally, we discuss **NiO.** Again, a filled oxygen $2p$-band and an empty nickel $4s$-band are involved. In order to form the nickel $3d$-bands required for conduction, a substantial overlap of the $3d$-wave functions would be required by quantum mechanics. Band structure calculations show, however, that these interactions do not take place. Instead, *deep-lying localized electron states* in the forbidden band close to the upper edge of the valence band are observed. Thus, no $3d$-band conduction can take place, which results in stoichiometric NiO being an insulator. Nonstoichiometry (obtained by removing some nickel atoms, thus creating cation vacancies) causes NiO to become a p-type semiconductor.

9.4. Amorphous Materials (Metallic Glasses)

Before we discuss electrical conduction in amorphous materials, we need to clarify what the term *amorphous* means in the present context. Strictly speaking, *amorphous* implies the random arrangement of atoms, the absence of any periodic symmetry, or the absence of any crystalline structure. One could compare the random distribution of atoms with the situation in a gas, as seen in an instantaneous picture. Now, such a completely random arrangement of atoms is seldom found even in liquids, much less in solids. In actuality, the relative positions of nearest neighbors surrounding a given atom in an amorphous solid are almost identical to the positions in crystalline solids because of the ever-present binding forces between the atoms. In short, the atomic order in amorphous materials is restricted to the nearest neighbors. Amorphous materials exhibit, therefore, only *short-range order*. In contrast to this, the exact positions of the atoms that are farther apart from a given central atom cannot be predicted. This is particularly the case when various kinds of stacking orders, i.e., if polymorphic modifications, are possible. As a consequence one observes atomic disorder at long range. The term *amorphous solid* should therefore be used *cum grano salis*. We empirically define materials to be amorphous when their diffraction patterns consist of diffuse rings, rather than sharply defined Bragg rings, as are characteristic for polycrystalline solids.

So far we have discussed **positional disorder** only as it might be found in pure materials. If more than one component is present in a material, a second type of disorder is possible: The individual species might be randomly distributed over the lattice sites; i.e., the species may not be alternately positioned as is the case for, say, sodium and chlorine atoms in NaCl. This random distribution of species is called **compositional disorder**.

The best-known representative of an amorphous solid is window glass, whose major components are silicon and oxygen. Glass is usually described as a supercooled liquid.

Interestingly enough, many elements and compounds that are generally known to be crystalline under equilibrium conditions can also be obtained in the nonequilibrium amorphous state by applying rapid solidification techniques, i.e., by utilizing cooling rates of about 10^5 K/s. These cooling rates can be achieved by fast quenching, melt spinning, vapor deposition, sputtering, radiation damage, filamentary casting in continuous operation, spark-processing, etc. The degree of amorphousness (or, the degree of short range order) may be varied by the severity of the quench. The resulting **metallic glasses**, or *glassy metals*, have unusual electrical, mechanical, opti-

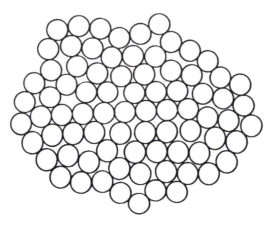

Figure 9.11. Two-dimensional schematic representation of a dense random packing of hard spheres (Bernal model).

cal, magnetic, and corrosion properties and are therefore of considerable interest. Amorphous semiconductors (consisting, e.g., of Ge, Si, GeTe, etc.) have also received substantial attention because they are relatively inexpensive to manufacture, have unusual switching properties, and have found applications in inexpensive photovoltaic cells.

We now turn to the atomic structure of amorphous metals and alloys. They have essentially nondirectional bonds. Thus, the short-range order does not extend beyond the nearest neighbors. The atoms must be packed together tightly, however, in order to achieve the observed density. There are only a limited number of ways of close packing. One way of arranging the atoms in amorphous metals is depicted by the *dense random packing of hard spheres* model (Fig. 9.11). This **Bernal model** is considered as the ideal amorphous state. No significant regions of crystalline order are present. In transition metal–metalloid compounds (such as Ni–P) it is thought that the small metalloid atoms occupy the holes which occur as a consequence of this packing (**Bernal–Polk model**).

The atoms in amorphous semiconductors, on the other hand, do not arrange themselves in a close-packed manner. Atoms of group IV elements are, as we know, covalently bound. They are often arranged in a *continuous random network* with correlations in ordering up to the third or fourth nearest neighbors (Fig. 9.12(b) and (c)). Amorphous pure silicon contains numerous *dangling bonds* similar to those found in crystalline silicon in the presence of vacancies (Fig. 9.12(a)).

Since amorphous solids have no long-range crystal symmetry, we can no longer apply the Bloch theorem, which led us in Section 4.4 from the distinct energy levels for isolated atoms to the broad quasi-continuous bands for crystalline solids. Thus, the calculation of electronic structures for amorphous metals and alloys has to use alternate techniques, e.g., the **cluster**

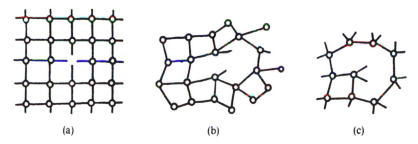

(a)   (b)   (c)

Figure 9.12. Defects in crystalline and amorphous silicon. (a) Monovacancy in a crystalline semiconductor; (b) one and (c) two dangling bonds in a continuous random network of an amorphous semiconductor. (Note the deviations in the interatomic distances and bond angles.)

model approach. This method has been utilized to calculate the electronic structure of amorphous Zr–Cu (which is a representative of a noble metal–transition metal metallic glass). A series of clusters were assumed which exhibit the symmetry of the close-packed lattices fcc (as for Cu) and hcp[15] (as for Zr). The energy level diagram depicted in Fig. 9.13 shows two distinct "bands" of levels. The lower band consists primarily of copper d-levels, while the upper band consists mainly of zirconium d-levels. A sort of gap separates the two bands of levels. Even though the concept of quasi-continuous energy bands is no longer meaningful for amorphous solids, the density of states concept still is, as can be seen in Fig. 9.13. We notice that the Fermi energy

---

[15] Hexagonal close-packed.

Figure 9.13. Schematic representation of the molecular orbital energy level diagram and the density of states curves for Zr–Cu clusters. The calculated density of states curves agree reasonably well with photoemission experiments.

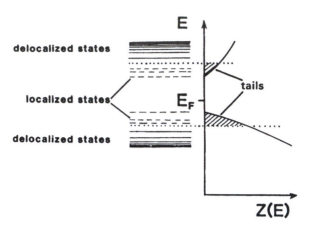

Figure 9.14. Localized and delocalized states and density of states $Z(E)$ for amorphous semiconductors. Note the band tails, which are caused by the localized states.

is located in the upper part of the zirconium levels. Further, we observe partially filled electron states. This has two interesting consequences. First, we expect *metal-like* conduction. Second, $Z(E)$ near E_F is small, which suggests relatively small values for the conductivity (see (7.26)). Indeed, σ for Cu–Zr is comparable to that of poor metallic conductors (i.e., approximately 5×10^3 $1/\Omega$ cm).

The electrical resistivity of many metallic glasses (such as $Pd_{80}Si_{20}$ or $Fe_{32}Ni_{36}Cr_{14}P_{12}B_6$[17]) stays constant over a fairly wide temperature range, up to the temperature which marks the irreversible transition from the amorphous into the crystalline state. This makes these alloys attractive as resistance standards. The mean free path for electrons in metallic glasses is estimated to be about 1 nm.

The energy level diagrams and the density of states curves for **amorphous semiconductors** are somewhat different from those for amorphous metals. Because of the stronger binding forces which exist between the atoms in covalently bound materials, the valence electrons are tightly bound, or *localized*. As a consequence, the density of states for the localized states extends into the "band gap" (Fig. 9.14). This may be compared to the localized impurity states in doped crystalline semiconductors, which are also located in the band gap. Thus, we observe density of states *tails*. These tails may extend, for some materials, so far into the gap that they partially overlap. In general, however, the density of electron and hole states for the localized levels is very small.

The electrical conductivity for amorphous semiconductors, σ_A, depends,

---

[17] METGLAS 2826A, trademark of Allied Chemical.

as usual (8.13), on the density of carriers, N_A, and the mobility of these carriers, μ_A:

$$\sigma_A = N_A e \mu_A. \tag{9.7}$$

The density of carriers in amorphous semiconductors is extremely small, because all electrons are, as said before, strongly bound (localized) to their respective nuclei. Likewise, the mobility of the carriers is small because the absence of a periodic lattice causes substantial incoherent scattering. As a consequence, the room-temperature conductivity in amorphous semiconductors is generally very low (about 10^{-7} $1/\Omega$ cm).

Some of the localized electrons might occasionally acquire sufficient thermal energy to overcome barriers which are caused by potential wells of variable depth and hop to a neighboring site. Thus, the conduction process in amorphous semiconductors involves a (temperature-dependent) activation energy, Q_A, which leads to an equation similar to (9.5), describing a so-called variable-range hopping

$$\sigma_A = \sigma_0 \exp\left[-\left(\frac{Q_A(T)}{k_B T}\right)\right]. \tag{9.8}$$

Equation (9.8) states that the conductivity in amorphous semiconductors increases exponentially with increasing temperature, because any increase in thermal energy provides additional free carriers.

The application of amorphous silicon for *photovoltaic devices* (see Section 8.7.6) will be discussed briefly in closing because of its commercial, as well as scientific, significance. If silicon is deposited out of the gas phase on relatively cold ($<500°C$) substrates (utilizing silane or sputtering), a structure as shown in Fig. 9.12 (b) and (c) results. Doping is virtually not possible in this condition since any free charge carriers recombine immediately with the dangling bonds. However, hydrogen, if added during deposition and incorporated into the solid, neutralizes the unsaturated valencies (and reduces internal strain in the lattice network). This results in **hydrogenated amorphous silicon**, which is, in its properties, quite comparable to crystalline silicon. Doping can be accomplished during deposition. This way, semiconducting materials can be produced which vary in their conductivity between 10^{-11} and 10^{-2} Ω^{-1} cm^{-1} depending on doping (see Fig. 7.1). Commercial flat-plate solar cells of this type have an efficiency of about 8% compared to 14% efficiency for commercial single-crystal silicon technology. The price (and the consumption of power during manufacturing) is, however, only one-half of that for crystalline silicon, mainly because of the simpler way of deposition (see Section 8.7.6).

The understanding of amorphous metals, alloys, and semiconductors is still in its infancy. Future developments in this field should be followed with a great deal of anticipation because of the potentially significant applications which might arise in the years to come.

9.4.1. Xerography

Xerography (from the Greek "dry writing") or **electrophotography** is an important application of amorphous semiconductors such as amorphous selenium or amorphous silicon, etc. Such a material, when deposited on a cylindrically shaped metallic substrate, constitutes the *photoreceptor* drum, as shown in Fig. 9.15.

Before copying, the photoreceptor is electrostatically charged by means of a corona wire to which a high voltage is applied (Step 1). Amorphous semi-conductors are essentially insulators (see above) which hold this electric charge reasonably well, as long as they are kept in the dark. If, however, light which has been reflected from the document to be copied falls on the photoreceptor, electron–hole pairs are formed, causing the photoreceptor to become conducting. This process discharges the affected parts on the drum, creating a latent image on the photoreceptor, i.e., a pattern consisting of charged and neutral areas. At the next step, electrostatically charged and pigmented polymer particles (called *toner*) are brought into contact with the drum. The toner clings to the charged areas only. Commonly, a two-component toner is utilized; one part consists of magnetically soft particles. They form brush-type chains under the influence of a magnetic field which is

Figure 9.15. Schematic representation of the electrophotography process. The individual steps are explained in the text.

caused by permanent magnets that are rotated inside a cylinder (see Fig. 9.15, Step 3). Eventually, the toner on the photoreceptor is electrostatically transferred to a piece of paper by properly corona-charging the back of the paper. Finally, the toner is fused to the paper by heat. A cleaning and photodischarging process prepares the photoreceptor drum for the next cycle.

Laser printers use the same principle. To create the latent image, the laser light is periodically scanned across the rotating photoreceptive drum by means of a rotating multisurface mirror. The spectral sensitivity of the amorphous semiconductor has to be matched to the wavelength of the laser light. Amorphous silicon (maximal photosensitivity near 700 nm) in conjunction with a helium–neon laser (see Table 13.1) is a usable combination.

9.5. Dielectric Properties

Insulators (also often called dielectric materials) possess a number of additional important electrical properties that make them useful in the electronics industry. They will be explained in this section.

When a voltage is momentarily applied to two parallel metal plates which are separated by a distance, L, as shown in Figure 9.16, then the resulting electric charge essentially remains on these plates even after the voltage has been removed (at least as long as the air is dry). This ability to store an electric charge is called **capacitance**, C, which is defined to be the charge, q, per unit applied voltage, V, that is:

$$C = \frac{q}{V},\qquad(9.9)$$

where C is given in coulombs per volt, or farad (see Appendix 4). Understandably, the capacitance is higher the larger the area, A, of the plates and

Figure 9.16. Two metal plates, separated by a distance, L, can store electric energy after having been charged momentarily by a battery.

Table 9.1. DC dielectric constants of some materials

Potassium tantalate niobate	6000	
Barium titanate (BaTiO$_3$)	4000	Ferroelectric
Potassium Niobate (KNbO$_3$)	700	
Rochelle salt (NaKC$_4$H$_4$O$_6 \cdot$ 4H$_2$O)	170	
Water	81.1	
Acetone	20	
Silicon	11.8	
GaAs	10.9	
Marble	8.5	
Soda-lime-glass	6.9	
Porcelain	6.0	
Epoxy	4.0	
Fused silica	4.0	Dielectric
Nylon 6,6	4.0	
PVC	3.5	
Ice	3.0	
Amber	2.8	
Polyethylene	2.3	
Paraffin	2.0	
Air	1.000576	

the smaller the distance, L, between them. Further, the capacitance depends on the material that may have been inserted between the plates. The experimental observations lead to

$$C = \varepsilon \varepsilon_0 \frac{A}{L},$$ (9.10)

where

$$\varepsilon = \frac{C}{C_{\text{vac}}}$$ (9.11)

determines the magnitude of the added storage capability. It is called the (unitless) **dielectric constant** (or occasionally the *relative permittivity*, ε_r). ε_0 is a universal constant having the value of 8.85×10^{-12} farad per meter (F/m), or As/Vm, and is known by the name **permittivity of empty space** (or of vacuum). Some values for the dielectric constant are given in Table 9.1. The dielectric constant of empty space is set to be 1, whereas ε of air and many other gases is nearly 1. The dielectric constant is frequency dependent.

We now need to explain why the capacitance increases when a piece of a dielectric material is inserted between two conductors [see Eq. (9.10)]. For this, one has to realize that, under the influence of an external electric field, the negatively charged electron cloud of an atom becomes displaced with respect to its positively charged core; compare Figure 9.17(a) with 9.17(b).

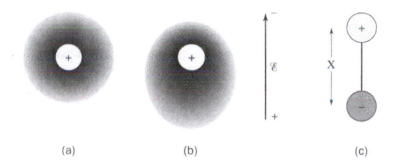

(a) (b) (c)

Figure 9.17. An atom is represented by a positively charged core and a surrounding, negatively charged, electron cloud (a) in equilibrium and (b) in an external electric field. (c) Schematic representation of an electric dipole as, for example, created by separation of the negative and positive charges by an electric field, as seen in (b).

As a result, a dipole is created, which has an **electric dipole moment**

$$p = q \cdot x, \tag{9.12}$$

where x is the separation between the positive and the negative charge as depicted in Figure 9.17(c). (The dipole moment is generally a vector pointing from the negative to the positive charge.) The process of dipole formation (or alignment of already existing dipoles) under the influence of an external electric field that has an electric field strength, $\mathscr{E}$, is called **polarization**. Dipole formation of all involved atoms within a dielectric material causes a charge redistribution so that the surface nearest to the positive capacitor plate is negatively charged (and vice versa), see Figure 9.18(a). As a consequence, electric field lines within a dielectric are created which are opposite in direction to the external field lines. Effectively, the electric field lines within a dielectric material are weakened due to polarization, as depicted in Figure 9.18(b). In other words, the **electric field strength** in a material,

$$\mathscr{E} = \frac{\mathscr{E}_{vac}}{\varepsilon}, \tag{9.13}$$

is reduced by inserting a dielectric between two capacitor plates.

Within a dielectric material the electric field strength, $\mathscr{E}$, is replaced by the **dielectric displacement**, D (also called the *surface charge density*), that is,

$$D = \varepsilon \varepsilon_0 \mathscr{E} = \frac{q}{A}. \tag{9.14}$$

The dielectric displacement is the superposition of two terms:

$$D = \varepsilon_0 \mathscr{E} + P, \tag{9.15}$$

where P is called the **dielectric polarization**, that is, the induced electric dipole moment per unit volume [Figures 9.18 (c and d)]. The units for D and

Figure 9.18. Schematic representation of two capacitor plates between which a dielectric material is inserted. (a) Induction of electric dipoles of opposite charge. (b) Weakening of the electric field *within* the dielectric material [Eq. (9.13)]. (c) The direction of the polarization vector is from the negative induced charge to the positive induced charge see Fig. 9.17(b). (d) The dielectric displacement, D, within the dielectric material is the sum of $\varepsilon_0 \mathscr{E}$ and P [Eq. (9.15)].

P are $C\ m^{-2}$; see Eq. (9.14). (D, $\mathscr{E}$, and P are generally vectors.) In summary, the polarization is responsible for the increase in charge density (q/A) above that for vacuum.

The mechanism just described is known by the name **electronic polarization**. It occurs in all dielectric materials that are subjected to an electric field. In ionic materials, such as the alkali halides, an additional process may occur, which is called **ionic polarization**. In short, cations and anions are somewhat displaced from their equilibrium positions under the influence of an external field and thus give rise to a net dipole moment. Finally, many materials already possess permanent dipoles that can be *aligned* in an external electric field. Among them are water, oils, organic liquids, waxes, amorphous polymers, polyvinylchloride, and certain ceramics, such as barium titanate ($BaTiO_3$). This mechanism is termed *orientation polarization*, or **molecular polarization**. All three polarization processes are additive if applicable; see Figure 9.19.

Most capacitors are used in alternating electric circuits. This requires the dipoles to reorient quickly under a rapidly changing electric field. Not all polarization mechanisms respond equally quick to an alternating electric field. For example, many molecules are relatively sluggish in reorientation. Thus, molecular polarization breaks down already at relatively low frequencies; see Figure 9.19. In contrast, electronic polarization responds quite rapidly to an alternating electric field even at frequencies up to 10^{16} Hz.

At certain frequencies a substantial amount of the excitation energy is absorbed and transferred into heat. This process is called *dielectric loss*. It is

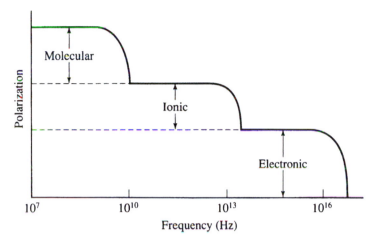

Figure 9.19. Schematic representation of the polarization as a function of excitation frequency for different polarization mechanisms.

imperative to know the frequency for dielectric losses for a given material so that the respective device is not operated in this range.

9.6. Ferroelectricity, Piezoelectricity, and Electrostriction

Ferroelectric materials, such as barium titanate, exhibit spontaneous polarization without the presence of an external electric field. Their dielectric constants may be orders of magnitude larger than those of dielectrics (see Table 9.1). Thus, they are quite suitable for the manufacturing of small-sized, highly efficient capacitors. Most of all, however, ferroelectric materials retain their polarization even after an external electric field has been removed. Specifically, if a ferroelectric is exposed to a strong electric field, $\mathscr{E}$, its permanent dipoles become increasingly aligned with the external field direction until eventually all dipoles are parallel to $\mathscr{E}$ and saturation of the polarization, P_s, has been achieved, as depicted in Figure 9.20. Once the external field has been withdrawn, a **remanent polarization**, P_r, remains which can only be removed by inverting the electric field until a coercive field, $\mathscr{E}_c$, is reached (Figure 9.20). By further increasing the reverse electric field, parallel orientation of the dipoles in the opposite direction is achieved. Finally, when reversing the field once more, a complete **hysteresis loop** is obtained, as depicted in Figure 9.20. Therefore, ferroelectrics can be utilized for memory devices in computers, etc. The area within a hysteresis loop is proportional to the energy per unit volume that is dissipated once a full field cycle has been completed.

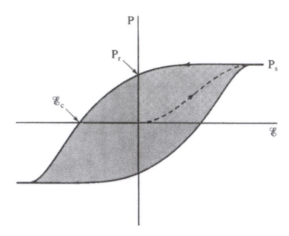

Figure 9.20. Schematic representation of a hysteresis loop for a *ferroelectric* material in an electric field. Compare to Figure 15.6.

It should be emphasized at this point that ferroelectrics do *not* contain iron, as the name might suggest. Instead, the name is derived from the similarity of some properties of ferroelectric substances to those of ferromagnetic materials such as iron. In other words, ferroelectricity is the electric analogue to ferromagnetism, which will be discussed in Section 15.1.3.

A critical temperature, called the **Curie temperature**, exists, above which the ferroelectric effects are destroyed and the material becomes dielectric. Typical Curie temperatures range from $-200°C$ for strontium titanate to at least $640°C$ for $NaNbO_3$.

The question remains to be answered, why do certain materials such as $BaTiO_3$ possess spontaneous polarization? This can be explained by knowing that in the tetragonal crystal structure of $BaTiO_3$, the negatively charged oxygen ions and the positively charged Ti^{4+} ion are slightly displaced from their symmetrical positions, as depicted in Figure 9.21. This results in a permanent ionic dipole moment along the c-axis within the unit cell. A large number of such dipoles line up in clusters (also called *domains*); see Figure 9.22. In the virgin state, the polarization directions of the individual domains are, however, randomly oriented, so that the material has no net polarization. An external field eventually orients the dipoles of the favorably oriented domains parallel to $\mathscr{E}$. Specifically, those domains in which the dipoles are already nearly parallel to $\mathscr{E}$ grow at the expense of unfavorably oriented domains.

By heating $BaTiO_3$ above its Curie temperature (120°C), the tetragonal unit cell transforms into a cubic cell whereby the ions now assume symmetric positions. Thus, no spontaneous alignment of dipoles remains, and $BaTiO_3$ becomes dielectric.

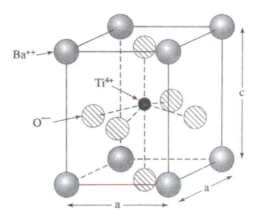

Figure 9.21. Tetragonal crystal structure of barium titanate at room temperature. Note the upward displacement of the Ti^{4+} ion in the center compared to the downward displacement of all surrounding O^{2-} ions. $a = 0.398$ nm; $c = 0.403$ nm.

If pressure is applied to a ferroelectric material, such as $BaTiO_3$, a change in the just-mentioned polarization may occur, which results in a small voltage across the sample. Specifically, the slight change in dimensions causes a variation in bond lengths between cations and anions. This effect is called **piezoelectricity**.[17] It is found in a number of materials, such as quartz (however, much weaker than in $BaTiO_3$), ZnO, and complicated ceramic compounds such as $PbZrTiO_6$. Piezoelectricity is utilized in devices that are designed to convert mechanical strain into electricity. Those devices are called **transducers**. Applications include strain gages, microphones, sonar detectors, and phonograph pickups, to mention a few.

The inverse mechanism, in which an electric field produces a change in dimensions in a ferroelectric material, is called **electrostriction**. An earphone

---

[17] *Piezo* (latin) = pressure.

Figure 9.22. Schematic representation of spontaneous alignments of electric dipoles within a domain and random alignment of the dipole moments of several domains in a ferroelectric material such as $BaTiO_3$. Compare to Figure 15.9.

utilizes such a device. Probably the most important application, however, is the quartz crystal resonator, which is used in electronic devices as a *frequency selective element*. Specifically, a periodic strain is applied to a quartz crystal by an alternating electric field, which excites this crystal to vibrations. These vibrations are monitored, in turn, by piezoelectricity. If the applied frequency coincides with the natural resonance frequency of the molecules, then amplification occurs. In this way, very distinct frequencies are produced, which are utilized for clocks or radio frequency signals.

Problems

1. Calculate the mobility of the oxygen ions in UO_2 at 700 K. The diffusion coefficient of O^{2-} at this temperature is 10^{-13} cm^2/s. Compare this mobility with electron or hole mobilities in semiconductors (see Appendix 4). Discuss the difference! (*Hint*: O^{2-} has two charges!)

2. Calculate the number of vacancy sites in an ionic conductor in which the metal ions are the predominant charge carriers. Assume a room-temperature ionic conductivity of 10^{-17} 1/Ω cm and an ionic mobility of 10^{-17} m^2/V s. Does the calculated result make sense? Discuss how the vacancies might have been introduced into the crystal.

3. Calculate the activation energy for ionic conduction for a metal ion in an ionic crystal at 300 K. Take $D_0 = 10^{-3}$ m^2/s and $D = 10^{-17}$ m^2/s.

4. Calculate the ionic conductivity at 300 K for an ionic crystal. Assume 6×10^{20} Schottky defects per cubic meter, an activation energy of 0.8 eV and $D_0 = 10^{-3}$ m^2/s.

5. Show that $\mathscr{E} = \mathscr{E}_{vac}/\varepsilon$ [Eq. (9.13)] by combining Eqs. (7.3), (9.9), and (9.11) and their equivalents for vacuum.

6. Show that the dielectric polarization is $P = (\varepsilon - 1)\varepsilon_0\mathscr{E}$. What values do P and D have for vacuum?

7. Show that $\varepsilon\varepsilon_0\mathscr{E} = q/A$ [Eq. (9.14)] by combining some pertinent equations.

Suggestions for Further Reading (Part II)

N.W. Ashcroft and N.D. Mermin, *Solid State Physics*, Hold, Rinehart and Winston, New York (1976).

A.R. Blythe, *Electrical Properties of Polymers*, Cambridge University Press, Cambridge (1979).

M.H. Bordsky, ed., *Amorphous Semiconductors*, Springer-Verlag, Berlin, (1979).

I. Brodie and J. Muray, *The Physics of Microfabrication*, Plenum Press, New York (1982).

R.H. Bube, *Electronic Properties of Crystalline Solids*, Academic Press, New York (1974).

R.H. Bube, *Electrons in Solids*, 3rd ed., Academic Press, New York (1992).

J. Czochralski, Z. Physik. Chem. *92*, 219 (1918).

W.C. Dash, J. Appl. Phys. *30*, 459 (1959).

J.R. Ferraro, and J.M. Williams, *Introduction to Synthetic Electrical Conductors*, Academic Press, Orlando (1987).

S.K. Ghandi, *The Theory and Practice of Microelectronics*, Wiley, New York (1968).

S.K. Ghandi, *VLSI Fabrication Principles*, Wiley, New York (1983).

N.J. Grant and B.C. Giessen, eds., *Rapidly Quenched Metals*, Second International Conference. The Massachusetts Institute of Technology, Boston, MA (1976).

C.R.M. Grovenor, *Materials for Semiconductor Devices*, The Institute of Metals (1987).

L.L. Hench and J.K. West, *Principles of Electronic Ceramics*, Wiley, New York (1990).

B.H. Kear, B.C. Giessen, and K.M. Cohen, eds., *Rapidly Solidified Amorphous and Crystalline Alloys*, North-Holland, Amsterdam (1982).

C. Kittel, *Introduction to Solid State Physics*, 7th ed., Wiley, New York (1996).

J. Mort and G. Pfister, eds., *Electronic Properties of Polymers*, Wiley, New York (1982).

G.W. Neudeck and R.F. Pierret, eds., *Modular Series on Solid State Devices*, Vols. I–VI, Addison-Wesley, Reading, MA (1987).

M.A. Omar, *Elementary Solid State Physics*, Addison-Wesley, Reading, MA (1978).

D.A. Seanov ed., *Electrical Properties of Polymers*, Academic Press, New York, (1982).

B.G. Streetman, *Solid State Electronic Devices*, 2nd ed., Prentice-Hall, Englewood Cliffs, NJ (1980).

S.M. Sze, *Physics of Semiconductor Devices*, 2nd ed., Wiley, New York (1981).

S.M. Sze, *Semiconductor Devices, Physics and Technology*, Wiley, New York (1985).

H.E. Talley and D.G. Daugherty, *Physical Principles of Semiconductor Devices*, Iowa University Press, Ames, IA (1976).

L.H. Van Vlack, *Physical Ceramics for Engineers*, Addison-Wesley, Reading, MA (1964).

C.A. Wert and R.M. Thompson, *Physics of Solids*, 2nd ed., McGraw-Hill, New York (1970).

PART III

OPTICAL PROPERTIES
OF MATERIALS

CHAPTER 10

The Optical Constants

10.1. Introduction

The most apparent properties of metals, their luster and their color, have been known to mankind since metals were known. Because of these properties, metals were already used in ancient times for mirrors and jewelry. The color was utilized 4000 years ago by the ancient Chinese as a guide to determine the composition of the melt of copper alloys: the hue of a preliminary cast indicated whether the melt, from which bells or mirrors were to be made, already had the right tin content.

The German poet Goethe was probably the first one who explicitly spelled out 200 years ago in his *Treatise on Color* that *color* is not an absolute property of matter (such as the resistivity), but requires a living being for its perception and description. Applying Goethe's findings, it was possible to explain qualitatively the color of, say, gold in simple terms. Goethe wrote: "If the color *blue* is removed from the spectrum, then blue, violet, and green are missing and red and yellow remain." Thin gold films are bluish–green when viewed in transmission. These colors are missing in reflection. Consequently, gold appears reddish–yellow.

This chapter treats the optical properties from a completely different point of view. Measurable quantities such as the index of refraction or the reflectivity and their spectral variations are used to characterize materials. In doing so, the term "color" will almost completely disappear from our vocabulary. Instead, it will be postulated that the interactions of light with the valence electrons of a material are responsible for the optical properties. As in previous chapters, where an understanding of the electrical properties was attempted, an atomistic model and later a quantum mechanical treatment will be employed. Thus, the electron theory of metals, as introduced in the first six chapters, will serve as a foundation.

Light comprises only an extremely small segment of the entire electromagnetic spectrum, which ranges from radio waves, via microwaves, infrared, visible, ultraviolet, and X-rays, to γ rays, as depicted in Figure 10.1. Many of the considerations that will be advanced in this chapter are therefore also valid for other wavelength ranges, e.g., for radio waves or X-rays.

At the beginning of this century the study of the interactions of light with matter (black body radiation, etc.) laid the foundations for quantum theory. Today, optical methods are among the most important tools for elucidating the electron structure of matter. Most recently, a number of optical devices, such as lasers, photodetectors, waveguides, light-emitting diodes, flat-panel displays, etc., have gained considerable technological importance. They are used in communication, fiber optics, medical diagnostics, night viewing, solar applications, optical computing, or for other optoelectronic purposes. Traditional utilizations of optical materials for windows, antireflection coatings, lenses, mirrors, etc., should be likewise mentioned. All taken, it is well justified to spend a major part of this book on the optical properties of materials.

Before we start our discourse, we need to define the optical constants. We make use of some elements of physics.

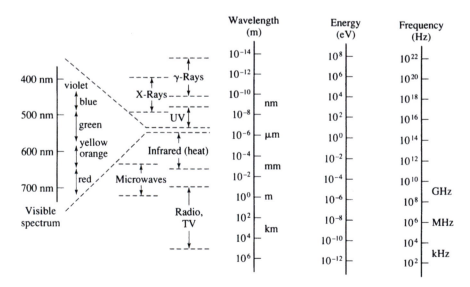

Figure 10.1. The spectrum of electromagnetic radiation. Note the small segment of this spectrum that is visible to human eyes.

10.2. Index of Refraction, n

When light passes from an optically "thin" into an optically dense medium, one observes that in the dense medium, the angle of refraction, β, (i.e., the angle between the refracted light beam and a line perpendicular to the surface) is smaller than the angle of incidence, α see Fig. 10.2. This well-known phenomenon is used for the definition of the refractive power of a material and is called **Snell's law**,

$$\frac{\sin \alpha}{\sin \beta} = \frac{n_{med}}{n_{vac}} = n. \tag{10.1}$$

Commonly, the **index of refraction** for vacuum, n_{vac}, is arbitrarily set to be unity. The refraction is caused by the different velocities, c, of the light in the two media,

$$\frac{\sin \alpha}{\sin \beta} = \frac{c_{vac}}{c_{med}}. \tag{10.2}$$

Thus, if light passes from vacuum into a medium, we find

$$n = \frac{c_{vac}}{c_{med}} = \frac{c}{v}. \tag{10.3}$$

The magnitude of the refractive index depends on the wavelength of the incident light. This property is called **dispersion**. In metals, the index of refraction varies, in addition, with the angle of incidence. This is particularly true when n is small.

In summary, when light passes from vacuum into a medium, its velocity as well as its wavelength, λ, decrease in order to keep the frequency, and thus, the energy, constant.

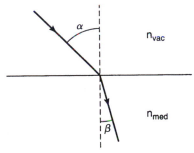

Figure 10.2. Refraction of a light beam when traversing the boundary from an optically thin medium into an optically denser medium.

10.3. Damping Constant, k

Metals damp the intensity of light in a relatively short distance. Thus, to characterize the optical properties of metals, an additional materials constant is needed.

We make use of the electromagnetic wave equation, which mathematically describes the propagation of light in a medium. The derivation of this wave equation from the well-known Maxwell equations does not further our understanding of the optical properties. (The interested reader can find the derivation in specialized texts.[1])

For simplification, we consider a plane-polarized wave that propagates along the positive z-axis and which vibrates in the x-direction (Fig. 10.3). We neglect possible magnetic effects. For this special case, the electromagnetic wave equation reads[2]

$$c^2 \frac{\partial^2 \mathscr{E}_x}{\partial z^2} = \varepsilon \frac{\partial^2 \mathscr{E}_x}{\partial t^2} + \frac{\sigma}{\varepsilon_0} \frac{\partial \mathscr{E}_x}{\partial t}, \qquad (10.4)$$

where $\mathscr{E}_x$ is the x-component of the electric field strength,[3] ε is the dielectric constant,[4] σ is the (a.c.) conductivity and ε_0 is a constant, called the permittivity of empty space (see Appendix 4). The solution to (10.4) is commonly

---

[1] For example, R.E. Hummel, *Optische Eigenschaften von Metallen und Legierungen*, Springer-Verlag, Berlin (1971).

[2] See also Appendix 1.

[3] We use for the electric field strength the symbol $\mathscr{E}$ to distinguish it from the energy.

[4] See Section 9.5.

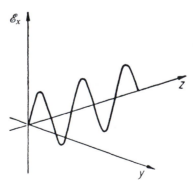

Figure 10.3. Plane-polarized wave which propagates in the positive z-direction and vibrates in the x-direction.

achieved by using the following trial solution:

$$\mathscr{E}_x = \mathscr{E}_0 \exp\left[i\omega\left(t - \frac{zn}{c}\right)\right], \tag{10.5}$$

where $\mathscr{E}_0$ is the maximal value of the electric field strength and $\omega = 2\pi\nu$ is the angular frequency. Differentiating (10.5) once with respect to time, and twice with respect to time *and* z, and inserting these values into (10.4) yields

$$\hat{n}^2 = \varepsilon - \frac{\sigma}{\varepsilon_0\omega}i = \varepsilon - \frac{\sigma}{2\pi\varepsilon_0\nu}i. \tag{10.6}$$

Equation (10.6) leads to an important result: The index of refraction is generally a complex number, as inspection of the right-hand side of (10.6) indicates. We denote for clarity the complex index of refraction by $\hat{n}$. As is true for all complex quantities, the complex index of refraction consists of a real and an imaginary part,

$$\hat{n} = n_1 - in_2. \tag{10.7}$$

In the literature, the imaginary part of the index of refraction, n_2, is often denoted by "k" and (10.7) is then written as

$$\hat{n} = n - ik. \tag{10.8}$$

We will call n_2 or k the **damping constant**. (In some books n_2 or k is named the *absorption constant*. We will not follow this practice because the latter term is extremely misleading. Other authors call k the *attenuation index* or the *extinction coefficient*, which we will not use either in this context.)

Squaring (10.8) yields, together with (10.6),

$$\hat{n}^2 = n^2 - k^2 - 2nki = \varepsilon - \frac{\sigma}{2\pi\varepsilon_0\nu}i. \tag{10.9}$$

Equating individually the real and imaginary parts of (10.9) yields two important relations between electrical and optical constants,

$$\varepsilon = n^2 - k^2, \tag{10.10}$$

$$\sigma = 4\pi\varepsilon_0 nk\nu. \tag{10.11}$$

Let us return to (10.9). The right-hand side is the difference between two dielectric constants (a real one and an imaginary one). Thus, the left side must be a dielectric constant too, and (10.9) may be rewritten as

$$\hat{n}^2 = n^2 - k^2 - 2nik \equiv \hat{\varepsilon} = \varepsilon_1 - i\varepsilon_2. \tag{10.12}$$

Equating individually the real and imaginary parts in (10.12) yields

$$\varepsilon_1 = n^2 - k^2 \tag{10.13}$$

and (with (10.11))

$$\varepsilon_2 = 2nk = \frac{\sigma}{2\pi\varepsilon_0 v}. \tag{10.14}$$

Similarly as above, ε_1 and ε_2 are called the real and the imaginary parts of the complex dielectric constant, $\hat{\varepsilon}$, respectively. (ε_1 in (10.13) is identical to ε in (10.10).) ε_2 is often called the *absorption product* or, briefly, the **absorption**.

We consider a special case: For insulators ($\sigma \approx 0$) it follows from (10.11) that $k \approx 0$ (see also Table 10.1). Then (10.10) reduces to $\varepsilon = n^2$ (**Maxwell relation**).

From (10.10), (10.11), (10.13), and (10.14) one obtains

$$n^2 = \frac{1}{2}\left(\sqrt{\varepsilon^2 + \left(\frac{\sigma}{2\pi\varepsilon_0 v}\right)^2} + \varepsilon\right) = \frac{1}{2}(\sqrt{\varepsilon_1^2 + \varepsilon_2^2} + \varepsilon_1), \tag{10.15}$$

$$k^2 = \frac{1}{2}\left(\sqrt{\varepsilon^2 + \left(\frac{\sigma}{2\pi\varepsilon_0 v}\right)^2} - \varepsilon\right) = \frac{1}{2}(\sqrt{\varepsilon_1^2 + \varepsilon_2^2} - \varepsilon_1). \tag{10.16}$$

It should be emphasized that (10.10)–(10.16) are only valid if ε, σ, n, and k are measured at the same wavelength, because these "constants" are wavelength dependent. For small frequencies, however, the d.c. values for ε and σ can be used with good approximation, as will be shown later. Finally, it should be noted that the above equations are only valid for optically isotropic media; otherwise ε becomes a tensor.

We return now to (10.5) in which we replace the index of refraction by the complex index of refraction (10.8). This yields

$$\mathscr{E}_x = \mathscr{E}_0 \exp\left[i\omega\left(t - \frac{z(n - ik)}{c}\right)\right], \tag{10.17}$$

which may be rewritten to read

$$\mathscr{E}_x = \mathscr{E}_0 \underbrace{\exp\left[-\frac{\omega k}{c}z\right]}_{\text{Damped amplitude}} \cdot \underbrace{\exp\left[i\omega\left(t - \frac{zn}{c}\right)\right]}_{\text{Undamped wave}}. \tag{10.18}$$

Equation (10.18) is now the complete solution of the wave equation (10.4). It represents a damped wave and expresses that in matter the amplitude decreases exponentially with increasing z (Fig. 10.4). The constant k determines how much the amplitude decreases, i.e., k expresses the degree of damping of the light wave. We understand now why k is termed the *damping constant*.

The result which we just obtained is well known to electrical engineers.

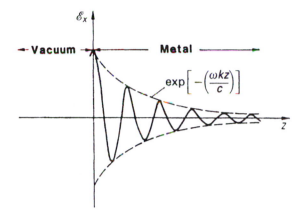

Figure 10.4. Modulated light wave. The amplitude decreases exponentially in an optically dense material. The decrease is particularly strong in metals, but less intense in dielectric materials, such as glass.

They observe that at high frequencies the electromagnetic waves are conducted only on the outer surface of a wire. They call this phenomenon the (*normal*) *skin effect*.

10.4. Characteristic Penetration Depth, W, and Absorbance, α

The field strength, $\mathscr{E}$, is hard to measure. Thus, the intensity, I, which can be measured effortlessly with light sensitive devices (such as a photodetector, see Section 8.7.6) is commonly used. The intensity equals the square of the field strength. Thus, the damping term in (10.18) may be written as

$$I = \mathscr{E}^2 = I_0 \exp\left(-\frac{2\omega k}{c} z\right).\tag{10.19}$$

We define a **characteristic penetration depth**, W, as that distance at which the intensity of the light wave, which travels through a material, has decreased to $1/e$ or 37% of its original value, i.e., when

$$\frac{I}{I_0} = \frac{1}{e} = e^{-1}.\tag{10.20}$$

This definition yields, in conjunction with (10.19),

$$z = W = \frac{c}{2\omega k} = \frac{c}{4\pi v k} = \frac{\lambda}{4\pi k}.\tag{10.21}$$

Table 10.1 presents values for k and W for some materials obtained by using light having $\lambda = 600$ nm.

The inverse of W is sometimes called the (exponential) *attenuation* or the **absorbance**, which is, by making use of (10.21), (10.14), and (10.11),

$$\alpha = \frac{4\pi k}{\lambda} = \frac{2\pi\varepsilon_2}{\lambda n} = \frac{\sigma}{nc\varepsilon_0} = \frac{2\omega k}{c}. \tag{10.21a}$$

It is measured, for example, in cm^{-1}. The **energy loss** per unit length (given for example in decibels, dB, per centimeter) is obtained by multiplying the absorbance, α, with 4.34, see Problem 13.6. (1 dB $= 10 \log I/I_0$.)

10.5. Reflectivity, R, and Transmittance, T

Metals are characterized by a large reflectivity. This stems from the fact that light penetrates a metal only a short distance, as shown in Fig. 10.4 and Table 10.1. Thus, only a small part of the impinging energy is converted into heat. The major part of the energy is reflected (in some cases close to 99%, see Table 10.1). In contrast to this, visible light penetrates into glass much

Table 10.1. Optical constants for some materials ($\lambda = 600$ nm)

	n	k	$W(nm)$	$R\%^b$
Metals				
Copper	0.14	3.35	14.2	95.6
Silver	0.05	4.09	11.7	98.9
Gold	0.21	3.24	14.7	92.9
Aluminum	0.97	6.0	7.9	90.3
Ceramics				
Silica glass (Vycor)	1.46	a		3.50
Soda-lime glass	1.51	a		4.13
Dense flint glass	1.75	a		7.44
Quartz	1.55	a	3×10^8	4.65
Al_2O_3	1.76	a		7.58
Polymers				
Polyethylene	1.51	a		4.13
Polystyrene	1.60	a		5.32
Polytetrafluoroethylene	1.35	a		2.22
Semiconductors				
Silicon	3.94	0.025	1,910	35.42
GaAs	3.91	0.228	209	35.26

[a]The damping constant for dielectrics is about 10^{-7}.
[b]The reflection is considered to have occurred on one reflecting surface only.

farther than into metals, i.e., approximately seven orders of magnitude more, see Table 10.1. As a consequence, very little light is reflected by glass. Nevertheless, a piece of glass about one or two meters thick eventually dissipates a substantial part of the impinging light into heat. (In practical applications, one does not observe this large reduction in light intensity because windows are as a rule only a few millimeters thick.) It should be noted that typical window panes reflect the light on the front as well as on the back surface.

The ratio between the reflected intensity, I_R, and the incoming intensity, I_0, of the light serves as a definition for the **reflectivity**:

$$R = \frac{I_R}{I_0}. \tag{10.22}$$

Quite similarly, one defines the ratio between the transmitted intensity, I_T, and the impinging light intensity as the **transmissivity**, or **transmittance**:

$$T = \frac{I_T}{I_0}. \tag{10.22a}$$

Experiments have shown that for insulators, R depends solely on the index of refraction. For perpendicular incidence one finds

$$R = \frac{(n-1)^2}{(n+1)^2}. \tag{10.23}$$

This equation can also be derived from the Maxwell equations.

We know already that n is generally a complex quantity. By definition, however, R has to remain real. Thus, the modulus of R becomes

$$R = \left| \frac{\hat{n} - 1}{\hat{n} + 1} \right|^2, \tag{10.24}$$

which yields

$$R = \frac{(n - ik - 1)}{(n - ik + 1)} \cdot \frac{(n + ik - 1)}{(n + ik + 1)} = \frac{(n-1)^2 + k^2}{(n+1)^2 + k^2} \tag{10.25}$$

(**Beer equation**). The reflectivity is a unitless materials constant and is often given in percent of the incoming light (see Table 10.1). R is, like the index of refraction, a function of the wavelength of the light.

The reflectivity is also a function of ε_1 and ε_2. We shall derive this relationship by performing a few transformations. Equation (10.25) is rewritten as

$$R = \frac{n^2 + k^2 + 1 - 2n}{n^2 + k^2 + 1 + 2n}, \tag{10.26}$$

(1) $n^2 + k^2 = \sqrt{(n^2 + k^2)^2} = \sqrt{n^4 + 2n^2k^2 + k^4}$

$= \sqrt{n^4 - 2n^2k^2 + k^4 + 4n^2k^2} = \sqrt{(n^2 - k^2)^2 + 4n^2k^2}$

$= \sqrt{\varepsilon_1^2 + \varepsilon_2^2},$ (10.27)

(2) $2n = \sqrt{4n^2} = \sqrt{2(n^2 + k^2 + n^2 - k^2)} = \sqrt{2(\sqrt{\varepsilon_1^2 + \varepsilon_2^2} + \varepsilon_1)}.$ (10.28)

Inserting (10.27) and (10.28) into (10.26) provides

$$R = \frac{\sqrt{\varepsilon_1^2 + \varepsilon_2^2} + 1 - \sqrt{2(\sqrt{\varepsilon_1^2 + \varepsilon_2^2} + \varepsilon_1)}}{\sqrt{\varepsilon_1^2 + \varepsilon_2^2} + 1 + \sqrt{2(\sqrt{\varepsilon_1^2 + \varepsilon_2^2} + \varepsilon_1)}}.$$ (10.29)

10.6. Hagen–Rubens Relation

Our next task is to find a relationship between reflectivity and conductivity. For small frequencies (i.e., $v < 10^{13}$ s^{-1}) the ratio $\sigma/2\pi\varepsilon_0 v$ for metals is very large, that is, $\sigma/2\pi\varepsilon_0 \approx 10^{17}$ s^{-1}. With $\varepsilon \approx 10$ we obtain

$$\frac{\sigma}{2\pi\varepsilon_0 v} \approx \frac{10^{17}}{10^{13}} \gg \varepsilon.$$ (10.30)

Then (10.15) and (10.16) reduce to

$$n^2 \approx \frac{\sigma}{2\pi\varepsilon_0 v} \approx k^2.$$ (10.31)

The reflectivity may now be rewritten by combining the slightly modified equation (10.26) with (10.31) to read

$$R = \frac{n^2 + 2n + 1 + k^2 - 4n}{n^2 + 2n + 1 + k^2} = 1 - \frac{4n}{2n^2 + 2n + 1}.$$ (10.32)

If $2n + 1$ is neglected as small compared to $2n^2$ (which can be done only for small frequencies for which n is much larger than 1), then (10.32) reduces by using (10.31) to

$$R = 1 - \frac{2}{n} = 1 - 4\sqrt{\frac{v}{\sigma}\pi\varepsilon_0}.$$ (10.33)

Finally, we set $\sigma = \sigma_0$ (d.c. conductivity) which is again only permissible for small frequencies, i.e., in the infrared region of the spectrum. This yields the

Hagen–Rubens equation,

$$R = 1 - 4\sqrt{\frac{v}{\sigma_0}\pi\varepsilon_0}, \tag{10.34}$$

which states that in the infrared (IR) region metals with large electrical conductivity are good reflectors. This equation was found empirically by Hagen and Rubens from reflectivity measurements in the IR and was derived theoretically by Drude. As stated above, the Hagen–Rubens relation is only valid at frequencies below 10^{13} s^{-1} or, equivalently, at wavelengths larger than about 30 μm.

Problems

1. Complete the intermediate steps between (10.5) and (10.6).

2. Calculate the conductivity from the index of refraction and the damping constant for copper (0.14 and 3.35, respectively; measurement at room temperature and $\lambda = 0.6$ μm). Compare your result with the conductivity of copper (see Appendix 4). You will notice a difference between these conductivities by several orders of magnitude. Why? (Compare only the same units!)

3. Express n and k in terms of ε and σ (or ε_1 and ε_2) by using $\varepsilon = n^2 - k^2$ and $\sigma = 4\pi\varepsilon_0 nk v$. (Compare with (10.15) and (10.16).)

4. The intensity of Na light passing through a gold film was measured to be about 15% of the incoming light. What is the thickness of the gold film? ($\lambda = 589$ nm; $k = 3.2$. Note: $I = \mathcal{E}^2$.)

5. Calculate the reflectivity of silver and compare it with the reflectivity of flint glass ($n = 1.59$). Use $\lambda = 0.6$ μm.

6. Calculate the characteristic penetration depth in aluminum for Na light ($\lambda = 589$ nm; $k = 6$).

7. Derive the Hagen–Rubens relation from (10.29). (Hint: In the IR region $\varepsilon_2^2 \gg \varepsilon_1^2$ can be used. Justify this approximation.)

8. The transmissivity of a piece of glass of thickness $d = 1$ cm was measured at $\lambda = 589$ nm to be 89%. What would the transmissivity of this glass be if the thickness were reduced to 0.5 cm?

CHAPTER 11

Atomistic Theory of the Optical Properties

11.1. Survey

In the preceding chapter, the optical constants and their relationship to electrical constants were introduced by employing the "continuum theory." The continuum theory considers only macroscopic quantities and interrelates experimental data. No assumptions are made about the structure of matter when formulating equations. Thus, the conclusions which have been drawn from the empirical laws in Chapter 10 should have general validity as long as nothing is neglected in a given calculation. The derivation of the Hagen–Rubens equation has served as an illustrative example for this.

The validity of equations derived from the continuum theory is, however, often limited to frequencies for which the atomistic structure of solids does not play a major role. Experience shows that the atomistic structure does not need to be considered in the far infrared (IR) region. Thus, the Hagen–Rubens equation reproduces the experimental results of metals in the far IR quite well. It has been found, however, that proceeding to higher frequencies (i.e., in the near IR and visible spectrum), the experimentally observed reflectivity of metals decreases faster than predicted by the Hagen–Rubens equation (Fig. 11.1(a)). For the visible and near IR region an atomistic model needs to be considered to explain the optical behavior of metals. Drude did this important step at the turn of the 20th century. He postulated that some electrons in a metal can be considered to be free, i.e., they can be separated from their respective nuclei. He further assumed that the free electrons can be accelerated by an external electric field. This preliminary Drude model was refined by considering that the moving electrons collide with certain metal atoms in a nonideal lattice.

The free electrons are thought to perform periodic motions in the alter-

(a)

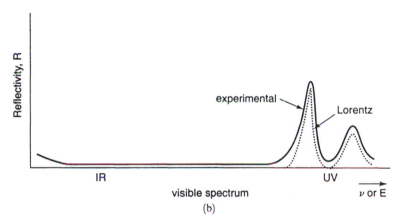

(b)

Figure 11.1. Schematic frequency dependence of the reflectivity of (a) metals, (b) dielectrics, experimentally (solid line) and according to three models.

nating electric field of the light. These vibrations are restrained by the above-mentioned interactions of the electrons with the atoms of a nonideal lattice. Thus, a **friction force** is introduced, which takes this interaction into consideration. The calculation of the frequency dependence of the optical constants is accomplished by using the well-known equations for vibrations, whereby the interactions of electrons with atoms are taken into account by a damping term which is assumed to be proportional to the velocity of the electrons. The free electron theory describes, to a certain degree, the dispersion of the optical constants of metals quite well. This is schematically shown in Fig. 11.1(a), in which the spectral dependence of the reflectivity is plotted for a specific case. The Hagen–Rubens relation reproduces the experimental findings only up to 10^{13} s^{-1}. In contrast to this, the Drude theory correctly reproduces the spectral dependence of R even in the visible spectrum. Proceeding to yet higher frequencies, however, the experimentally found reflectivity eventually rises and then decreases again. Such an **absorption band** cannot be explained by the Drude theory. For its interpretation, a new concept needs to be applied.

Lorentz postulated that the electrons should be considered to be bound to their nuclei and that an external electric field displaces the positive charge of an atomic nucleus against the negative charge of its electron cloud. In other words, he represented each atom as an electric dipole. Retracting forces were thought to occur which try to eliminate the displacement of charges. Lorentz postulated further that the centers of gravity of the electric charges are identical if no external forces are present. However, if one shines light onto a solid, i.e., if one applies an alternating electric field to the atoms, then the dipoles are thought to perform forced vibrations. Thus, a dipole is considered to behave similarly as a mass which is suspended on a spring, i.e., the equations for a harmonic oscillator may be applied. An oscillator is known to absorb a maximal amount of energy when excited near its resonance frequency (Fig. 11.2). The absorbed energy is thought to be dissipated mainly by diffuse radiation. Figure 11.2 resembles an absorption band as shown in Fig. 11.1.

Forty or fifty years ago, many scientists considered the electrons in metals to behave at low frequencies as if they were free and at higher frequencies as if they were bound. In other words, electrons in a metal under the influence of light were described to behave as a series of classical free electrons and a series of classical harmonic oscillators. Insulators and semiconductors, on the other hand, were described by harmonic oscillators only, see Fig. 11.1(b).

We shall now treat the optical constants of materials by applying the above-mentioned theories.

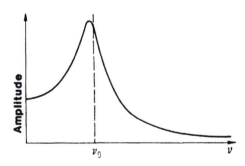

Figure 11.2. Frequency dependence of the amplitude of a harmonic oscillator that is excited to perform forced vibrations, assuming weak damping. ν_0 is the resonance frequency.

11.2. Free Electrons Without Damping

We consider the simplest case at first and assume that the free electrons are excited to perform forced but undamped vibrations under the influence of an external alternating field, i.e., under the influence of light. As explained in Section 11.1, the damping of the electrons is thought to be caused by collisions between electrons and atoms of a nonideal lattice. Thus, we neglect in this section the influence of lattice defects. For simplicity, we treat the one-dimensional case because the result obtained this way does not differ from the general case. Thus, we consider the interaction of plane-polarized light with the electrons. The momentary value of the field strength of a plane-polarized light wave is given by

$$\mathscr{E} = \mathscr{E}_0 \exp(i\omega t), \tag{11.1}$$

where $\omega = 2\pi\nu$ is the angular frequency, t is the time, and $\mathscr{E}_0$ is the maximal value of the field strength. The equation describing the motion of an electron that is excited to perform forced, harmonic vibrations under the influence of light is (see Appendix 1 and (7.6))

$$m\frac{d^2x}{dt^2} = e\mathscr{E} = e\mathscr{E}_0 \exp(i\omega t), \tag{11.2}$$

where e is the electron charge, m is the electron mass, and $e \cdot \mathscr{E}$ is the modulus of the excitation force. The stationary solution of this vibrational equation is obtained by forming the second derivative of the trial solution $x = x_0 \exp(i\omega t)$ and inserting it into (11.2). This yields

$$x = -\frac{e\mathscr{E}}{m4\pi^2\nu^2}. \tag{11.3}$$

The vibrating electrons carry an electric dipole moment, which is the product of the electron charge, e, and displacement, x, see (9.12). The polarization, P, is defined to be the sum of the dipole moments of all N_f free electrons per cubic centimeter:

$$P = exN_f. \tag{11.4}$$

The dielectric constant can be calculated from polarization and electric field strength by combining (9.14) and (9.15):

$$\varepsilon = 1 + \frac{P}{\varepsilon_0 \mathscr{E}}. \tag{11.5}$$

Inserting (11.3) and (11.4) into (11.5) yields

$$\hat{\varepsilon} = 1 - \frac{e^2 N_f}{4\pi^2\varepsilon_0 m\nu^2}. \tag{11.6}$$

(It is appropriate to use in the present case the *complex* dielectric constant, see below.) The dielectric constant equals the square of the index of refraction, n, (see (10.12)). Equation (11.6) thus becomes

$$\hat{n}^2 = 1 - \frac{e^2 N_f}{4\pi^2 \varepsilon_0 m v^2}. \qquad (11.7)$$

We consider two special cases:

(a) For small frequencies, the term $e^2 N_f / 4\pi^2 \varepsilon_0 m v^2$ is larger than one. Then $\hat{n}^2$ is negative and $\hat{n}$ imaginary. An imaginary $\hat{n}$ means that the real part of $\hat{n}$ disappears. Equation (10.25) becomes, for $n = 0$,

$$R = \frac{(n-1)^2 + k^2}{(n+1)^2 + k^2} = \frac{1 + k^2}{1 + k^2} = 1,$$

i.e., the reflectivity is 100% (see Fig. 11.3).

(b) For large frequencies (UV light), the term $e^2 N_f / 4\pi^2 \varepsilon_0 m v^2$ becomes smaller than one. Thus, $\hat{n}^2$ is positive and $\hat{n} \equiv n$ real (but smaller than one). The reflectivity for real values of $\hat{n}$, i.e., for $k = 0$, becomes

$$R = \frac{(n-1)^2}{(n+1)^2},$$

i.e., the material is essentially transparent for these wavelengths (and perpendicular incidence) and therefore behaves optically like an insulator, see Fig. 11.3.

We define a characteristic frequency, v_1, often called the **plasma frequency**, which separates the reflective region from the transparent region (Fig. 11.3). The plasma frequency can also be deduced from (11.6) or (11.7). We observe in these equations that $e^2 N_f / 4\pi^2 \varepsilon_0 m$ must have the unit of the square of a frequency, which we define to be v_1. This yields

$$v_1^2 = \frac{e^2 N_f}{4\pi^2 \varepsilon_0 m}. \qquad (11.8)$$

Because of (11.8) we conclude from (11.6) that the dielectric constant becomes zero at the plasma frequency. $\hat{\varepsilon} = 0$ is the condition for a **plasma**

Figure 11.3. Schematic frequency dependence of an alkali metal according to the free electron theory without damping. v_1 is the plasma frequency.

Table 11.1. Plasma Frequencies and Effective Numbers of Free Electrons for Some Alkali Metals.

Metal	Li	Na	K	Rb	Cs
v_1 (10^{14} s^{-1}), observed	14.6	14.3	9.52	8.33	6.81
v_1 (10^{14} s^{-1}), calculated	19.4	14.3	10.34	9.37	8.33
λ_1 nm ($= c/v_1$), observed	150	210	290	320	360
N_{eff} [free electrons/atom]	0.57	1.0	0.8	0.79	0.67

oscillation, i.e., a fluidlike oscillation of the entire electron gas. We will discuss this phenomenon in detail in Section 13.2.2.

The alkali metals behave essentially as shown in Fig. 11.3. They are transparent in the near UV and reflect the light in the visible region. This result indicates that the s-electrons[5] of the outer shell of the alkali metals can be considered to be free.

Table 11.1 contains some measured, as well as some calculated, plasma frequencies. For the calculations, applying (11.8), *one* free electron per atom was assumed. This means that N_f was set equal to the number of atoms per volume, N_a. (The latter quantity is obtained by using

$$N_a = \frac{N_0 \cdot \delta}{M}, \tag{11.9}$$

where N_0 is the Avogadro constant, $\delta =$ density, and $M =$ atomic mass.)

We note in Table 11.1 that the calculated and the observed values for v_1 are only identical for sodium. This may be interpreted to mean that only in sodium does exactly *one* free electron per atom contribute to the electron gas. For other metals an "effective number of free electrons" is commonly introduced, which is defined to be the ratio between the observed and calculated v_1^2 values:

$$\frac{v_1^2 \text{ (observed)}}{v_1^2 \text{ (calculated)}} = N_{eff}. \tag{11.10}$$

The **effective number of free electrons** is a parameter of great interest, because it is contained in a number of nonoptical equations (such as the Hall constant, electromigration, superconductivity, etc.). Since for most metals the plasma frequency, v_1, cannot be measured as readily as for the alkalis, another avenue for determining N_{eff} has to be found. For reasons which will become clear later, N_{eff} can be obtained by measuring n and k in the red or IR spectrum (i.e., in a frequency range without absorption bands, Fig. 11.1) and by applying

$$N_{eff} = \frac{(1 - n^2 + k^2)v^2 4\pi^2 \varepsilon_0 m}{e^2}. \tag{11.10a}$$

---

[5] See Appendix 3.

Equation (11.10a) follows by combining (11.6) with (10.10) and replacing N_f by N_{eff}.

11.3. Free Electrons With Damping (Classical Free Electron Theory of Metals)

The simple reflectivity spectrum as depicted in Fig. 11.3 is seldom found for metals. We need to refine our model. We postulate that the motion of electrons in metals is damped. More specifically, we postulate that the velocity is reduced by collisions of the electrons with atoms of a nonideal lattice. Lattice defects may be introduced into a solid by interstitial atoms, vacancies, impurity atoms, dislocations, grain boundaries, or thermal motion of the atoms.

To take account of the damping, we add to the vibration equation (11.2) a damping term, $\gamma(dx/dt)$, which is proportional to the velocity (See Appendix 1 and (7.7)):

$$m\frac{d^2x}{dt^2} + \gamma\frac{dx}{dt} = e\mathscr{E} = e\mathscr{E}_0\exp(i\omega t). \tag{11.11}$$

We determine first the damping factor, γ. For this we write a particular solution of (11.11) which is obtained by assuming that the electrons drift under the influence of a steady or slowly varying electric field (see Section 7.3) with a velocity $v' = $ const. through the crystal. (The drift velocity of the electrons, which is caused by an external field, is superimposed on the random motion of the electrons.) The damping is depicted to be a friction force which counteracts the electron motion. $v' = $ const. yields

$$\frac{d^2x}{dt^2} = 0. \tag{11.12}$$

By using (11.12), Equation (11.11) becomes

$$\frac{e\mathscr{E}}{\gamma} = \frac{dx}{dt} = v'. \tag{11.13}$$

The drift velocity is

$$v' = \frac{j}{eN_f} \tag{11.14}$$

(see (7.4)), where j is the current density (i.e., that current which passes through an area of one square centimeter). N_f is the number of free electrons per cubic centimeter. The current density is connected with the d.c. conductivity, σ_0, and the field strength, $\mathscr{E}$, by Ohm's law (7.2),

$$j = \sigma_0\mathscr{E}. \tag{11.15}$$

Inserting (11.14) and (11.15) into (11.13) yields

$$\gamma = \frac{N_f e^2}{\sigma_0}. \tag{11.16}$$

Thus, (11.11) becomes

$$m\frac{d^2x}{dt^2} + \frac{N_f e^2}{\sigma_0}\frac{dx}{dt} = e\mathscr{E} = e\mathscr{E}_0 \exp(i\omega t). \tag{11.17}$$

We note that the damping term in (11.17) is inversely proportional to the conductivity, i.e., proportional to the resistivity. This result makes sense.

The stationary solution of (11.17) is obtained, similarly as in Section 11.2, by differentiating the trial solution $x = x_0 \exp(i\omega t)$ by the time, and inserting first and second derivatives into (11.17), which yields

$$-m\omega^2 x + \frac{N_f e^2}{\sigma_0} x\omega i = \mathscr{E}e. \tag{11.18}$$

Rearranging (11.18) provides

$$x = \frac{\mathscr{E}}{\dfrac{N_f e\omega}{\sigma_0}i - \dfrac{m\omega^2}{e}}. \tag{11.19}$$

Inserting (11.19) into (11.4) yields the polarization,

$$P = \frac{eN_f\mathscr{E}}{\dfrac{N_f e\omega}{\sigma_0}i - \dfrac{m\omega^2}{e}}. \tag{11.20}$$

With (11.20) and (11.5) the complex dielectric constant becomes

$$\hat{\varepsilon} = 1 + \frac{P}{\varepsilon_0\mathscr{E}} = 1 + \frac{1}{\dfrac{2\pi\varepsilon_0 v}{\sigma_0}i - \dfrac{m4\pi^2\varepsilon_0}{N_f e^2}v^2}. \tag{11.21}$$

The term $N_f e^2/m4\pi^2\varepsilon_0$ is set, as in (11.8), equal to v_1^2, which reduces (11.21) to

$$\hat{\varepsilon} = 1 + \frac{1}{\dfrac{2\pi\varepsilon_0 v}{\sigma_0}i - \dfrac{v^2}{v_1^2}} = 1 + \frac{v_1^2}{iv\dfrac{2\pi\varepsilon_0 v_1^2}{\sigma_0} - v^2}. \tag{11.22}$$

The term $2\pi\varepsilon_0 v_1^2/\sigma_0$ in (11.22) has the unit of a frequency. Thus, for abbreviation, we define a **damping frequency**

$$v_2 = \frac{2\pi\varepsilon_0 v_1^2}{\sigma_0} = 2\pi\varepsilon_0 v_1^2\rho_0. \tag{11.23}$$

Table 11.2. Resistivities and Damping Frequencies for Some Metals.

Metal	Li	Na	K	Rb	Cs	Cu	Ag	Au
ρ_0 ($\mu\Omega$ cm)[a]	8.55	4.2	6.15	12.5	20	1.67	1.59	2.35
v_2 (10^{12} s^{-1})	10.1	4.8	3.1	4.82	5.15	4.7	4.35	5.9

[a] *Handbook of Chemistry and Physics*, 1977; room-temperature values.

(Table 11.2 lists values for v_2 which were calculated using experimental ρ_0 and v_1 values.) Now (11.22) becomes

$$\hat{\varepsilon} = 1 + \frac{v_1^2}{ivv_2 - v^2},\tag{11.24}$$

where $\hat{\varepsilon}$ is, as usual, identical to $\hat{n}^2$,

$$(\hat{n})^2 = n^2 - 2nki - k^2 = 1 - \frac{v_1^2}{v^2 - vv_2 i}.\tag{11.25}$$

Multiplying the numerator and denominator of the fraction in (11.25) by the complex conjugate of the denominator $(v^2 + vv_2 i)$ allows us to equate individually real and imaginary parts. This provides the Drude equations for the optical constants,

$$n^2 - k^2 = \varepsilon_1 = 1 - \frac{v_1^2}{v^2 + v_2^2}\tag{11.26}$$

and

$$2nk = \varepsilon_2 = \frac{v_2}{v} \frac{v_1^2}{v^2 + v_2^2},\tag{11.27}$$

with the characteristic frequencies

$$v_1 = \sqrt{\frac{e^2 N_f}{4\pi^2 \varepsilon_0 m}}\tag{11.8}$$

and

$$v_2 = \frac{2\pi \varepsilon_0 v_1^2}{\sigma_0}.\tag{11.23}$$

The absorption, ε_2, and the polarization, ε_1, are plotted in Figs. 11.4 and 11.5 as a function of frequency, making use of (11.27) and (11.26).

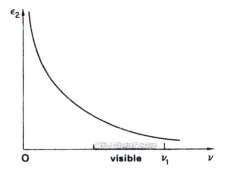

Figure 11.4. The absorption, $\varepsilon_2 = 2nk$, versus frequency, ν, according to the free electron theory (schematic).

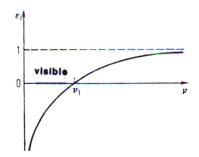

Figure 11.5. The polarization, $\varepsilon_1 = n^2 - k^2$, as a function of frequency according to the Drude theory (schematic).

11.4. Special Cases

For the *UV, visible*, and *near IR* regions, the frequency varies between 10^{14} and 10^{15} s^{-1}. The average damping frequency, ν_2, is 5×10^{12} s^{-1} (Table 11.2). Thus, $\nu^2 \gg \nu_2^2$. Equation (11.27) then reduces to

$$\varepsilon_2 = \frac{\nu_2}{\nu} \frac{\nu_1^2}{\nu^2}. \tag{11.28}$$

With $\nu \approx \nu_1$ (Table 11.1) we obtain

$$\varepsilon_2 \approx \frac{\nu_2}{\nu}. \tag{11.29}$$

Equation (11.29) confirms that ε_2 plotted versus the frequency yields a hyperbola with ν_2 as parameter (Fig. 11.4). For *very small frequencies*

$(v^2 \ll v_2^2)$, we may neglect v^2 in the denominator of (11.27). This yields, with (11.23),

$$nkv = \frac{\sigma}{4\pi\varepsilon_0} = \frac{1}{2}\frac{v_1^2}{v_2} = \frac{\sigma_0}{4\pi\varepsilon_0}. \tag{11.30}$$

Thus, in the far IR the a.c. conductivity, σ, and the d.c. conductivity, σ_0, may be considered to be identical. We have already made use of this condition in Section 10.6. In general, however, σ is *not* identical to the d.c. conductivity, σ_0. (The same is true for the dielectric constant, ε.)

11.5. Reflectivity

The reflectivity of metals is calculated using (10.29) in conjunction with (11.26) and (11.27), see Fig. 11.6. We notice that the experimental behavior for not-too-high frequencies (Fig. 11.1) is essentially reproduced. See also in this context the experimentally obtained reflectivities in Figs. 13.7, 13.10, and 13.12. For higher frequencies, however, we need to resort to a model different from the one discussed so far. This will be done in the next chapter.

Figure 11.6. Calculated spectral reflectivity for a metal using the exact Drude equation (solid line), and the Hagen–Rubens equation (10.34) using $v_1 = 2 \times 10^{15}$ s^{-1} and $v_2 = 3.5 \times 10^{12}$ s^{-1}.

Figure 11.7. An atom is represented as a positively charged core and a surrounding, negatively charged electron cloud (a) in equilibrium and (b) in an external electric field.

11.6. Bound Electrons (Classical Electron Theory of Dielectric Materials)

The preceding sections have shown that the optical properties of metals can be described and calculated quite well in the low-frequency range by applying the free electron theory. We mentioned already that this theory has its limits at higher frequencies, at which we observe that light is absorbed and reflected by metals as well as by nonmetals in a narrow frequency band. To interpret these absorption bands, Lorentz postulated that the electrons are bound to their respective nuclei. He assumed that under the influence of an external electric field, the positively charged nucleus and the negatively charged electron cloud are displaced with respect to each other (Fig. 11.7). An electrostatic force tries to counteract this displacement. For simplicity, we describe the negative charge of the electrons to be united in one point. Thus, we describe the atom in an electric field as consisting of a positively charged core which is bound quasielastically to *one* electron (electric dipole, Fig. 11.8). A bound electron, thus, may be compared to a mass which is

Figure 11.8. Quasi-elastic bound electron in an external electric field (harmonic oscillator).

suspended from a spring. Under the influence of an alternating electric field (i.e., by light), the electron is thought to perform forced vibrations. For the description of these vibrations, the well-known equations of mechanics dealing with a harmonic oscillator may be applied. This will be done now.

We first consider an isolated atom, i.e., we neglect the influence of the surrounding atoms upon the electron. An external electric field with force

$$e\mathscr{E} = e\mathscr{E}_0 \exp(i\omega t) \tag{11.31}$$

periodically displaces an electron from its rest position by a distance x. This displacement is counteracted by a restoring force, $\kappa \cdot x$, which is proportional to the displacement, x. Then, the vibration equation becomes (see Appendix 1)

$$m\frac{d^2x}{dt^2} + \gamma'\frac{dx}{dt} + \kappa x = e\mathscr{E}_0 \exp(i\omega t). \tag{11.32}$$

The factor κ is the *spring constant*, which determines the binding strength between the atom and electron. Each vibrating dipole (e.g., an antenna) loses energy by radiation. Thus, $\gamma'(dx/dt)$ represents the damping of the oscillator by radiation ($\gamma' =$ damping parameter). The stationary solution of (11.32) for weak damping is (see Appendix 1)

$$x = \frac{e\mathscr{E}_0}{\sqrt{m^2(\omega_0^2 - \omega^2)^2 + \gamma'^2\omega^2}} \exp[i(\omega t - \phi)], \tag{11.33}$$

where

$$\omega_0 = 2\pi\nu_0 = \sqrt{\frac{\kappa}{m}} \tag{11.34}$$

is called the **resonance frequency** of the oscillator, i.e., that frequency at which the electron vibrates freely without an external force. ϕ is the phase difference between forced vibration and the excitation force of the light wave. It is defined to be (see Appendix 1)

$$\tan\phi = \frac{\gamma'\omega}{m(\omega_0^2 - \omega^2)} = \frac{\gamma'\nu}{2\pi m(\nu_0^2 - \nu^2)}. \tag{11.35}$$

As in the previous sections, we calculate the optical constants starting with the polarization, P, which is the product of the dipole moment, $e \cdot x$, of *one* dipole times the number of all dipoles (oscillators), N_a. As before, we assumed *one* oscillator per atom. Thus, N_a is identical to the number of atoms per unit volume. We obtain

$$P = exN_a. \tag{11.36}$$

Inserting (11.33) yields

$$P = \frac{e^2 N_a \mathscr{E}_0 \exp[i(\omega t - \phi)]}{\sqrt{m^2(\omega_0^2 - \omega^2)^2 + \gamma'^2\omega^2}}. \tag{11.37}$$

With

$$\exp[i(\omega t - \phi)] = \exp(i\omega t) \cdot \exp(-i\phi) \qquad (11.38)$$

we obtain

$$P = \frac{e^2 N_a \mathscr{E}}{\sqrt{m^2(\omega_0^2 - \omega^2)^2 + \gamma'^2 \omega^2}} \exp(-i\phi), \qquad (11.39)$$

which yields with (11.5) and (10.12)

$$\hat{\varepsilon} = n^2 - k^2 - 2nki = 1 + \frac{e^2 N_a}{\varepsilon_0 \sqrt{m^2(\omega_0^2 - \omega^2)^2 + \gamma'^2 \omega^2}} \exp(-i\phi). \qquad (11.40)$$

Equation (11.40) becomes with[6]

$$\exp(-i\phi) = \cos\phi - i\sin\phi, \qquad (11.41)$$

$$n^2 - k^2 - 2nki = 1 + \frac{e^2 N_a}{\varepsilon_0 \sqrt{m^2(\omega_0^2 - \omega^2)^2 + \gamma'^2 \omega^2}} \cos\phi$$

$$- i \frac{e^2 N_a}{\varepsilon_0 \sqrt{m^2(\omega_0^2 - \omega^2)^2 + \gamma'^2 \omega^2}} \sin\phi. \qquad (11.42)$$

The trigonometric terms in (11.42) are replaced, using (11.35), as follows:

$$\cos\phi = \frac{1}{\sqrt{1 + \tan^2\phi}} = \frac{m(\omega_0^2 - \omega^2)}{\sqrt{m^2(\omega_0^2 - \omega^2)^2 + \gamma'^2 \omega^2}}, \qquad (11.43)$$

$$\sin\phi = \frac{\tan\phi}{\sqrt{1 + \tan^2\phi}} = \frac{\gamma'\omega}{\sqrt{m^2(\omega_0^2 - \omega^2)^2 + \gamma'^2 \omega^2}}. \qquad (11.44)$$

Separating the real and imaginary parts in (11.42) finally provides the optical constants

$$\varepsilon_1 = n^2 - k^2 = 1 + \frac{e^2 m N_a(\omega_0^2 - \omega^2)}{\varepsilon_0 [m^2(\omega_0^2 - \omega^2)^2 + \gamma'^2 \omega^2]},$$

that is,

$$\boxed{\varepsilon_1 = 1 + \frac{e^2 m N_a(v_0^2 - v^2)}{\varepsilon_0 [4\pi^2 m^2(v_0^2 - v^2)^2 + \gamma'^2 v^2]},} \qquad (11.45)$$

---

[6] See Appendix 2.

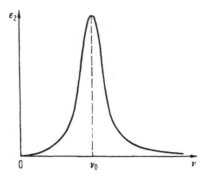

Figures 11.9 and 11.10. Frequency dependence of polarization, $\varepsilon_1 = n^2 - k^2$, and absorption, $\varepsilon_2 = 2nk$, as calculated with (11.45) and (11.46), respectively, using characteristic values for N_a and γ'.

and

$$\varepsilon_2 = 2nk = \frac{e^2 N_a \gamma' \omega}{\varepsilon_0 [m^2(\omega_0^2 - \omega^2)^2 + \gamma'^2 \omega^2]},$$

or

$$\varepsilon_2 = \frac{e^2 N_a \gamma' v}{2\pi\varepsilon_0 [4\pi^2 m^2(v_0^2 - v^2)^2 + \gamma'^2 v^2]}. \qquad (11.46)$$

The frequency dependencies of ε_1 and ε_2 are plotted in Figs. 11.9 and 11.10. Figure 11.9 resembles the dispersion curve for the index of refraction as it is experimentally obtained for dielectrics. Figure 11.10 depicts the absorption product, ε_2, in the vicinity of the resonance frequency, v_0, (absorption band) as experimentally observed for dielectrics. Equations (11.45) and (11.46) reduce to the Drude equations for $v_0 \rightarrow 0$ (no oscillators).

*11.7. Discussion of the Lorentz Equations for Special Cases

11.7.1. High Frequencies

We observe in Fig. 11.10 that ε_2 approaches zero at high frequencies and far away from any resonances (absorption bands). In the same frequency region, $\varepsilon_1 = n^2 - k^2$ and, thus, essentially n, assumes the constant value 1 (Fig. 11.9). This is consistent with experimental observations that X-rays are not refracted and are not absorbed by many materials. (Note, however, that highly energetic X-rays interact with the *inner* electrons, i.e., they may be absorbed by the K, L,..., etc. electrons. Metals are, therefore, opaque for high-energetic X-rays).

11.7.2. Small Damping

We consider the case for which the radiation-induced energy loss of the oscillator is very small. Then, γ' is small. With $\gamma'^2 v^2 \ll 4\pi^2 m^2 (v_0^2 - v^2)^2$ (which is only valid for $v \neq v_0$), equation (11.45) reduces to

$$\varepsilon_1 = n^2 - k^2 = 1 + \frac{e^2 N_a}{4\pi^2 \varepsilon_0 m(v_0^2 - v^2)}. \tag{11.47}$$

Figure 11.11 depicts a sketch of (11.47). We observe that for small damping, ε_1 (and thus essentially n^2) approaches infinity near the resonance frequency. A dispersion curve such as Fig. 11.11 is indeed observed for many dielectrics (glass, etc.).

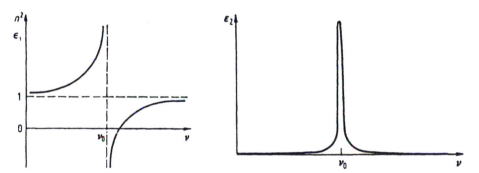

Figures 11.11 and 11.12. The functions $\varepsilon_1 (n^2)$ and ε_2, respectively, versus frequency according to the bound electron theory for the special case of small damping.

11.7.3. Absorption Near v_0

Electrons absorb most energy from light at the resonance frequency, i.e., ε_2 has a maximum near v_0. For small damping, the absorption band becomes an absorption line (see Fig. 11.12). Inserting $v = v_0$ into (11.46) yields

$$\varepsilon_2 = \frac{e^2 N_a}{2\pi \varepsilon_0 \gamma' v_0}, \tag{11.48}$$

which shows that the absorption becomes large for small damping (γ').

11.7.4. More Than One Oscillator

At the beginning of Section 11.6 we assumed that *one* electron is quasielastically bound to a given nucleus; in other words, we assumed *one* oscillator per atom. This assumption is certainly a gross simplification, as one can deduce from the occurrence of multiple absorption bands in experimental optical spectra. Thus, each atom has to be associated with a number of i oscillators, each having an oscillator strength, f_i. The ith oscillator vibrates with its resonance frequency, ν_{0i}. The related damping constant is γ_i'. (This description has its equivalent in the mechanics of a system of mass points having one basic frequency and higher harmonics.) If all oscillators are taken into account, (11.45) and (11.46) become

$$\varepsilon_1 = n^2 - k^2 = 1 + \frac{e^2 m N_a}{\varepsilon_0} \sum_i \frac{f_i(\nu_{0i}^2 - \nu^2)}{4\pi^2 m^2(\nu_{0i}^2 - \nu^2)^2 + \gamma_i'^2 \nu^2}, \quad (11.49)$$

$$\varepsilon_2 = 2nk = \frac{e^2 N_a}{2\pi\varepsilon_0} \sum_i \frac{f_i \nu \gamma_i'}{4\pi^2 m^2(\nu_{0i}^2 - \nu^2)^2 + \gamma_i'^2 \nu^2}. \quad (11.50)$$

Equations (11.49) and (11.50) reduce for weak damping (see above) to

$$\varepsilon_1 = n^2 - k^2 \approx n^2 = 1 + \frac{e^2 N_a}{4\pi^2 \varepsilon_0 m} \sum_i \frac{f_i}{\nu_{0i}^2 - \nu^2}, \quad (11.51)$$

$$\varepsilon_2 = 2nk = \frac{e^2 N_a}{8\pi^3 \varepsilon_0 m^2} \sum_i \frac{f_i \nu \gamma_i'}{(\nu_{0i}^2 - \nu^2)^2}. \quad (11.52)$$

11.8. Contributions of Free Electrons and Harmonic Oscillators to the Optical Constants

In the previous section, we ascribed two different properties to the electrons of a solid. In Section 11.4 we postulated that N_f electrons move freely in metals under the influence of an electric field and that this motion is damped by collisions of the electrons with vibrating lattice atoms and lattice defects. In Section 11.6 we postulated that a certain number of electrons are quasi-elastically bound to N_a atoms which are excited by light to perform forced vibrations. The energy loss was thought to be by radiation.

The optical properties of metals may be described by postulating a certain number of free electrons and a certain number of harmonic oscillators. Both the free electrons and the oscillators contribute to the polarization. Thus, the equations for the optical constants may be rewritten, by combining (11.26),

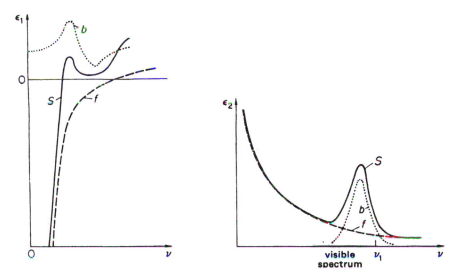

Figures 11.13 and 11.14. Frequency dependence of ε_1 and ε_2 according to (11.53) and (11.54). ($i = 1$). f = free electron theory; b = bound electron theory; S = summary curve (schematic).

(11.27), (11.49), and (11.50),

$$\varepsilon_1 = 1 - \frac{v_1^2}{v^2 + v_2^2} + \frac{e^2 m N_a}{\varepsilon_0} \sum_i \frac{f_i(v_{0i}^2 - v^2)}{4\pi^2 m^2 (v_{0i}^2 - v^2)^2 + \gamma_i'^2 v^2}, \qquad (11.53)$$

$$\varepsilon_2 = 2nk = \frac{v_2}{v} \frac{v_1^2}{v^2 + v_2^2} + \frac{e^2 N_a}{2\pi\varepsilon_0} \sum_i \frac{f_i v \gamma_i'}{4\pi^2 m^2 (v_{0i}^2 - v^2)^2 + \gamma_i'^2 v^2}. \qquad (11.54)$$

Figures 11.13 and 11.14 depict schematically the frequency dependence of ε_1 and ε_2 as obtained by using (11.53) and (11.54). These figures also show the contributions of free and bound electrons on the optical constants. The experimentally found frequency dependence of ε_1 and ε_2 resembles these calculated spectra quite well. We will elaborate on this in Chapter 13, in which experimental results are presented.

Problems

1. Calculate the reflectivity of sodium in the frequency ranges $v > v_1$ and $v < v_1$ using the theory for free electrons without damping. Sketch R versus frequency.

2. The plasma frequency, v_1, can be calculated for the alkali metals by assuming *one* free electron per atom, i.e., by substituting for N_f the number of atoms per unit volume (atomic density, N_a). Calculate v_1 for potassium and lithium.

3. Calculate N_{eff} for sodium and potassium. For which of these two metals is the assumption of *one* free electron per atom justified?

4. What is the meaning of the frequencies v_1 and v_2? In which frequency ranges are they situated compared to visible light?

5. Calculate the reflectivity of gold at $v = 9 \times 10^{12}$ s^{-1} from its conductivity. Is the reflectivity increasing or decreasing at this frequency when the temperature is increased? Explain.

6. Calculate v_1 and v_2 for silver (0.5×10^{23} free electrons per cubic centimeter).

7. The experimentally found dispersion of NaCl is as follows:

λ [μm]	0.3	0.4	0.5	0.7	1	2	5
n	1.607	1.568	1.552	1.539	1.532	1.527	1.519

Plot these results along with calculated values obtained by using the equations of the "bound electron theory" assuming small damping. Let

$$\frac{e^2 N_a}{4\pi^2 \varepsilon_0 m} = 1.81 \times 10^{30} \text{ s}^{-2} \quad \text{and} \quad v_0 = 1.47 \times 10^{15} \text{ s}^{-1}.$$

8. The optical properties of an absorbing medium can be characterized by various sets of parameters. One such set is the index of refraction and the damping constant. Explain the physical significance of those parameters, and indicate how they are related to the complex dielectric constant of the medium. What other sets of parameters are commonly used to characterize the optical properties? Why are there always "sets" of parameters?

9. Describe the damping mechanisms for free electrons and bound electrons.

10. Why does it make sense that we assume *one* free electron per atom for the alkali metals?

11. Derive the Drude equations from (11.45) and (11.46) by setting $v_0 \rightarrow 0$.

12. Calculate the effective number of free electrons per cubic centimeter and per atom for silver from its optical constants ($n = 0.05$ and $k = 4.09$ at 600 nm). (*Hint*: Use the free electron mass.) How many free electrons per atom would you expect? Does the result make sense? Why may we use the free electron theory for this wavelength?

13. *Computer problem.* Plot (11.26), (11.27), and (10.29) for various values of v_1 and v_2. Start with $v_1 = 2 \times 10^{15}$ s^{-1} and $v_2 = 3.5 \times 10^{12}$ s^{-1}.

14. *Computer problem.* Plot (11.45), (11.46), and (10.29) for various values of N_a, γ', and v_0. Start with $v_0 = 1.5 \times 10^{15}$ s^{-1} and $N_a = 2.2 \times 10^{22}$ cm^{-3} and vary γ' between 100 and 0.1.

15. *Computer problem.* Plot (11.51), (11.52), and (10.29) by varying the parameters as in the previous problems. Use one, two or three oscillators. Try to "fit" an experimental curve such as the ones in Figs. 13.10 or 13.11.

CHAPTER 12

Quantum Mechanical Treatment of the Optical Properties

12.1. Introduction

We assumed in the preceding chapter that the electrons behave like particles. This working hypothesis provided us (at least for small frequencies) with equations which reproduce the optical spectra of solids reasonably well. Unfortunately, the treatment had one flaw: For calculation and interpretation of the infrared (IR) absorption we used the concept that electrons in metals are free; whereas the absorption bands in the visible and ultraviolet (UV) spectrum could only be explained by postulating harmonic oscillators. From the classical point of view, however, it is not immediately evident why the electrons should behave freely at low frequencies and respond as if they would be bound at higher frequencies. An unconstrained interpretation for this is only possible by applying wave mechanics. This will be done in the present chapter. We make use of the material presented in Chapters 5 and 6.

12.2. Absorption of Light by Interband and Intraband Transitions

When light (photons) having sufficiently large energy impinges on a solid, the electrons in this crystal are thought to be excited into a higher energy level, provided that unoccupied higher energy levels are available. For these transitions the total momentum of electrons and photons must remain constant (conservation of momentum). For optical frequencies, the momentum of a photon, and thus its wave vector $\mathbf{k}_{phot} = p/h$ (see (4.7)), is much smaller than that of an electron. Thus, $\mathbf{k}_{phot}$ is much smaller than the diameter of the

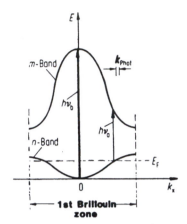

Figure 12.1. Electron bands and direct interband transitions in a reduced zone. (Compare with Fig. 5.4.)

Brillouin zone (Fig. 12.1). Electron transitions at which **k** remains constant (vertical transitions) are called "**direct interband transitions.**" Optical spectra for metals are dominated by direct interband transitions.

Another type of interband transition is possible however. It involves the absorption of a light quantum under participation of a *phonon* (lattice vibration quantum, see Chapter 20). To better understand these "**indirect interband transitions**" (Fig. 12.2) we have to know that a phonon can only absorb very small energies, but is able to absorb a large momentum comparable to that of an electron. During an indirect interband transition, the excess momentum (i.e., the wave number vector) is transferred to the lattice (or

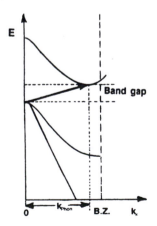

Figure 12.2. Indirect interband transition. (The properties of phonons are explained in Chapter 20.)

is absorbed from the lattice). In other words, a phonon is exchanged with the solid. Indirect interband transitions may be disregarded for the interpretation of metal spectra, because they are generally weaker than direct transitions by two or three orders of magnitude. They are only observed in the absence of direct transitions. In the case of semiconductors, however, and for the interpretation of photoemission, indirect interband transitions play an important role.

We now make use of the simplified model depicted in Fig. 12.1 and consider direct interband transitions from the n to the m band. The smallest photon energy in this model is absorbed by those electrons whose energy equals the Fermi energy, E_F, i.e., by electrons which already possess the highest possible energy at $T = 0$ K. This energy is marked in Fig. 12.1 by $h v_a$. Similarly, $h v_b$ is the *largest* energy, which leads to an interband transition from the n to the m band. In the present case, a variety of interband transitions may take place between the energy interval $h v_a$ and $h v_b$.

Interband transitions are also possible by skipping one or more bands, which occur by involving photons with even larger energies. Thus, a multitude of absorption bands are possible. These bands may partially overlap.

As an example for interband transitions in an actual case, we consider the band diagram for copper. In Fig. 12.3, a portion of Fig. 5.22 is shown, i.e., the pertinent bands around the L-symmetry point are depicted. The interband transition having the smallest possible energy difference is shown to occur between the upper d-band and the Fermi energy. This smallest energy is called the "**threshold energy for interband transitions**" (or the "**fundamental edge**") and is marked in Fig. 12.3 by a solid arrow. We mention in passing that this transition, which can be stimulated by a photon energy of 2.2 eV, is

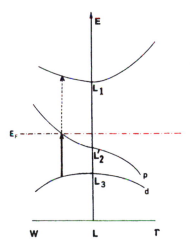

Figure 12.3. Section of the band diagram for copper (schematic). Two pertinent interband transitions are shown with arrows. The smallest possible interband transition occurs from a filled d-state to an unfilled state just above the Fermi energy.

responsible for the red color of copper. At slightly higher photon energies, a second transition takes place, which originates *from* the Fermi energy. It is marked in Fig. 12.3 by a dashed arrow. Needless to say, many more transitions are possible. They can take place over a wide range in the Brillouin zone. This will become clearer in Chapter 13 when we return to the optical spectra of materials and their interpretation.

We now turn to another photon-induced absorption mechanism. Under certain conditions photons may excite electrons into a higher energy level *within the same band*. This occurs with participation of a phonon, i.e., a lattice vibration quantum. We call such a transition, appropriately, an **intraband transition** (Fig. 12.4). It should be kept in mind, however, that because of the Pauli principle, electrons can only be excited into *empty* states. Thus, intraband transitions are mainly observed in metals because metals have unfilled electron bands. We recognize, however, that semiconductors with high doping levels or which are kept at high temperatures may likewise have partially filled conduction bands.

*Intra*band transitions are equivalent to the behavior of *free* electrons in classical physics, i.e., to the "classical infrared absorption." Insulators and semiconductors have no classical infrared absorption because their bands are either completely filled or completely empty (except at high temperatures and due to doping). This explains why some insulators (such as glass) are transparent in the visible spectrum. The largest photon energy, E_{max}, that can be absorbed by means of an intraband transition corresponds to an excitation from the lower to the upper band edge, see Fig. 12.4. All energies smaller than E_{max} are absorbed continuously.

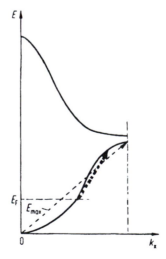

Figure 12.4. Intraband transitions. The largest energy that can be absorbed by intraband transitions is obtained by projecting the arrow marked "E_{max}" onto the energy axis.

In summary, at low photon energies, *intra*band transitions (if possible) are the prevailing absorption mechanism. Intraband transitions are not quantized and occur essentially in metals only. Above a critical light energy *inter*band transitions set in. Only certain energies or energy intervals are absorbed in this case. The onset of this absorption mechanism depends on the energy difference between the bands in question. *Interband* transitions occur in metals as well as in insulators or semiconductors. They are analogous to optical excitations in solids with *bound* electrons. In an intermediate frequency range, interband as well as intraband transitions may take place (see Figure 11.1).

12.3. Optical Spectra of Materials

Optical spectra are the principal means to obtain experimentally the band gaps and energies for interband transitions. For isolated atoms and ions, the absorption and emission spectra are known to be extremely sharp. Thus, absorption and emission energies for atoms can be determined with great accuracy. The same is basically true for molecular spectra. In contrast to this, the optical spectra of solids are rather broad. This stems from the high particle density in solids and from the interatomic interactions, which split the atomic levels into quasi-continuous bands. The latter extend through the three-dimensional momentum space of a Brillouin zone.

A further factor has to be considered, too. Plain *reflection* spectra of solids are, in general, not too useful for the deduction of transition energies, mainly because R is a rather involved function of ε_1 and ε_2 (see (10.29)). Thus, ε_2 (i.e., absorption) spectra are often utilized instead. The characteristic features in the ε_2-spectra of solids stem from discontinuities in the energy profile of the density of states. However, relatively sharp features in ε_2-spectra are superimposed on noncharacteristic transitions from other parts of the Brillouin zone. In other words, the ε_2-spectra derive their shape from a summation over extended, rather than localized, regions in the Brillouin zone. Modulated optical spectra (see Section 13.1.3) separate the small contributions stemming from points of high symmetry (such as the centers and edges of a Brillouin zone) from the general, much larger background. This will become clearer in the next chapter.

*12.4. Dispersion

To calculate the behavior of electrons in a periodic lattice we used, in Section 4.4, the periodic potential shown in Fig. 4.9. We implied at that time that the potential does not vary with time. This proposition needs to be dropped when the interaction of light with a solid is considered. The alternating electric field of the light which impinges on the solid perturbs the potential field of the lattice periodically. Thus, we need to add to the potential energy a correction term, the so-called perturbation potential, V',

$$V = V_0 + V' \tag{12.1}$$

(V_0 = unperturbed potential energy). It goes without saying that this perturbational potential oscillates with the frequency, ν, of the light.

We consider, as always, plane-polarized light. The momentary value of the field strength, $\mathscr{E}$, is

$$\mathscr{E} = A \cos \omega t, \tag{12.2}$$

where A is the maximal value of the field strength. Then, the perturbation potential (potential energy of the perturbation, or force times displacement x) is

$$V' = e\mathscr{E}x = eA \cos(\omega t) \cdot x. \tag{12.3}$$

Since the potential now varies with time, we need to make use of the time-dependent Schrödinger equation (3.8),

$$\nabla^2 \Psi - \frac{2m}{\hbar^2} V\Psi - \frac{2im}{\hbar} \frac{\partial \Psi}{\partial t} = 0, \tag{12.4}$$

which reads, with (12.1) and (12.3),

$$\nabla^2 \Psi - \frac{2m}{\hbar^2} (V_0 + eAx \cos \omega t)\Psi - \frac{2im}{\hbar} \frac{\partial \Psi}{\partial t} = 0. \tag{12.5}$$

Our goal is to calculate the optical constants from the polarization, in a similar way as it was done in Sections 11.2, 11.3, and 11.6. We have to note, however, the following: In wave mechanics, the electron is not considered to be a point, but instead is thought to be "smeared" about the space $d\tau$. The locus of the electron in classical mechanics is thus replaced by the probability, $\Psi\Psi^*$, of finding an electron in space (see (2.12)). The classical polarization

$$P = Nex$$

(11.4) is replaced in wave mechanics by

$$P = Ne \int x\Psi\Psi^* \, d\tau. \tag{12.6}$$

We seek to find a solution Ψ of the *perturbed Schrödinger equation* (12.5) and calculate from that the *norm* $\Psi\Psi^*$; then, by using (12.6) we can calculate the polarization P. The equation for the optical constants thus obtained is given in (12.31).

The detailed calculation of this approach will be given below. The first step is to transform the space- *and* time-dependent Schrödinger equation into a Schrödinger equation that is only space-dependent. The perturbed

Schrödinger equation (12.5) is rewritten, using the Euler equation[7] $\cos \rho = \frac{1}{2}(e^{i\rho} + e^{-i\rho})$, as

$$\nabla^2 \Psi - \frac{2m}{\hbar^2} V_0 \Psi - \frac{2im}{\hbar} \frac{\partial \Psi}{\partial t} = \frac{2m}{\hbar^2} eAx \frac{1}{2}[e^{i\omega t} + e^{-i\omega t}]\Psi. \qquad (12.7)$$

Now, the left side of (12.7) has the form of the unperturbed Schrödinger equation (12.4). We assume that the perturbation is very small. Then, we can insert in the perturbation term (right side of (12.7)) the expression (3.4), and get

$$\Psi_i^0(x, y, z, t) = \psi_i^0(x, y, z)e^{i\omega_i t} \qquad (12.8)$$

for the unperturbed ith eigenfunction. This yields

$$\nabla^2 \Psi - \frac{2m}{\hbar^2} V_0 \Psi - \frac{2im}{\hbar} \frac{\partial \Psi'}{\partial t} = \frac{m}{\hbar^2} eAx\psi_i^0[e^{i(\omega_i + \omega)t} + e^{i(\omega_i - \omega)t}]. \qquad (12.9)$$

The right-hand side will be contracted to simplify the calculation:

$$\nabla^2 \Psi - \frac{2m}{\hbar^2} V_0 \Psi - \frac{2im}{\hbar} \frac{\partial \Psi}{\partial t} = \frac{m}{\hbar^2} eAx\psi_i^0 e^{i(\omega_i \pm \omega)t}. \qquad (12.10)$$

To solve (12.10), we seek a trial solution which consists of an unperturbed solution and two terms with the angular frequencies $(\omega_i + \omega)$ and $(\omega_i - \omega)$:

$$\Psi = \Psi_i^0 + \psi_+ e^{i(\omega_i + \omega)t} + \psi_- e^{i(\omega_i - \omega)t}. \qquad (12.11)$$

This trial solution is condensed as before

$$\Psi = \Psi_i^0 + \psi_\pm e^{i(\omega_i \pm \omega)t}. \qquad (12.12)$$

Equation (12.12) is differentiated twice with respect to space and once with respect to time, and the results are inserted into (12.10). This yields

$$\underline{\nabla^2 \Psi_i^0} + \nabla^2 \psi_\pm e^{i(\omega_i \pm \omega)t} - \underline{\frac{2m}{\hbar^2} V_0 \Psi_i^0} - \frac{2m}{\hbar^2} V_0 \psi_\pm e^{i(\omega_i \pm \omega)t}$$

$$\underline{- \frac{2im}{\hbar} \frac{\partial \Psi_i^0}{\partial t}} + \frac{2m}{\hbar}(\omega_i \pm \omega)\psi_\pm e^{i(\omega_i \pm \omega)t} = \frac{m}{\hbar^2} eAx\psi_i^0 e^{i(\omega_i \pm \omega)t}. \qquad (12.13)$$

The underlined terms in (12.13) vanish according to (12.4) if Ψ_i^0 is the solution to the unperturbed Schrödinger equation. In the remaining terms, the exponential factors can be cancelled, which yields, with $\hbar\omega = h\nu = E$,

$$\nabla^2 \psi_\pm + \frac{2m}{\hbar^2} \psi_\pm (E_i \pm h\nu - V_0) = \frac{m}{\hbar^2} eAx\psi_i^0. \qquad (12.14)$$

In writing (12.14) we have reached our first goal, i.e., to obtain a time-independent, perturbed Schrödinger equation. We solve this equation with a

---

[7] See Appendix 2.

procedure that is common in perturbation theory. We develop the function $x\psi_i^0$ (the right side of (12.14)) in a series of eigenfunctions

$$x\psi_i^0 = a_{1i}\psi_1^0 + a_{2i}\psi_2^0 + \cdots + a_{ni}\psi_n^0 + \cdots = \sum a_{ni}\psi_n^0, \tag{12.15}$$

multiply (12.15) by ψ_n^{0*}, and integrate over the entire space $d\tau$. Then, due to $\int \psi\psi^* \, d\tau = 1$ (3.15) and $\int \psi_m \psi_n^* \, d\tau = 0$ (for $m \neq n$), we obtain

$$\int x\psi_i^0 \psi_n^{0*} \, d\tau = a_{1i} \underbrace{\int \psi_1^0 \psi_n^{0*} \, d\tau}_{0} + \cdots + a_{ni} \underbrace{\int \psi_n^0 \psi_n^{0*} \, d\tau}_{1} + \cdots = a_{ni}. \tag{12.16}$$

Similarly, we develop the function $\psi_\pm$ in a series of eigenfunctions

$$\psi_\pm = \sum b_{\pm n}\psi_n^0. \tag{12.17}$$

Inserting (12.15) and (12.17) into (12.14) yields

$$\sum b_{\pm n}\left(\underline{\nabla^2\psi_n^0 + \frac{2m}{\hbar^2}E_i\psi_n^0} \pm \frac{2m}{\hbar^2}h\nu\psi_n^0 - \frac{2m}{\hbar^2}V_0\psi_n^0\right) = \frac{m}{\hbar^2}eA\sum a_{ni}\psi_n^0. \tag{12.18}$$

Rewriting the unperturbed time-independent Schrödinger equation (3.1) yields

$$\nabla^2\psi_n^0 - \frac{2m}{\hbar^2}V_0\psi_n^0 = -\frac{2m}{\hbar^2}E_n\psi_n^0. \tag{12.19}$$

Equation (12.19) shows that the underlined terms in (12.18) may be equated to the right side of (12.19). Thus, (12.18) may be rewritten as

$$\frac{2m}{\hbar^2}\sum \psi_n^0 b_{\pm n}(E_i - E_n \pm h\nu) = \frac{2m}{\hbar^2}\frac{eA}{2}\sum \psi_n^0 a_{ni}. \tag{12.20}$$

Comparing the coefficients in (12.20) yields, with

$$E_i - E_n = E_{ni} = h\nu_{ni}, \tag{12.21}$$

the following expression:

$$b_{\pm n} = \frac{eA a_{ni}}{2(E_i - E_n \pm h\nu)} = \frac{eA a_{ni}}{2h(\nu_{ni} \pm \nu)}. \tag{12.22}$$

Now we are able to determine the functions ψ_+ and ψ_- by using (12.22) and (12.17). We insert these functions together with (3.4) into the trial solution (12.11) and obtain a solution for the time-dependent, perturbed Schrödinger equation (12.5)

$$\Psi = \psi_i^0 e^{i\omega_i t} + \frac{1}{2h}\sum eA a_{ni}\psi_n^0\left[\frac{e^{i(\omega_i+\omega)t}}{\nu_{ni} + \nu} + \frac{e^{i(\omega_i-\omega)t}}{\nu_{ni} - \nu}\right], \tag{12.23}$$

and thus

$$\Psi^* = \psi_i^{0*} e^{-i\omega_i t} + \frac{1}{2h} \sum eA a_{ni}^* \psi_n^{0*} \left[\frac{e^{-i(\omega_i+\omega)t}}{\nu_{ni}+\nu} + \frac{e^{-i(\omega_i-\omega)t}}{\nu_{ni}-\nu} \right]. \tag{12.24}$$

In order to write the polarization (12.6) we have to form the product $\Psi\Psi^*$. As can be seen from (12.23) and (12.24), this calculation yields time-dependent as well as time-independent terms. The latter ones need not be considered here, since they provided only an additive constant to the polarization (light scattering). The time-dependent part of the norm $\Psi\Psi^*$ is

$$\Psi\Psi^* = \frac{eA}{2h} \left[\sum a_{ni}^* \psi_n^{0*} \psi_i^0 \underbrace{\left(\frac{e^{-i\omega t}}{\nu_{ni}+\nu} + \frac{e^{i\omega t}}{\nu_{ni}-\nu} \right)}_{Q} \right.$$

$$\left. + \sum a_{ni} \psi_n^0 \psi_i^{0*} \underbrace{\left(\frac{e^{i\omega t}}{\nu_{ni}+\nu} + \frac{e^{-i\omega t}}{\nu_{ni}-\nu} \right)}_{R} \right]. \tag{12.25}$$

To simplify, we abbreviate the terms in parentheses by Q and R, respectively. The polarization (12.6) is then

$$P = \frac{Ne^2 A}{2h} \left[\sum a_{ni}^* Q \underbrace{\int x\psi_n^{0*}\psi_i^0 \, d\tau}_{a_{ni}} + \sum a_{ni} R \underbrace{\int x\psi_n^0 \psi_i^{0*} \, d\tau}_{a_{ni}^*} \right], \tag{12.26}$$

which reduces, with (12.16),

$$\int x\psi_i^0 \psi_n^{0*} \, d\tau = a_{ni}$$

and

$$a_{ni} \cdot a_{ni}^* = |a_{ni}|^2 \equiv a_{ni}^2 \tag{12.27}$$

to

$$P = \frac{Ne^2 A}{2h} \sum a_{ni}^2 (Q + R). \tag{12.28}$$

A numerical calculation applying the above-quoted Euler equation yields

$$Q + R = \frac{2\nu_{ni}e^{-i\omega t}}{\nu_{ni}^2 - \nu^2} + \frac{2\nu_{ni}e^{i\omega t}}{\nu_{ni}^2 - \nu^2} = \frac{4\nu_{ni}\cos\omega t}{\nu_{ni}^2 - \nu^2}, \tag{12.29}$$

which gives, with (12.2),

$$P = \frac{Ne^2 \mathscr{E}}{\pi h} \sum a_{ni}^2 \frac{\nu_{ni}}{\nu_{ni}^2 - \nu^2}. \tag{12.30}$$

Finally, we make use of (10.13) and (11.5) and obtain with, (12.30),

$$\varepsilon_1 = n^2 - k^2 = 1 + \frac{Ne^2}{\varepsilon_0 \pi \hbar} \sum a_{ni}^2 \frac{v_{ni}}{v_{ni}^2 - v^2}. \tag{12.31}$$

Equation (12.31) is the sought-after relation for the optical properties of solids, obtained by wave mechanics. It is similar in form to the classical dispersion equation (11.51). A comparison of classical and quantum mechanical results might be helpful to better understand the meaning of the empirically introduced oscillator strength, f_i. We obtain

$$f_i = \frac{4\pi m}{\hbar} a_{ni}^2 v_{ni}. \tag{12.32}$$

We know that $h v_{ni}$ is that energy which an electron absorbs when it is excited from the n-band into the i-band (e.g., the m-band). Thus, the resonance frequency, v_{oi}, of the ith oscillator introduced in Section 11.7.4 is replaced in wave mechanics by a frequency, v_{ni}, that corresponds to an allowed electron transition from the nth into the ith band. Furthermore, we see from (12.16) that a_{ni} is proportional to the probability of an electron transition from the nth into the ith band. The oscillator strength, f_i, is, therefore, essentially the probability for a certain interband transition.

Problems

1. What information can be gained from the quantum mechanical treatment of the optical properties of metals which cannot be obtained by the classical treatment?

2. What can we conclude from the fact that the spectral reflectivity of a metal (e.g., copper) has "structure"?

3. Below the reflection spectra for two materials A and B are given.
   (a) What type of material belongs to reflection spectrum A, what type to B? (Justify). Note the scale difference!
   (b) For which colors are these (bulk) materials transparent?
   (c) What is the approximate threshold energy for interband transitions for these materials?
   (d) For which of the materials would you expect intraband transitions in the infrared region? (Justify.)
   (e) Why do these intraband transitions occur in this region?

4. What is the smallest possible energy for interband transitions for aluminum? (*Hint*: Consult the band diagram in Fig. 5.21.)

5. Are intraband transitions possible in semiconductors at high temperatures?

CHAPTER 13

Applications

13.1. Measurement of the Optical Properties

The measurement of the optical properties of solids is simple in principle, but can be involved in practice. This is so because many bulk solids (particularly metals) are opaque, so that the measurements have to be taken in reflection. Light penetrates about 10 nm into a metal (see Table 10.1). As a consequence, the optical properties are basically measured near the surface, which is susceptible to oxidation, deformation (polishing), or contamination by adsorbed layers. One tries to alleviate the associated problems by utilizing ultrahigh vacuum, vapor deposition, sputtering, etc. Needless to say, the method by which a given sample was prepared may have an effect on the numerical value of its optical properties.

Let us assume that the surface problems have been resolved. Then, still another problem remains. The most relevant optical properties, namely, n, k, ε_1, ε_2, and the energies for interband transitions cannot be easily deduced by simply measuring the reflectivity, i.e., the ratio between reflected and incident intensity. Thus, a wide range of techniques have been developed in the past century to obtain the above-mentioned parameters. Only three methods will be briefly discussed here. It should be mentioned, however, that thirty or forty other techniques could be easily presented. They all have certain advantages for some specific applications and disadvantages for others. The reader who is not interested in the measurement of optical properties may skip the next three sections for the time being and return to them at a later time.

*13.1.1. Kramers–Kronig Analysis (Dispersion Relations)

This method was very popular in the 1960s and involves the measurement of the reflectivity over a wide spectral range. A relationship exists between real and imaginary terms of any complex function, which enables one to calculate *one* component of a complex quantity if the other one is known. In the present case, one calculates the phase jump, δ', (between the reflected and incident ray) from the reflectivity, R, which was measured at a given frequency, v. This is accomplished by the Kramers–Kronig relation,

$$\delta'(v_x) = \frac{1}{\pi} \int_0^\infty \frac{d \ln \rho}{dv} \ln \left| \frac{v + v_x}{v - v_x} \right| dv, \tag{13.1}$$

where

$$\rho = \sqrt{R} = \sqrt{\frac{I_R}{I_0}} \tag{13.2}$$

is obtained from the reflected intensity, I_R, and the incident intensity, I_0, of the light. The optical constants are calculated by applying

$$n = \frac{1 - \rho^2}{1 + \rho^2 + 2\rho \cos \delta'} \tag{13.3}$$

and

$$k = \frac{2\rho \sin \delta'}{1 + \rho^2 + 2\rho \cos \delta'}. \tag{13.4}$$

Equation (13.1) shows that the reflectivity should be known in the entire frequency range (i.e., between $v = 0$ and $v = \infty$). Since measured values can hardly be obtained for such a large frequency range, one usually extrapolates the reflectivity beyond the experimental region using theoretical or phenomenological considerations. Such an extrapolation would not cause a substantial error if one could assume that no interband transitions exist beyond the measured spectral range. This assumption is probably valid only on rare occasions. (For details, see specialized books listed at the end of Part III.)

*13.1.2. Spectroscopic Ellipsometry

This technique was developed in its original form at the turn of the 20th century. The underlying idea is as follows: If plane-polarized light impinges under an angle α on a metal, the reflected light is generally elliptically polarized. The analysis of this elliptically polarized light yields two parameters, the **azimuth** and the **phase difference**, from which the optical properties are calculated.

We consider plane-polarized light whose vibrational plane is inclined by 45° towards the plane of incidence (Fig. 13.1). This angle is called azimuth,

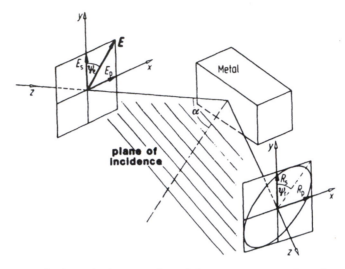

Figure 13.1. Reflection of plane-polarized light on a metal surface. (*Note*: In the figure $\mathscr{E}_{Rp} \equiv R_p$ and $\mathscr{E}_{Rs} \equiv R_s$.)

ψ_e, in contrast to the azimuth of the reflected light, ψ_r, which is defined as

$$\tan \psi_r = \frac{\mathscr{E}_{Rp}}{\mathscr{E}_{Rs}} \tag{13.5}$$

(see Fig. 13.1), where $\mathscr{E}_{Rp}$ and $\mathscr{E}_{Rs}$ are parallel and perpendicular components of the reflected electric field strength $|\mathscr{E}|$, i.e., the amplitudes of the reflected light wave.

In elliptically polarized light, the length and direction of the light vector is altered periodically. The tip of the light vector moves along a continuous screw, having the direction of propagation as an axis (Fig. 13.2(a)). The projection of this screw onto the $x - y$ plane is an ellipse (Fig. 13.1). Elliptically polarized light can be thought of as composed of two mutually perpendicular, plane-polarized waves, having a phase difference δ between them (expressed in fractions of 2π) (see Fig. 13.2(b)).

For the actual measurement of ψ_r and δ, one needs two polarizers (consisting of a birefringent material, which allows only plane-polarized light to pass), and a compensator (also consisting of birefringent material, which allows one to measure the phase difference δ; see Fig. 13.3). In Fig 13.4, the light reflected from a metal is represented by two light vectors pointing in the x- and y-directions, respectively. They have a phase difference δ between them. By varying the thickness of the birefringent materials in the compensator, one eventually accomplishes that the light which leaves the compensator is plane-polarized (i.e., $\delta = 0°$). The resultant vector, R_{res}, is then tilted by an angle, ψ_r, against the normal to the plane of incidence. One determines ψ_r by turning the analyzer to a position at which its axis is perpendicular to R_{res}. In short, δ and ψ_r are measured by simultaneously altering the thickness

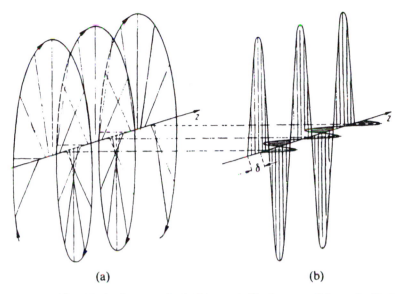

(a)   (b)

Figure 13.2. (a) Elliptically polarized light and (b) decomposition of elliptically polarized light into two mutually perpendicular plane-polarized waves with phase difference δ. Adapted from R.W. Pohl, *Optik und Atomphysik*. Springer-Verlag, Berlin (1958).

of the compensator and turning the analyzer until no light leaves the analyzer. It is evident that this method is cumbersome and time-consuming, particularly in cases in which an entire spectrum needs to be measured point by point. Thus, in recent years automated and computerized ellipsometers have been developed.

The optical constants are calculated using

$$n^2 = \tfrac{1}{2}\left[\sqrt{(a^2 - b^2 + \sin^2\alpha)^2 + 4a^2b^2} + a^2 - b^2 + \sin^2\alpha\right], \qquad (13.6)$$

$$k^2 = \tfrac{1}{2}\left[\sqrt{(a^2 - b^2 + \sin^2\alpha)^2 + 4a^2b^2} - (a^2 - b^2 + \sin^2\alpha)\right], \qquad (13.7)$$

Figure 13.3. Schematic of an **ellipsometer** (polarizer and analyzer are identical devices).

Figure 13.4. Vector diagram of light reflected from a metal surface. The vectors having solid arrowheads give the vibrational direction and magnitude of the light.

with

$$a = \frac{\sin \alpha \tan \alpha \cos 2\psi_r}{1 - \cos \delta \sin 2\psi_r} \tag{13.8}$$

and

$$b = -a \sin \delta \tan 2\psi_r. \tag{13.9}$$

Alternatively, one obtains, for the polarization ε_1 and absorption ε_2,

$$\varepsilon_1 = n^2 - k^2 = \sin^2\alpha \left[1 + \frac{\tan^2 \alpha (\cos^2 2\psi_r - \sin^2 2\psi_r \sin^2 \delta)}{(1 - \sin 2\psi_r \cos \delta)^2} \right], \tag{13.10}$$

$$\varepsilon_2 = 2nk = -\frac{\sin 4\psi_r \sin \delta \tan^2 \alpha \sin^2 \alpha}{(1 - \sin 2\psi_r \cos \delta)^2}. \tag{13.11}$$

*13.1.3. Differential Reflectometry

The information gained by differential reflectometry is somewhat different from that obtained by the aforementioned techniques. A "differential reflectogram" allows the direct measurement of the energies that electrons absorb from photons as they are raised into higher allowed energy states. The **differential reflectometer** measures the normalized difference between the reflectivities of two similar specimens which are mounted side by side (Fig. 13.5). For example, one specimen might be pure copper and the other copper with, say, 1% zinc. Unpolarized light coming from a monochromator

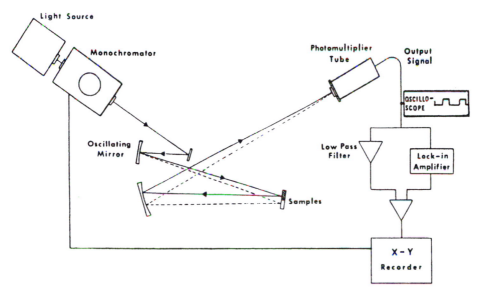

Figure 13.5. Schematic diagram of the differential reflectometer. (For clarity, the angle of incidence of the light beam impinging on the samples is drawn larger than it is in reality.) From R.E. Hummel, *Phys. Stat. Sol.* (a) **76**, 11 (1983).

is alternately deflected under near-normal incidence to one or the other sample by means of a vibrating mirror. The reflected light is electronically processed to yield $\Delta R/\bar{R} = 2(R_1 - R_2)/(R_1 + R_2)$. A complete differential reflectogram, i.e., a scan from the near IR through the visible into the near UV, is generated automatically and takes about two minutes. The main advantage of differential reflectometry over conventional optical techniques lies in its ability to eliminate any undesirable influences of oxides, deformations, windows, electrolytes (for corrosion studies), or instrumental parameters upon a differential reflectogram, owing to the differential nature of the technique. No vacuum is needed. Thus, the formation of a surface layer due to environmental interactions can be studied *in situ*. Finally, the data can be taken under near-normal incidence.

Differential reflectometry belongs to a family of techniques, called **modulation spectroscopy**, in which the derivative of the unperturbed reflectivity (or ε_2) with respect to an external parameter is measured. Modulation techniques restrict the action to so-called *critical points* in the band structure, i.e., they emphasize special electron transitions from an essentially featureless background. This background is caused by the allowed transitions at practically all points in the Brillouin zone. Most modulation techniques, such as differential reflectometry, wavelength modulation, thermoreflectance, or piezoreflectance, are *first-derivative* techniques (Fig. 13.6(a)). In semiconductor research (Section 13.6) another modulation technique, called electroreflectance, is often used, which provides the third derivative of R or ε_2. (It utilizes an alternating electric field which is applied to the semiconducting

Figure 13.6. Schematic representation of (a) the first derivative and (b) the third derivative of an ε_2-spectrum. The equivalent interband transitions at a so-called M_0 symmetry point are shown in the inserts. Adapted from D.E. Aspnes, *Surface Science* **37**, 418 (1973).

material during the reflection measurement.) The third derivative provides sharper and more richly structured spectra than the first-derivative techniques (Fig. 13.6(b)). In a first-derivative modulation spectrum, the lattice periodicity is retained, the optical transitions remain vertical and the interband transition energy changes with the perturbation (see insert of Fig. 13.6(a)). In electromodulation, the formerly sharp vertical transitions are spread over a finite range of initial and final momenta (see insert of Fig. 13.6(b)). A relatively involved line-shape analysis of electroreflectance spectra eventually yields the interband transition energies.

We shall make use of reflection, absorption, and first-derivative spectra in the sections to come.

13.2. Optical Spectra of Pure Metals

13.2.1. Reflection Spectra

The spectral dependence of the optical properties of metals was described and calculated in Chapter 11 by postulating that light interacts with a certain

Figure 13.7. Reflectivity spectrum for silver. Adapted from H. Ehrenreich et al., *IEEE Spectrum* **2**, 162 (1965). © 1965 IEEE.

number of *free electrons* and a certain number of *classical harmonic oscillators*, or equivalently, by *intraband* and *interband* transitions. In the present section we shall inspect experimental reflection data and see what conclusions can be drawn from these results with respect to the electron band structure.

Figure 13.7 depicts the spectral reflectivity for silver. From this diagram, the optical constants (i.e., the real and imaginary parts of the complex dielectric constant, $\varepsilon_1 = n^2 - k^2$ and $\varepsilon_2 = 2nk$) have been calculated by means of a Kramers–Kronig analysis (Section 13.1.1). Comparing Fig. 13.8 with

Figure 13.8. Spectral dependence of ε_1 and ε_2 for silver. ε_1 and ε_2 were obtained from Fig. 13.7 by a Kramers–Kronig analysis. Adapted from H. Ehrenreich et al., *IEEE Spectrum* **2**, 162 (1965). © 1965 IEEE.

Fig. 11.5 shows that for small photon energies, i.e., for $E < 3.8$ eV, the spectral dependences of ε_1 and ε_2 have the characteristic curve shapes for free electrons. In other words, the optical properties of silver can be described in this region by the concept of free electrons. Beyond 3.8 eV, however, the spectral dependences of ε_1 and ε_2 deviates considerably from the free electron behavior. In this range, classical oscillators, or equivalently, interband transitions, need to be considered.

Now it is possible to separate the contributions of free and bound electrons in ε_1- and ε_2-spectra. For this, one *fits* the theoretical ε_2 to the experimental ε_2 curves in the low-energy region. The theoretical spectral dependence of ε_2 is obtained by the Drude equation (11.27). An "effective mass" and the damping frequency, v_2, are used as adjustable parameters. With these parameters, the free electron part of ε_1 (denoted by ε_1^f) is calculated in the *entire* spectral range by using (11.26). Next, ε_1^f is subtracted from the experimental ε_1, which yields the bound electron contribution, ε_1^b. Figure 13.9 depicts an absorption band thus obtained, which resembles a calculated absorption band quite well (Fig. 11.9).

We now turn to the optical spectra for copper (Figs. 13.10 and 13.11). We notice immediately one important feature: Copper possesses an absorption

Figure 13.9. Separation of ε_1 for silver into ε_1^f (free electrons) and ε_1^b (bound electrons). Adapted from H. Ehrenreich et al., *IEEE Spectrum* **2**, 162 (1965). © 1965 IEEE.

Figure 13.10. Reflectivity spectrum for copper. Adapted from H. Ehrenreich et al., *IEEE Spectrum* **2**, 162 (1965). © 1965 IEEE.

band in the *visible* spectrum, which is, as already mentioned, responsible for the characteristic color of copper. We defined above a threshold energy at which interband transitions set in. In copper, the threshold energy is about 2.2 eV (Fig. 13.11), which is assigned to the *d*-band $\rightarrow E_F$ transition near the *L*-symmetry point. (This is marked by an arrow in Fig. 5.22.) Another peak is observed at slightly above 4 eV, which is ascribed to interband transitions *from* the Fermi energy near the *L*-symmetry point, as depicted in Figs. 12.3 and 5.22.

Figure 13.11. Spectral dependence of ε_1 and ε_2 for copper. ε_1 and ε_2 were obtained from Fig. 13.10 by a Kramers–Kronig analysis. Adapted from H. Ehrenreich et al., *IEEE Spectrum* **2**, 162 (1965). © 1965 IEEE.

Figure 13.12. Reflection spectrum for aluminum. Adapted from H. Ehrenreich et al., *IEEE Spectrum* **2**, 162 (1965). © 1965 IEEE.

As a final example, we inspect the reflection spectrum of aluminum. Figures 13.12 and 13.13 show that the spectral dependences of ε_1 and ε_2 resemble those shown in Fig. 11.5, except in the small energy region around 1.5 eV. Thus, the behavior of aluminum may be described essentially by the free electron theory. This free electron-like behavior of aluminum can also be deduced from its band structure (Fig. 5.21), which has essential characteristics of free electron bands for fcc metals (Fig. 5.20). Interband transitions which contribute to the ε_2-peak near 1.5 eV occur between the W_2' and W_1 symmetry points and the closely spaced and almost parallel Σ_3 and Σ_1 bands. A small contribution stems from the $W_3 \rightarrow W_1$ transition near 2 eV.

Figure 13.13. Spectral dependence of ε_1 and ε_2 for aluminum. Adapted from H. Ehrenreich et al., *IEEE Spectrum* **2**, 162 (1965). © 1965 IEEE.

*13.2.2. Plasma Oscillations

We postulate now that the free electrons of a metal interact electrostatically, thus forming an electron "**plasma**" that can be excited by light of proper photon energy to collectively perform fluidlike oscillations. These plasma oscillations are quantized. One quantum of plasma oscillations is called a "**plasmon**". This plasma possesses, just as an oscillator, a resonance frequency, often called the plasma frequency. We already introduced in Section 11.2 the plasma frequency, v_1, and noted that the dielectric constant, $\hat{\varepsilon}$, becomes zero at v_1. Thus, (10.12) reduces to

$$\hat{\varepsilon} = \varepsilon_1 - i\varepsilon_2 = 0, \tag{13.12}$$

from which we conclude that at the plasma frequency ε_1 as well as ε_2 must be zero. Experience shows that oscillations of the electron plasma already occur when ε_1 and ε_2 are close to zero.

The frequency dependence of the imaginary part of the reciprocal dielectric constant peaks at the **plasma frequency**, as we will see shortly. We write

$$\frac{1}{\hat{\varepsilon}} = \frac{1}{\varepsilon_1 - i\varepsilon_2} = \frac{\varepsilon_1 + i\varepsilon_2}{\varepsilon_1^2 + \varepsilon_2^2} = \frac{\varepsilon_1}{\varepsilon_1^2 + \varepsilon_2^2} + i\frac{\varepsilon_2}{\varepsilon_1^2 + \varepsilon_2^2}. \tag{13.13}$$

The imaginary part of the reciprocal dielectric constant, i.e.,

$$\operatorname{Im}\frac{1}{\hat{\varepsilon}} = \frac{\varepsilon_2}{\varepsilon_1^2 + \varepsilon_2^2}, \tag{13.14}$$

is called the "**energy loss function**" which is large for $\varepsilon_1 \to 0$ and $\varepsilon_2 < 1$, i.e., at the plasma frequency. We will now inspect the energy loss functions for some metals. We begin with aluminum because its behavior may well be interpreted by the free electron theory. We observe in Fig. 13.14 a pronounced maximum of $\operatorname{Im}(1/\hat{\varepsilon})$ near 15.2 eV. The real part of the dielectric constant

Figure 13.14. Energy loss function for aluminum. Adapted from H. Ehrenreich et al., *IEEE Spectrum* **2**, 162 (1965). © 1965 IEEE.

Figure 13.15. Energy loss function for silver. Adapted from H. Ehrenreich et al., *IEEE Spectrum* **2**, 162 (1965). © 1965 IEEE.

(ε_1) is zero at this frequency and ε_2 is small (see Fig. 13.13). Thus, we conclude that aluminum has a plasma resonance at 15.2 eV.

Things are slightly more complicated for silver. Here, the energy loss function has a steep maximum near 4 eV (Fig. 13.15), which cannot be solely attributed to free electrons, since ε_1^f is only zero at 9.2 eV (see Fig. 13.9). The plasma resonance near 4 eV originates by cooperation of the d- as well as the conduction electrons. The loss function for silver has another, but much weaker, resonance near 7.5 eV. This maximum is essentially caused by the conduction electrons, but is perturbed by interband transitions which occur at higher energies.

The reflection spectrum for silver (Fig. 13.7) can now be completely interpreted. The sharp decrease in R near 4 eV by almost 99% within a fraction of an electron volt is caused by a weakly damped plasma resonance. The sudden increase, only 0.1 eV above the plasma resonance, takes place because of interband transitions that commence at this energy. Such a dramatic change in optical constants is unparalleled.

13.3. Optical Spectra of Alloys

It was demonstrated in the previous sections that knowledge of the spectral dependence of the optical properties contributes to the understanding of the electronic structure of metals. We will now extend our discussion to alloys. Several decades ago, N.F. Mott suggested that when a small amount of metal A is added to a metal B, the Fermi energy would simply assume an average value, while leaving the electron bands of the solvent intact. It was eventually recognized, however, that this "rigid-band model" needed some modification and that the electron bands are somewhat changed for an alloy. We use copper–zinc as an example. Figure 13.16 shows a series of differen-

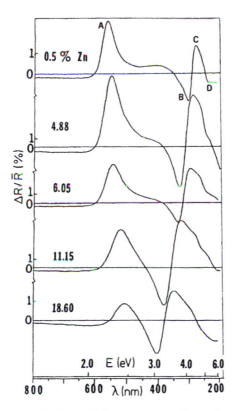

Figure 13.16. Experimental differential reflectograms for various copper–zinc alloys. The parameter on the curves is the average zinc concentration of the two alloys in at.%. The curve marked 0.5%, e.g., resulted by scanning the light beam between pure copper and a Cu–1% Zn alloy. Peaks A and D are designated as ε_2-type structures (Fig. 11.10) whereas features B and C belong to an ε_1-type structure (Fig. 11.9). From R.J. Nastasi-Andrews and R.E. Hummel, *Phys. Rev. B* **16**, 4314 (1977).

tial reflectograms (see Section 13.1.3) from which the energies for interband transitions, E_T, can be taken. Peak A represents the threshold energy for interband transitions, which can be seen to shift to higher energies with increasing zinc content. E_T is plotted in Fig. 13.17 as a function of solute (X). Essentially, a linear increase in E_T with increasing X is observed. The threshold energy for copper has been identified in Section 12.2 to be associated with electron transitions from the upper d-band to the conduction band, just above the Fermi surface (see Fig. 12.3). The rise in energy difference between the upper d-band and Fermi level, caused by solute additions, can be explained in a first approximation by suggesting a rise in the Fermi energy which results when extra electrons are introduced into the copper matrix from the higher-valent solute. Gallium, which has three valence electrons, would thus raise the Fermi energy more than zinc, which is indeed observed in Fig. 13.17. The slope of the $E_T = f(X)$ curve in Fig. 13.17 for

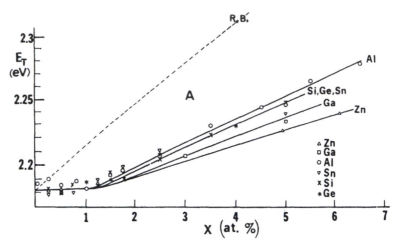

Figure 13.17. Threshold energies, E_T, for interband transitions for various copper-based alloys as a function of solute content. The E_T values are taken from differential reflectograms similar to those shown in Fig. 13.16. The rigid band line (R.B.) for Cu–Zn is added for comparison. From R.J. Nastasi-Andrews and R.E. Hummel, *Phys. Rev. B* **16**, 4314 (1977).

zinc (as well as for other solutes) is considerably smaller than that predicted by the rigid band model. This suggests that the d-bands are likewise raised with increasing solute content and/or that the Fermi level is shifted up much less than anticipated. Band calculations substantiate this suggestion. They reveal that upon solute additions to copper, the d-bands become narrower (which results from a reduction in Cu–Cu interactions) and that the d-bands are lifted up as a whole. Furthermore, the calculations show that solute additions to copper cause a rise in E_F and a downward shift of the bottom of the s-band. Figure 13.18 reflects these results. Because of the lowering of the bottom of the s-band (Γ_1 in Fig. 5.22), the Fermi energy rises much less than predicted had E_{Γ_1} remained constant.

An unexpected characteristic of all $E_T = f(X)$ curves is that the threshold energy for interband transitions, E_T, does not vary appreciably for solute concentrations up to slightly above 1 at.% (Fig. 13.17). Friedel predicted just this type of behavior and related it to "screening" effects. He argued that for the first few atomic percent solute additions to copper, the additional charge from the higher-valent solute is effectively screened and the copper matrix behaves as if the impurities were not present. The matrix remains essentially unperturbed as long as the impurities do not mutually interact.

The differential reflectograms shown in Fig. 13.16 suggest two additional pieces of structure, one of which corresponds to feature 'D' near 5 eV and is assigned to electron transitions from the *lower* d-bands to the Fermi surface. This interband transition is *not* shown in Fig. 13.18 because of its large energy, which is beyond the scale of this figure. An E_T versus X plot for peak 'D' resembles Fig. 13.17.

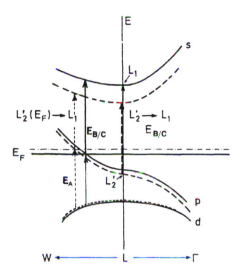

Figure 13.18. Schematic band structure near L for copper (solid lines) and an assumed dilute copper-based alloy (dashed lines). Compare with Figs. 12.3 and 5.22.

The third transition in the chosen energy region occurs at about 4 eV and involves the structural features 'B' and 'C'. The associated transition energy is seen to decrease with increasing solute content (Fig. 13.19). Features 'B' and 'C' are ascribed to transitions near the L-symmetry point, originating near the Fermi energy and terminating at the conduction band. It can be seen in Fig. 13.18 that the transition energy just mentioned is smaller for copper-based alloys than for pure copper, quite in agreement with the ex-

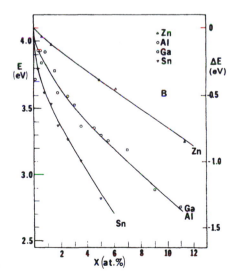

Figure 13.19. Energy of peak B for various dilute copper-based alloys. From R.J. Nastasi-Andrews, and R.E. Hummel, *Phys. Rev. B* **16**, 4314 (1977).

perimental findings. The reader is asked at this point to compare Figs. 13.10 and 13.11 with Fig. 13.16 and see how different optical techniques complement each other in revealing the electronic structure of solids.

*13.4. Ordering

It was shown in Section 7.5.3 that the resistivity decreases when solute atoms of an alloy are periodically arranged on the regular lattice sites. Thus, we conclude that ordering has an effect on the electronic structure and hence on the optical properties of alloys. The best way to study ordering is to compare two specimens of the same alloy when one of them is ordered and the other is in the disordered state. This way, peaks occur in a differential optical spectrum whenever the ordered state causes *extra* interband transitions comparable to superlattice lines in X-ray spectroscopy. As an example, Fig. 13.20 depicts an optical spectrum for the intermetallic phase Cu_3Au. We note several transitions, among them an ε_2-type structure with a peak energy at 2.17 eV and an ε_1-type structure with a transition energy around 3.6 eV (median between 3.29 eV and 3.85 eV, see Fig. 11.9). We shall explain them by referring to Fig. 13.21, which depicts the first Brillouin zone of the disordered fcc lattice in which a simple cubic Brillouin zone, representing the superlattice, is inscribed. The $\Gamma - X$ direction of the fcc Brillouin zone is bisected by the face of the cubic Brillouin zone at the point $\bar{X}$. The point X is then thought to be folded back to the point Γ. A new transition from the d-bands (e.g., at Γ_{12}) to the point X_4' (unfolded) can now take place (see Fig. 5.22). Folding along $\Gamma - M - K$, and possibly along other directions, explains the other transitions.

Figure 13.20. Differential reflectogram of (long-range) ordered versus disordered Cu_3Au. From R.E. Hummel, *Phys. Stat. Sol.* (*a*) **76**, 11 (1983).

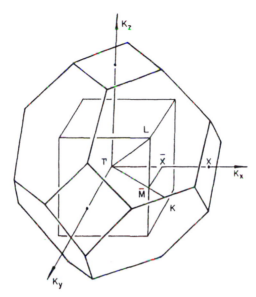

Figure 13.21. First Brillouin zone of an fcc lattice with inscribed Brillouin zone representing a cubic primitive superlattice.

Short-range ordering shows comparatively smaller effects than long-range ordering (Fig. 13.22). The reflectivity difference between ordered and disordered alloys is about 3% for long-range ordering compared to 0.5% in the case of short-range ordering. Still, even in the latter case, a superlattice transition is observed, which is attributed to the periodic arrangement of solute atoms in small domains (about 1–2 nm in diameter).

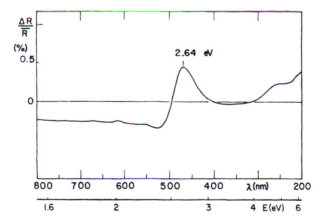

Figure 13.22. Differential reflectogram of (short-range) ordered versus disordered Cu–17 at.% Al. From J.B. Andrews, R.J. Andrews, and R.E. Hummel, *Phys. Rev. B* **22**, 1837 (1980).

Interestingly enough, optical investigations provide a further piece of information, which enables us to look upon the short-range ordered state from a different perspective. It has been observed that certain peaks in a differential reflectogram shift due to ordering, exactly as they would do when a solute is added to a solvent (see Section 13.3). From this we conclude that in the short-range ordered state, the interaction between dissimilar atoms is slightly larger than that for similar atoms.

*13.5. Corrosion

Studies of the optical properties have been used for many decades for the investigation of environmentally induced changes of surfaces. Optical studies are *nondestructive*, simple, and allow the investigation of oxides during their formation. No vacuum is required, in contrast to many other surface techniques. We use as an example the electrochemical corrosion of copper in an aqueous solution. A copper disc is divided into two parts that are electrically insulated from each other by a thin polymer film. One half is held electrically at the protective potential (as reference) and the other at the corrosion potential. No artifacts from the electrolyte, the corrosion cell window, or the metal substrate are experienced since the only difference in the light path of a differential reflectometer is the corrosion film itself. Figure 13.23 depicts a series of differential reflectograms demonstrating the evolution of Cu_2O on a

Figure 13.23. Differential reflectograms depicting the *in situ* evolution of Cu_2O on a copper substrate in a buffered electrolyte of pH 9. One sample half was held potentiostatically at -200 mV (SCE) for various times, the other at the protective potential (-500 mV (SCE)). From R.E. Hummel, *Phys. Stat. Sol. (a)* **76**, 11 (1983).

copper substrate. We observe that the peak height near 3.25 eV, and thus the corrosion film thickness, initially grows rapidly. The growth rate slows down as the film becomes thicker. The growth kinetics has been observed to obey a logarithmic relationship.

13.6. Semiconductors

Intrinsic semiconductors have, at low temperatures, a completely filled valence band and an empty conduction band (see Chapter 8). Consequently, no *intra*band transition, or classical infrared (IR) absorption, is possible at low temperatures. Thus, the optical behavior of an intrinsic semiconductor is similar to that of an insulator, i.e., it is transparent in the low energy (far IR) region. Once the energy of the photons is increased and eventually reaches the gap energy, then the electrons are excited from the top of the valence band to the bottom of the conduction band. The semiconductor becomes opaque like a metal (see Fig. 13.24). The onset for *inter*band transitions is thus determined by the gap energy, which characteristically has values between 0.2 eV and 3.5 eV (see Table 8.1 and Appendix 4). The corresponding wavelength lies in the near IR or visible region.

The reader certainly knows from Chapter 8 that silicon is the most important semiconductor material. It is therefore quite appropriate at this point to look at the absorption spectrum of Si, Fig 13.25. The situation is, however, not as simple as just explained, because Si is a so-called "indirect–band gap material". By inspecting its band diagram (see Fig. 5.23 or Fig. 13.26) we notice that the maximum of the valence band and the minimum of

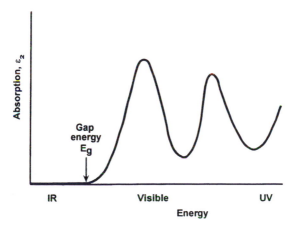

Figure 13.24. Schematic representation of the absorption spectrum of an intrinsic, direct–band gap semiconductor. The material is transparent below the gap energy and opaque above E_g.

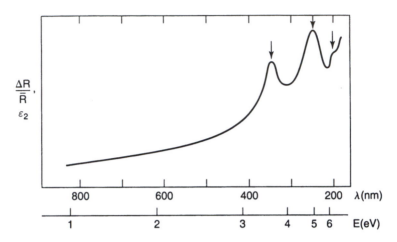

Figure 13.25. Differential reflectogram of silicon (after R.E. Hummel and W. Xi). $\Delta R/\bar{R}$ is essentially the absorption, ε_2, as explained in Section 13.1.3

the conduction band are not at the same point in k-space. Vertical transitions are thus not permissible (or have only a very small probability) at energies below about 3.4 eV. Accordingly, we observe in the optical spectrum depicted in Fig 13.25 three distinct absorption peaks, which are known by the designations $L_3' \rightarrow L_1$ (3.4 eV), Σ (4.2 eV), and $L_3' \rightarrow L_3$ (5.6 eV) (see Fig. 5.23). These peaks are all caused by direct interband transitions in specific areas of k-space.

Nevertheless, indirect transitions between the top of the valence band and the bottom of the conduction band may be possible to a limited degree provided the necessary momentum (wave vector **k**) is furnished by a phonon (see Fig. 13.26). We have already discussed phonon-assisted transitions in Section 12.2 and explained there that indirect interband transitions are particularly observed in the absence of direct transitions. Indirect interband transitions are generally quite weak.

Our discussion of the optical spectra of semiconductors is not complete by considering only direct or indirect interband transitions. Several other absorption mechanisms may occur. It has been observed, for example, that the absorption spectra for semiconductors show a structure for photon energies slightly *below* the gap energy (Fig. 13.27(a)). Frenkel explained this behavior by postulating that a photon may excite an electron so that it remains in the vicinity of its nucleus, thus forming an electron–hole pair, called an **exciton**. Electrons and holes are thought to be bound together by electrostatic forces and revolve around their mutual center of mass. The electrons may hop through the crystal and change their respective partners. This motion can also be described as an exciton wave. One depicts the excitons by introducing "exciton levels" into the forbidden band (Fig. 13.27(b)). They are separated from the conduction band by the "binding energy", E_x, whose position can be calculated by an equation similar to (4.18a) (see also

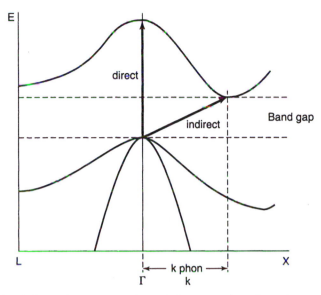

Figure 13.26. Schematic representation of direct versus indirect interband absorptions in Si. In the case of an indirect transition, a phonon needs to be additionally absorbed. Compare to Fig. 5.23 and 12.2.

Problem 8/10):

$$E_x = -\frac{m^* e^4}{(4\pi\varepsilon_0)^2 2n^2\hbar^2\varepsilon^2},$$ (13.15)

where n is an integer, m^* is the effective mass of the exciton (which is the average of m_e and m_h), and ε is the a.c. dielectric constant. E_x is character-

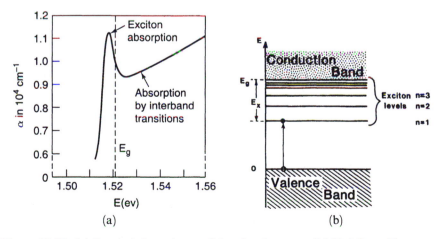

Figure 13.27. (a) Spectral dependence of the absorbance, α, (10.21a) for gallium arsenide at 21 K. Adapted from M.D. Sturge, *Phys. Rev.* **127**, 768 (1962). (b) Schematic representation of exciton energy levels and an exciton in a semiconductor (or insulator).

istically about 0.01 eV. The exciton levels are broadened by interactions with impurities or phonons.

Finally, **extrinsic semiconductors** have, as we know, donor or acceptor states near the conduction or the valence band, respectively (Section 8.3). At sufficiently high temperatures, optical transitions from and to these states can take place, which also cause weak absorption peaks below the gap energy.

It should be noted that the temperature slightly influences the absorption characteristics of a semiconductor. The change in gap energy is about -2×10^{-4} eV/K (see Appendix, and Equation (8.1)), which stems from an apparent broadening of valence and conduction levels with increasing temperature due to transitions with simultaneous emission and absorption of photons. Another temperature-enhanced effect should be considered, too. Once electrons have been excited from the valence into the conduction band (either by photons or thermal excitation), holes are present in the upper part of the valence bands. Then, photons having energies well below E_g can be absorbed by *intra*band transitions. These transitions are, however, relatively weak.

Optical absorption measurements are widely used in semiconductor research since they provide the most accurate way to determine the gap energies and the energies of the localized states. Measurements are normally performed at low temperatures so that the thermal excitations of the electrons do not mask the transitions to be studied. Optical measurements are capable of discriminating between direct and indirect transitions, based on the magnitude of the absorption peaks.

13.7. Insulators (Dielectric Materials and Glass Fibers)

As we know, insulators are characterized by completely filled valence bands and empty "conduction" bands. Thus, no *intra*band transitions, i.e., no classical IR absorption, takes place. Furthermore, the gap energy for insulators is fairly large (typically 5 eV or larger) so that *inter*band transitions do not occur in the IR and visible spectrum either. They take place, however, in the ultraviolet (UV) region. Third, *excitons* may be created, which cause absorption peaks somewhat below the gap energy. For example, the lowest energy for an exciton level (and thus for the first exciton absorption peak) for NaCl has been found to be at about 7 eV, i.e., in the vacuum UV region. (Other alkali halides have very similar exciton energies.) We suspect, therefore, that insulators are transparent from the far IR throughout the visible up to the UV region. This is indeed essentially observed. However, in the IR region a new absorption mechanism may take place which we have not yet discussed. It is caused by the light-induced vibrations of the lattice atoms,

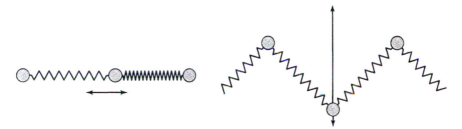

Figure 13.28. One-dimensional representations of possible vibration modes of atoms that have been excited by IR electromagnetic radiation (heat). Left: stretching mode, right: bending mode.

i.e., by the **excitation of phonons by photons**. We need to explain this in some more detail.

Let us first consider a **monatomic crystal** (one kind of atom). The individual atoms are thought to be excited by light of appropriate frequency to perform oscillations about their points of rest. Now, the individual atoms are surely not vibrating independently. They interact with their neighbors, which causes them to move simultaneously. For simplicity, we model the atoms to be interconnected by elastic springs, see Fig. 13.28. Thus, the interaction of light with the lattice can be mathematically represented in quite a similar manner to the one used when we discussed and calculated the classical *electron* theory of dielectric materials. (In Section 11.6 we represented *one* atom in an electric field as consisting of a positively charged core which is bound quasi-elastically to an electron.) A differential equation similar to (11.32) may be written for the present case as

$$m\frac{d^2x}{dt^2} + \gamma'\frac{dx}{dt} + \kappa x = e\mathscr{E}_0 \exp(i\omega t), \qquad (13.16)$$

which represents the oscillations of atoms under the influence of light whose excitation force is $e\mathscr{E}_0 \exp(i\omega t)$. As before, the factor $\kappa \cdot x$ is the restoring force that contains the displacement x and an interatomic force constant κ (i.e., a "spring constant," or a "binding strength" between the atoms). The damping of the oscillations is represented by the second term in (13.16). Damping is thought to be caused by interactions of the phonons with lattice imperfections, or with external surfaces of the crystal, or with other phonons. The oscillators possess one or several resonance frequencies ω_0, which depend on the mass of the atoms on the vibrational modes (see Fig. 13.28), and on the restoring force (see (11.34)). The solution of the differential equation (13.16) yields a spectral dependence of ε_1 and ε_2 that is very similar to that shown in Figs. 11.9–11.12.

The situation becomes slightly more complicated when **diatomic solids**, such as ionic crystals, are considered. In this case, two differential equations of the type of (13.16) need to be written. They have to be solved simultaneously. Actually, one needs to solve $2N$ coupled differential equations,

Figure 13.29. Spectral reflectivity of NaCl at room temperature in the far IR region.

where N is the number of unit cells in the lattice. The result is, however, qualitatively still the same. The resonance frequency for diatomic crystals is

$$\omega_0 = 2\kappa\left(\frac{1}{m_1} + \frac{1}{m_2}\right), \tag{13.17}$$

where m_1 and m_2 are the masses of the two ion species. Figure 13.29 depicts the spectral reflectivity of NaCl in the IR. Sodium chloride is transmissive between 0.04 eV and 7 eV. At the upper boundary energy, exciton absorption sets in.

Fused quartz (depending on the method of manufacturing) is essentially transparent between 0.29 eV and 6.9 eV (4.28 μm and 0.18 μm), having, however, two pronounced absorption peaks near 1.38 μm and 2.8 μm. Window glass has a similar transmission spectrum as fused quartz, with the exception that its UV cut-off wavelength is already near 0.38 μm (3.3 eV). In recently developed **sol–gel silica "glasses"** the absorption peaks near 1.38 μm and 2.8 μm are virtually suppressed, which causes this material to be transparent from 0.16 μm to 4 μm. The energy loss spectrum for the commercially important **borosilicate/phosphosilicate glass**, used for **optical fibers**, is shown in Fig. 13.30. We notice the aforementioned peak near 1.38 μm, which is caused by oscillations of OH^- ions. We shall refer to this spectrum in a later section.

A word should be added about the **opacity** of some dielectric materials, such as enamels, opal glasses, glazes, or porcelains, which should be transparent in the visible region according to our discussion above. This opacity is caused by the scattering of light on small particles which are contained in the matrix. Part of the light is diffusely transmitted and part of it is diffusely reflected. The larger the specular part of the reflected light, the higher the gloss. Very often, *opacifiers* are purposely added to a dielectric material to cause wanted effects. The particle size should be nearly the same as the wavelength

Figure 13.30. Energy loss spectrum of highly purified glass for fiber-optic applications which features a phosphosilicate core surrounded by a borosilicate cladding. The communication channels near 1.3 μm and 1.5 μm are marked.

of the light, and the index of refraction should be largely different from that of the material, to obtain maximal scattering.

13.8. Emission of Light

13.8.1. Spontaneous Emission

So far we have discussed only the *absorption* of light by matter. We learned that due to the interaction of photons with electrons, the electrons are excited into higher energy states. The present section deals with the *emission* of photons.

An electron, once excited, must eventually revert back into a lower, empty energy state. This occurs, as a rule, spontaneously within a fraction of a second and is accompanied by the emission of a photon and/or the dissipation of heat, that is, *phonons*. The emission of light due to reversion of electrons from a higher energy state is called **luminescence**. If the electron transition occurs within nanoseconds or faster, the process is called **fluorescence**. In some materials, the emission takes place after microseconds or milliseconds. This slower process is referred to as **phosphorescence**. A third process, called **afterglow**, which is even slower (seconds), occurs when excited electrons have been temporarily trapped, for example, in impurity states from which they eventually return after some time into the valence band. Commercially used phosphorescing materials consist, for example, of metal sulfides (such as ZnS), tungstates, oxides (such as ZnO having a surplus of Zn), and many organic substances. More expensive phosphors are rare earth elements, such as Eu^{3+} (red), Eu^{2+} (blue), or Tb.

Photoluminescence is observed when photons impinge on a material which in turn re-emits light of a lower energy. **Electroluminescing** materials emit light as a consequence of an applied voltage or electric field. **Cathodoluminescence**, finally, is the term which is used to describe light emission from a substance that has been showered by electrons of higher energy. All of these effects have commercial applications. For example, the inside walls of picture tubes of television sets are coated with a cathodoluminescing material, basically ZnS, which emits light when hit by electrons generated by a hot filament (cathode ray). Silver-doped ZnS yields blue, and Cu-doping yields green colors. In fluorescent lamps, the inside of a glass tube is covered with tungstates, silicates, rare earth elements, or halophosphates, which emit light as a consequence of bombardment with ultraviolet light that has been generated by a mercury glow discharge. The image generated in electron microscopes is made visible by a screen that consists of a *phosphor* such as ZnS. The same is true when X-rays or γ-rays need to be made visible.

Spontaneous light emission occurs also in common devices such as candles or incandescent light bulbs. In both of these cases, the electrons have been excited into higher energy states by heat energy (**thermoluminescence**). The larger the temperature, the higher the energy of the photons and the shorter their wavelength. For example, heating to about 700 °C yields a dark red color whereas heating near 1600 °C results in orange hues. At still higher temperatures, the emitted light appears to be white, since large portions of the visible spectrum are emitted. Spontaneous emission possesses none of the characteristic properties of laser light: the radiation is emitted through a wide-angle region in space, the light is phase *in*coherent (see below) and is often polychromatic (more than one wavelength).

Light-emitting diodes (LED's) have recently gained substantial importance as electroluminescing devices whose efficiencies have surpassed those of incandescent lamps. They can be manufactured to emit light throughout the entire visible spectrum. They are rugged, small and relatively inexpensive, and will probably dominate the lighting market soon. We shall devote Section 13.8.13 to this topic. Light-emitting devices for display purposes will be discussed in subsequent sections.

13.8.2. Stimulated Emission (Lasers)

A quite different type of light source is the laser, which is, among others, used for telecommunications (optical fiber networks), data storage (compact discs), laser printers, and grocery scanners. This section will explain how lasers work.

Let us consider two energy levels, E_1 and E_2, and let us assume for a moment that the higher energy level, E_2, contains more electrons than the lower level, E_1, i.e., let us assume a **population inversion** of electrons (Fig. 13.31(a)). We further assume that by some means (which we shall discuss in a moment)

E_2

E_1

(a)   (b)   (c)   (d)

Figure 13.31. Schematic representation of stimulated emission between two energy levels, E_2 and E_1. The dots symbolize electrons.

the electrons in E_2 are made to stay there for an appreciable amount of time. Nevertheless, one electron will eventually revert to the lower state. As a consequence, a photon with energy $E_{21} = h\nu_{21}$ is emitted (Fig. 13.31(b)). This photon might stimulate a second electron to *descend in step* to E_1, thus causing the emission of another photon which vibrates in phase with the first one. The two photons are consequently **phase coherent** (Fig. 13.31(c)). They might stimulate two more electrons to descend in step (Fig. 13.31(d)) and so on until an avalanche of photons is created. In short, stimulated emission of light occurs when electrons are forced by incident radiation to add more photons to an incident beam. The acronym LASER can now be understood; it stands for *light amplification by stimulated emission of radiation*.

Laser light is highly **monochromatic** because it is generated by electron transitions between two narrow energy levels. (As a consequence, laser light can be focused to a spot less than 1 μm in diameter.) Another outstanding feature of laser light is its strong **collimation**, i.e., the parallel emergence of light from a laser window. (The cross-section of a laser beam transmitted to the moon is only 3 km in diameter!) We understand the reason for the collimation best by knowing the physical setup of a laser.

The lasing material is embodied in a long narrow container called the cavity; the two faces at opposite ends of this cavity must be absolutely parallel to each other. One of the faces is silvered and acts as a perfect mirror, whereas the other face is partially silvered and thus transmits some of the light (Fig. 13.32). The laser light is reflected back and forth by these mirrors, thus increasing the number of photons during each pass. After the laser has

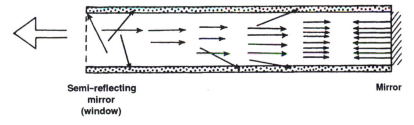

Semi-reflecting   Mirror
mirror
(window)

Figure 13.32. Schematic representation of a laser cavity and the buildup of laser oscillations. The stimulated emission eventually dominates over the spontaneous emission. The light leaves the cavity at the left side.

been started, the light is initially emitted in all possible directions (left part of Fig. 13.32). However, only photons that travel strictly parallel to the cavity axis will remain in action, whereas the photons traveling at an angle will eventually be absorbed by the cavity walls (center part in Fig. 13.32). A fraction of the photons escape through the partially transparent mirror. They constitute the emitted beam.

We now need to explain how the electrons arrive at the higher energy level, i.e., we need to discuss how they are *pumped* from E_1 into E_2. One of the methods is, of course, *optical pumping*, i.e., the absorption of light stemming from a polychromatic light source. (Xenon flashlamps for pulsed lasers, or tungsten–iodine lamps for continuously operating lasers, are often used for pumping. The lamp is either wrapped in helical form around the cavity, or the lamp is placed in one of the focal axes of a specularly reflecting elliptical cylinder, whereas the laser rod is placed along the second focal axis.) Other pumping methods involve collisions in an electric discharge, chemical reactions, nuclear reactions, or external electron beam injection.

The **pumping efficiency** is large if the bandwidth, δE, of the upper (and the lower) electron state is broad. This way, an entire frequency range (rather than a single wavelength) leads to excited electrons (Fig. 13.33(a)).

Next, we discuss how **population inversion** can be achieved. For this we need to quote Heisenberg's uncertainty principle,

$$\delta E \cdot \delta t \propto h, \qquad (13.18)$$

which states that the time span, δt, for which an electron remains at the higher energy level, E_2, is large when the bandwidth, δE, of E_2 is narrow. In other words, a sharp energy level (δE small, δt large) supports the population inversion, Fig. 13.33(b). On the other hand, a large pumping efficiency requires a large δE (Fig. 13.33(a)), which results in a small δt and a small population inversion. Thus, high pumping efficiency and large population inversion mutually exclude each other in a two-level configuration. In es-

Figure 13.33. Examples of possible energy states in a two-level configuration. (a) δE large, i.e., large pumping efficiency but little or no population inversion. (b) Potentially large population inversion (δt large) but small pumping efficiency. (*Note:* Two-level lasers do not produce a population inversion, because absorption and emission compensate each other.)

Figure 13.34. Three-level laser. The nonradiative, phonon-assisted decay is marked by a dashed line. Lasing occurs between levels E_2 and E_1. High pumping efficiency to E_3. High population inversion at E_2.

sence, a two-level configuration as depicted in Fig. 13.33 does not yield laser action.

The **three-level laser** (Fig. 13.34) provides improvement. There, the "pump band", E_3, is broad, which enables a good pumping efficiency. The electrons revert after about 10^{-14} s into an intermediate level, E_2, via a nonradiative, phonon-assisted process. Since E_2 is sharp and not strongly coupled to the ground state, the electrons remain much longer, i.e., for some microseconds or even milliseconds on this level. This provides the required population inversion.

An even larger population inversion is obtained using a **four-level laser**. In this configuration the energy level E_2 is emptied rapidly by electron transitions into a lower level, E_1 (Fig. 13.35). It should be added that some three- and four-level lasers have several closely spaced pumping bands, which, of course, increases the pumping efficiency.

The highest population inversion is achieved by adding **Q-switching**. For this method, the mirror in Fig. 13.32 is turned sideways during pumping to reduce stimulated emission, i.e., to build up a substantial population inversion. After some time, the mirror is turned back into its original vertical position, which results in a burst of light lasting 10–20 ns.

Figure 13.35. Four-level laser.

Laser materials cannot be created at will in, say, three- or four-level configurations. They can, however, be selected from hundreds of substances to suit a specific purpose. Laser materials include *crystals* (such as ruby), *glasses* (such as neodymium-doped glass), *gases* (such as helium, argon, xenon), *metal vapors* (such as cadmium, zinc, or mercury), *molecules* (such as carbon dioxide), or *liquids* (solvents which contain organic dye molecules). Table 13.1 lists the properties of some widely used lasers. We observe that many lasers emit their light in the red or IR spectrum. Exceptions are the He–Cd laser ($\lambda = 325$ nm), the argon laser ($\lambda = 520$ nm), the tunable dye lasers, and certain semiconductor lasers. Lasers can be operated in a continuous mode (CW), or, with a higher power output, in the pulsed mode. The power output varies over many orders of magnitude and it can even be increased if Q-switching is applied (specifically, from 10^{-9} to 10^{20} Watt). A few important laser types need special mention.

13.8.3. Helium–Neon Laser

A cavity about 2 mm in diameter is filled with 0.1 Torr Ne and 1 Torr He (Fig. 13.36(a)). A current that passes through the gas produces free electrons (and ions). The electrons are accelerated by the electric field and excite the helium gas by electron–atom collisions. Some of the helium levels are reso-

Figure 13.36. Helium–neon laser. (a) Schematic diagram of the laser cavity with Littrow prism to obtain preferred oscillation at one wavelength. (The end windows are inclined at the Brewster angle for which plane-polarized light suffers no reflection losses.) (b) Energy level diagram for helium and neon. The decay time for the *p*-states is ~10 ns; that of the *s*-states 100 ns. The letters on the energy levels represent the angular momentum quantum number; the number in front of the letters gives the value for the principal quantum number; and the superscripts represent the multiplicity (singlet, doublet, etc.), see Appendix 3.

Table 13.1. Properties of Some Common Laser Materials.

Type of laser	Wave-length(s) (nm)	Beam diver-gence (mrad)	Peak power output (W)	Comments
Ruby (Cr^{3+}-doped Al_2O_3)	694.3	10 5 0.5	CW:[a] $\sim$5 pulsed (1–3 ms): 10^6–10^8 Q-switched (10 ns): 10^9	Optically pumped three-level laser. Lasing occurs between Cr^{3+} levels. Low efficiency (0.1%). Historic device (1960).
Neodymium (Nd^{3+}-doped glass or YAG[b])	1,064	3–8	CW: 10^3 pulsed (0.1–1 μs): $\sim$$10^4$	Optically pumped four-level laser. High efficiency 2%.
HeNe	632.8 (1150; 3390)	1	10^{-3}–10^{-2}	See Fig. 13.36 and text. Most widely used.
HeCd (gas/metal vapor)	441.6 325 353.6		150 mW CW 100 mW 20 mW	Similarly pumped as HeNe laser. Used for high-speed laser printers, and writing data on photoresists for CD-ROMs. Efficiency: up to 0.02%.
Argon ion	488		$\sim$25 CW	0.1% Efficiency
CO_2	10,600; 9,600	2	CW: 10–1.5$\times$$10^4$ pulsed (10^2–10^3 ns): 10^5	High efficiency (20%). Lasing occurs between vibrational levels (Fig. 13.37).
Semiconductor GaAs	$\sim$870	250	Homojunction, pulsed: (10^2 ns) 10–30 Heterojunction, CW:	Small size, direct conversion of electrical energy into
GaAlAs[c]	$\sim$850	500	1–4×10^{-1}	optical energy. 10–55% efficiency. See Figs. 13.38 and 13.43.
Dye (organic dyes in solvents)	350–1000	3 10	CW: $\sim$$10^{-1}$ pulsed (6 ns) $\sim$$10^5$	Lasing occurs between vibrational sublevels of molecules. Tunable by Littrow prism (Fig. 13.36(a)).

[a] CW: Continuous wave.
[b] Yttrium aluminum garnet ($Y_3Al_2O_{15}$).
[c] See Fig. 13.40.

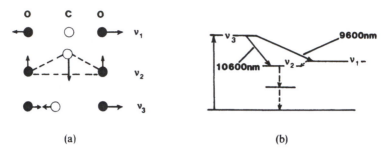

(a)   (b)

Figure 13.37. CO_2 laser. (a) Fundamental modes of vibration for a CO_2 molecule; v_1: symmetric stretching mode; v_2: bending mode; v_3: asymmetric stretching mode. (b) Energy level diagram for various vibrational modes.

nant with neon levels so that the neon gas also becomes excited by resonant energy transfer (Fig. 13.36(b)). This constitutes a very efficient pumping into the neon 2s- and 3s-levels. (Direct electron–neon collisions also contribute to the pumping). Lasing occurs between the neon s- and p-levels and produces three characteristic wavelengths. Suppression of two of the wavelengths is accomplished by multilayer dielectric mirrors, which provide a maximum reflectivity at the desired wavelength, or by a Littrow prism, as shown in Fig. 13.36(a).

13.8.4. Carbon Dioxide Laser

The CO_2 molecule possesses three fundamental modes of vibration, as shown in Fig. 13.37(a). The lasing occurs between these levels as shown in Fig. 13.37(b). Pumping is accomplished by electron–atom collisions (see above) and by a resonant energy transfer from N_2 molecules which are added to the CO_2. Nitrogen (and helium) greatly improves the pumping efficiency of the CO_2 laser, which is one of the most efficient and powerful lasers.

13.8.5. Semiconductor Laser

The "cavity" for this laser consists of heavily doped (10^{18} cm^{-3}) n- and p-type semiconductor materials such as GaAs. The energy band diagram for a p–n junction has been shown in Fig. 8.19 and is redrawn in Fig. 13.38(a) for the case of forward bias. We notice a population inversion of electrons in the depletion layer. Two opposite end faces of this p–n junction are made parallel and are polished or cleaved along crystal planes. The other faces are left untreated to suppress lasing in unwanted directions (Fig. 13.38(b)). A reflective coating of the window is usually not necessary since the reflectivity of the semiconductor is already 35%. The pumping occurs by direct injection of electrons and holes into the depletion region. Semiconductor lasers are small and can be quite efficient.

13.8.6. Direct–Versus Indirect–Band Gap Semiconductor Lasers

We need to discuss now whether or not *all* semiconducting materials are equally well suited for a laser. Indeed, they are not. Direct–band gap materials, such as GaAs, have a much higher quantum efficiency for the emission of light than indirect–band gap materials, such as silicon. This needs some explanation. Let us assume that an electron at the top of the valence band in silicon has absorbed energy, and has thus been excited (pumped) by means of a direct interband transition into the conduction band, as shown in Fig. 13.39. This *"hot"* electron quickly *thermalizes*, i.e., it reverts down within 10^{-14} s to the bottom of the conduction band in a nonradiative process, involving a phonon (to conserve momentum, see Section 12.2). In order to recombine finally with the left-behind hole in the valence band (by means of an indirect transition) a second phonon-assisted process has to take place. This requirement substantially reduces the probability for emission. The time interval which elapses before such a recombination takes place may be as much as 0.25 s, which is substantially longer than it would take for a direct recombination in a pumped semiconductor. Before this quarter of a second has passed, the electron and also the hole have already recombined through some other nonradiative means involving impurity states, lattice defects, etc. Thus, the electron in question is lost before a radiative emission occurs. This does not mean that indirect emissive transitions would never take place. In

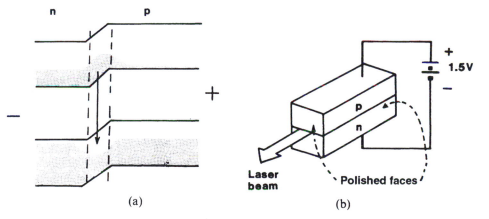

Figure 13.38. (a) Energy band diagram of a heavily doped, forward-biased semiconductor. (b) Schematic setup of a semiconductor laser.

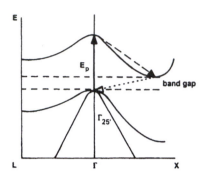

Figure 13.39. Direct interband transition pumping (E_p) and phonon-involved rever-sion-of a *hot* electron by indirect transitions for an indirect–band gap semiconductor such as silicon. (Compare with Figs. 5.23 and 12.2.)

fact, they do occur occasionally and have been observed, for example, in GaP, but with a very small quantum efficiency. Indirect–band gap semi-conductors therefore seem to be not suited for lasers. It should be added, however, that silicon, when made porous by anodically HF etching, has been observed to emit visible light. It is speculated that the etching creates an ar-ray of columns which act as fine quantum lines, and thus alter the electronic band structure of silicon to render it direct. Moreover, spark-processed Si emits quite efficiently in the blue and green spectral range and is extremely stable against high temperatures, laser light, and HF etchings. However, none of these Si-based materials have yielded a laser so far.

13.8.7. Wavelength of Emitted Light

The wavelength of a binary GaAs laser is about 0.87 μm. This is, however, not the most advantageous wavelength for telecommunication purposes be-cause glass attenuates light of this wavelength appreciably. By inspecting Fig. 13.30 we note that the optical absorption in glass is quite wavelength dependent, having minima in absorption at 1.3 μm and 1.55 μm. For-tunately, the band gap energy, i.e., the wavelength at which a laser emits light, can be adjusted to a certain degree by utilizing ternary or quaternary compound semiconductors (Fig. 13.40). Among them, $In_{1-x}Ga_xAs_yP_{1-y}$ plays a considerable role for telecommunication purposes, because the useful emission wavelengths of these compounds can be varied between 0.886 μm and 1.55 μm (which corresponds to gap energies from 1.4 eV to 0.8 eV). In other words, the above-mentioned desirable wavelengths of 1.3 μm and 1.55 μm can be conveniently obtained by utilizing a properly designed indium–gallium–arsenide–phosphide laser.

Red lasers are quite common. They are widely used, for example in laser printers, grocery scanners, and compact disc players. On the other hand,

Figure 13.40. Lattice constants, energy gaps, and emission wavelengths of some ternary and quaternary compound semiconductors at 300 K. The lines between the binary compounds denote ternaries. The cross-hatched lines indicate indirect interband transitions. Pure silicon is also added for comparison.

blue semiconductor lasers have been more of a problem to fabricate. This hurdle seems to have been partially overcome now by an InGaN laser that emits at 399 nm. It involves a two-dimensional matrix of surface-emitting lasers that are optically pumped at 367 nm by a nitrogen laser–pumped dye laser. In other words, this laser does not yet emit blue light by merely applying a voltage, as in the case of the red lasers.

A note on compound semiconductor fabrication needs to be inserted at this point. Semiconductor compounds are usually deposited out of the gaseous or liquid phase onto an existing semiconductor substrate, whereby a relatively close match of the lattice structure of substrate and layer has to be maintained. This process, in which the lattice structure of the substrate is continued into the deposited layer, is called "**epitaxial growth.**" The important point is that, in order to obtain a strain-free epitaxial layer, the lattice constants of the involved components have to be nearly identical. Figure 13.40 shows, for example, that this condition is fulfilled for GaAs and AlAs. These compounds have virtually identical lattice constants. A near-perfect lattice match can also be obtained for ternary $In_{0.53}Ga_{0.47}As$ on an InP substrate, see Fig. 13.40. In short, the critical parameters for designing lasers from compound semiconductors include the band gap energy, the similarities of the lattice constants of the substrate and active layer, the fact whether or not a direct–band gap material is involved, and the refractive indices of the core and cladding materials (see Section 13.8.9).

Finally, the emission wavelength depends on the temperature of operation, because the band gap decreases with increasing temperature (see Equation (8.1) and Appendix 4) according to the empirical equation

$$E_{gT} = E_{g0} - \frac{\xi T^2}{T + \theta_D},$$
(13.19)

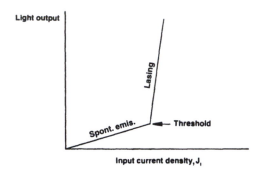

Figure 13.41. Schematic representation of the power output of a diode laser versus the pump current density. The threshold current density for a homojunction GaAs laser is on the order of 10^4 A/cm^2.

where E_{g0} is the band gap energy at $T = 0$ K, $\xi \approx 5 \times 10^{-4}$ eV/K, and θ_D is the Debye temperature (see Table 19.2 and Section 19.4), which is 204 K for GaAs.

13.8.8. Threshold Current Density

A few more peculiarities of semiconductor lasers will be added to deepen our understanding. Each diode laser has a certain power output characteristic which depends on the input current density, as depicted in Fig. 13.41. Applying low pumping currents results in predominantly *spontaneous* emission of light. The light is in this case incoherent and is not strongly monochromatic, i.e., the spectral line width is spread over several hundred Ångströms. However, when the current density increases above a certain threshold, population inversion eventually occurs. At this point, the stimulated emission (lasing) dominates over spontaneous emission and the laser emits a single wavelength having a line width of about 1 Å. Above the threshold the laser operates about one hundred times more efficiently than below the threshold. The electric vector vibrates perpendicular to the length axis of the cavity, i.e., the emitted light is plane-polarized. Additionally, standing waves are formed within the laser, which avoids destructive interference of the radiation. The distance between the two cavity faces must therefore be an integer multiple of half a wavelength.

13.8.9. Homojunction Versus Heterojunction Lasers

Lasers for which the *p*-type and *n*-type base materials are alike (e.g., GaAs) are called *homojunction* lasers. In these devices the photon distribution ex-

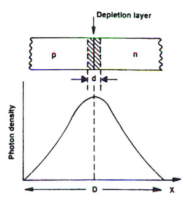

Figure 13.42. Schematic representation of the photon distribution in the vicinity of the depletion layer of a homojunction diode laser.

tends considerably beyond the electrically active region (in which the lasing occurs) into the adjacent inactive regions, as shown in Fig. 13.42. The total light-emitting layer, D, for GaAs is about 10 μm wide, whereas the depletion layer, d, i.e., the active region, might be as narrow as 1 μm. The photons that penetrate into the nonactive region do not stimulate further emission and thus reduce the quantum efficiency (which in the present case is about 10%). In essence, some of these photons are eventually absorbed and thus increase the temperature of the laser. The homojunction laser has therefore to be cooled or operated in a pulsed mode employing bursts of 100 ns duration, allowing for intermittent cooling times as long as 10^{-2} s. This yields peak powers of about 10–30 W.

Cooling or pulsing is not necessary for heterojunction lasers in which, for example, *two* junctions are utilized as depicted in Fig. 13.43. If the refractive

Figure 13.43. Schematic representation of a double heterojunction laser in which the active region consists of an *n*-doped GaAs layer.

index of the active region is larger than that of the neighboring areas, an "*optical waveguide structure*" is effectively achieved which confines the photon within the GaAs layer (total reflection!). This way, virtually no energy is absorbed in the nonactive regions. The threshold current density can be reduced to 400 A/cm^2. The quantum efficiency can reach 55% and the output power in continuous mode may be as high as 390 mW. The disadvantage of a double heterojunction laser is, however, its larger angular divergence of the emerging beam, which is between 20° and 40°.

13.8.10. Laser Modulation

For telecommunication purposes it is necessary to impress an a.c. signal on the output of a laser, i.e., to modulate directly the emerging light by, say, the speech. This can be accomplished, for example, by **amplitude modulation**, i.e., by biasing the laser initially above the threshold and then superimposing on this d.c. voltage an a.c. signal (Fig. 13.44). The amplitude of the emerging laser light depends on the slope of the power–current characteristic. Another possibility is **pulse modulation**, i.e., the generation of subnanosecond pulses having nanosecond spacings between them. (For digitalization, see Section 13.10.) This high-speed pulsing is possible because of the inherently short turn-on and turn-off times (10^{-10} s) of semiconductor lasers when initially biased just below the threshold current density. Finally, **frequency modulation** can be achieved by applying, perpendicularly to the diode junction, a periodic varying mechanical pressure (by means of a transducer), thus periodically altering the dielectric constant of the cavity. This way, modulation rates of several hundred megahertz have been achieved.

13.8.11. Laser Amplifier

The laser can also function as an optical amplifier, which is again used for telecommunication purposes. A weak optical signal enters a laser through one of its windows and there stimulates the emission of photons. The amplified signal leaves the other window after having passed the cavity only once. This **traveling-wave laser** is biased slightly below the threshold current in order to exclude spurious lasing not triggered by an incoming signal. Nevertheless, some photons are always spontaneously generated, which causes some background noise.

A new development is the **erbium-doped fiber amplifier**, which works quite similar to the above-mentioned traveling-wave laser. Erbium atoms, contained in lengths of a coiled glass fiber, are pumped to higher energies by an indium–gallium–arsenide–phosphide laser at a wavelength of 0.98 μm or

Figure 13.44. Amplitude modulation of a semiconductor laser: (a) input current–output power characteristic; (b) circuit diagram. (The d.c. power supply has to be electrically insulated from the a.c. source.)

1.48 μm. When a weakened signal enters one end of this erbium-doped fiber, the erbium atoms gradually transfer their energy to the incoming signal by stimulated emission, thus causing amplification. A mere 10 mW of laser power can thus achieve a gain of 30–40 dB. Networks which include fiber amplifiers, linked at certain distance intervals to cladded optical glass fibers (Fig. 13.30), have the potential of transmitting data at very high rates, e.g., 2.5 gigabits of information per second over more than 20,000 km. This is possible because fibers are able to support a large (but finite) number of channels. The advantage of erbium-doped optical fibers is that they do not interrupt the path of a light signal as conventional *"repeaters"* do (which convert light into an electric current, amplify the current, and then transform the electrical signal back into light).

13.8.12. Quantum Well Lasers

Quantum well lasers are the ultimate in miniaturization, as already discussed in Section 8.7.10. We have explained there that some unique properties are observed when device dimensions become comparable to the wavelength of electrons. In essence, when a thin (20 nm wide) layer of a *small–band gap* material (such as GaAs) is sandwiched between two *large–band gap* materials (such as AlGaAs), a similar energy configuration is encountered as known for an *electron in a box* (Fig. 8.33). Specifically, the carriers are confined in this case to a potential well having "infinitely" high walls. Then, as we know from Section 4.2, the formerly continuous conduction or valence bands reduce to discrete energy levels, see Fig. 13.45.

The light emission in a *quantum well laser* occurs as a result of electron transitions from these conduction band levels into valence levels. It goes almost without saying that the line width of the emitted light is small in this case, because the transitions occur between *narrow* energy levels. Further, the threshold current density for lasing (Fig. 13.41) is reduced by one order of magnitude, and the number of carriers needed for population inversion is likewise smaller.

If a series of large–and small–band gap materials are joined, thus forming a *multiple quantum well laser*, the gain is even further increased and the sta-

Figure 13.45. Band structure of a single quantum dot structure. See in this context Section 8.7.10 and Fig. 8.33(b).

bility of the threshold current toward temperature fluctuations is improved. The GaAs/GaAlAs combination yields an emission wavelength somewhat below 0.87 μm, whereas InGaAsP quantum well lasers emit light near 1.3 μm or 1.5 μm (depending on their composition). It appears to be challenging to eventually fabricate quantum wire or quantum dot lasers (see Section 8.7.10) which are predicted to have even lower threshold current densities and higher modulation speeds.

13.8.13. Light-Emitting Diodes (LED)

Light-emitting diodes are of great technical importance as inexpensive, rugged, small, and efficient light sources. The LED consists, like the semiconductor laser, of a forward biased $p - n$ junction. The above mentioned special facing procedures are, however, omitted during the manufacturing process. Thus, the LED does not operate in the lasing mode. The emitted light is therefore neither phase coherent nor collimated. It is, of course, desirable that the light emission occurs in the visible spectrum. Certain III–V compound semiconductors, such as $Ga_xAs_{1-x}P$, GaP, $Ga_xAl_{1-x}As$ (for red and yellow-green) and the newly discovered nitride-based compound semiconductors (for green and blue colors) fulfill this requirement. Their emission efficiencies (measured in lumens per watt) are at par or even better than those of unfiltered incandescent light bulbs and are almost one order of magnitude larger than certain color-filtered tungsten-filament lamps. All three basic colors necessary for covering the visible spectrum and also the infrared (450–1,500 nm) are now available with adequate intensities. Because of these properties, the lighting industry is currently undergoing a revolution that will lead to LED-based large flat-panel color displays, bright outdoor color signs, better projection television, full-color photographic printers, more efficient and particularly durable traffic lights, and even changes in home and office illumination.

In order to vividly demonstrate the spectral emission properties of LEDs a **chromaticity diagram** as shown in Fig. 13.46 is helpful. It is based on the peculiarities of the three types of cones in the human eye, which are sensitive for either blue, green, or red radiation. (The corresponding wavelengths mark the corners of the chromaticity diagram). A given "color" is represented by two parameters or percentages (x and y) in this graph, while the percentage of the third color is the difference between $x + y$ and 100%. Monochromatic light (such as from a laser) is depicted by a specific point on the perimeter of the graph. Any other hue is created by mixing the basic colors. When the spectral width of the light increases and the emission is therefore less pure, the color coordinates move towards the center. As an example, "white light", that is, the broad emission spectrum of a black-body radiator heated to very high temperatures (e.g., the sun), is described by a point in this diagram at which x as well as y are about 33%.

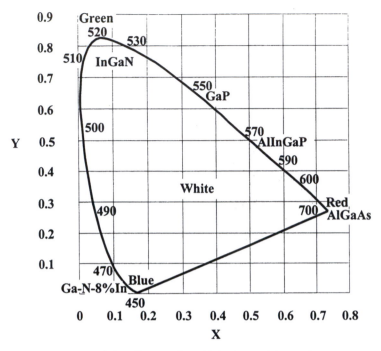

Figure 13.46. Chromaticity diagram in which the positions of some commercially available LEDs are shown.

Fig. 13.46 displays the color coordinates of some of the presently available electroluminescing compound semiconductors. Ga–N containing about 8% In is depicted to provide blue (470 nm) light. This color changes into green (520 nm) by adding successively larger amounts of indium to GaN. Al–Ga–As yields red hues (700 nm) and yellow-green light is emitted by Al–In–Ga–P (590 nm). Moreover, green- or blue-emitting LEDs, when covered by one or more appropriate phosphors, (see Section 13.8.1) can be made to vary their color according to the spectral emission of the phosphor. White-radiating LEDs are obtained by exciting suitable phosphors by the ultra-violet radiation from GaN. Alternatively, white-emitting LEDs are obtained by adding blue and yellow LED light.

A few technical details on nitride-based semiconductors shall be added. As mentioned above, LEDs require n-and p-type components to manufacture a diode. Si in the form of silane (SiH$_4$) is used for n-doping. On the other hand, incorporating Mg, followed by low-energy electron radiation, or thermal annealing (to activate the Mg-doped GaN) leads to p-type segments.

As already mentioned in Section 13.8.9 (Fig. 13.43), double hetero-structures increase the output power of LEDs. Specifically, incorporating Zn and Si dopants into an In–Ga–N active layer that is surrounded by Al–Ga–

N layers leads to output powers near 3 mW when the chip size is 3×3 mm^2. Further, a peak wavelength of 450 nm, and an external quantum efficiency of 5.4% (that is, the ratio of the number of photons produced to the number of injected electrons) are obtained. Forward currents of typically 20 mA are generally applied.

The devices consisting of indium–gallium nitrides are commonly deposited on sapphire (α-Al$_2$O$_3$) or silicon carbide (6H-SiC) substrates. Metal-organic chemical vapor deposition at 700 to 1,100°C, involving trimethyl gallium (Ga(CH$_3$)$_3$) and ammonia (NH$_3$) as gas sources, is generally used for growing epitaxial GaN films. InGaN is laid down by additions of trimethyl indium. The devices are contacted and eventually encapsulated in epoxy resins. The radiation leaves the device through a semitransparent metal contact on the top or through a transparent n-GaN contact on the substrate. In the case of GaAs, the light may leave the device through a window which has been etched through the metallic contact (*surface emitter*).

The "lifetimes" of LEDs are in excess of 50,000 hours, or more precisely, the time after which the light intensity has decreased to 50% of its original value is approximately 30,000 hours. (This compares to an average lifetime of 1,000 to 2,000 hours for a typical incandescent light bulb.) The failures are generally caused by a break-down of the contacts or the encapsulate, rather than of the semiconductor itself. The cost at present is about one US dollar for a blue- or green-emitting diode and less than 10 cents for red and orange LEDs.

A short note on other recently developed LED materials shall be added. Blue-emitting SiC has been investigated for some time, but its efficiency is orders of magnitude smaller than the above-described nitrides. This is mainly due to its indirect–band gap characteristics. ZnSe (a II–VI compound) has also blue and green emissions, but its lifetime is substantially reduced by the formation of structural defects. LEDs based on polymers with ionic materials as electron-injecting and hole-blocking layers have been demonstrated. Finally, as a matter of curiosity, a "light-emitting vegetable diode", utilizing a pickle, has been reported in the literature to emit yellow light (and an unpleasant smell).

13.8.14. Liquid Crystal Displays (LCDs)

Many consumer products need to communicate the processed information to their owners, such as in wrist watches, calculator read-outs, video cameras, video recorders (VCRs), automobile dashboards, etc. The display market is currently still dominated (70%) by traditional cathode ray tubes as known for TVs and many computer monitors. However, flat-panel displays are rapidly gaining ground. Among them, the liquid crystal display dominates with 85% of this market, having annual sales near 10 billion dollars world-

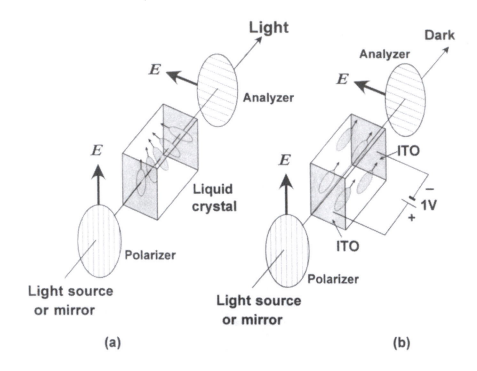

Figure 13.47. Schematic representation of a liquid crystal display unit (a) in the light-transmitting mode, (b) in the non-light-transmitting mode, caused by a potential that is applied to the end faces of the (twisted nematic) liquid crystal. Polarizer and analyzer are identical devices that allow the light (i.e., the electric field vector) to oscillate in only one direction as indicated by arrows (see also Section 13.1.2). The end faces of the liquid crystal–containing glass vessel are coated by transparent electrodes such as indium-tin-oxide (ITO), see Section 9.3.

wide. LCDs are "non-emissive devices", that is, they do not emit light by themselves, but rather depend on external illumination, as we will see momentarily.

LCDs contain peculiar viscous liquids whose rod-shaped molecules are arranged in a specifically ordered pattern. Each of these rod-shaped molecules has a strong electric dipole moment and can thus be oriented in an electric field, see Section 9.5. The viscous liquid is encapsulated in a glass container and is initially treated so that the molecules on one end are aligned at right angles to the ones on the other end, see Fig. 13.47(a). Moreover, the orientations of the molecules vary gradually from, say, a vertical to a horizontal array, as also depicted in Fig. 13.47(a). It is therefore called a "twisted nematic" type LCD.

If light which is polarized parallel to the aligned molecules of one end impinges on such a crystal, its electric vector will follow the twist of the molecules through the liquid crystal to the other end and emerges therefore on the opposite side with its polarization direction perpendicular to its original orientation. Since the analyzer that is placed behind the liquid crystal is oriented perpendicular to the polarizer, the emerging light beam is transmitted.

On the other hand, if a small voltage (about 1 volt) is applied to the conducting end faces of the liquid crystal, the molecules (dipoles) align parallel to the field direction and the light is therefore not caused to change its polarization direction, see Fig 13.47(b). Thus, since polarizer and analyzer are mutually perpendicular to each other, the light is blocked from transmission.

In practice, a mirror is placed behind the liquid crystal arrangement which reflects the ambient light back to the viewer if the LCD is in the transmission mode. Alternatively, the display can be illuminated from the back to allow dark readability.

The advantage of an LCD is that it is inexpensive and that it consumes very little energy (at least as long as no back-lighting is utilized). On the negative side, an LCD cannot be read in the dark or in dim light without back-lighting. Furthermore, it has a limited usable temperature range (20°C to 47°C), slow time response, and most of all, it has a very narrow viewing angle. LCDs are matrix addressed by applying voltages to rows and columns, similarly as depicted in Fig. 8.43.

13.8.15. Emissive Flat-Panel Displays

Electroluminescent devices utilize a thin phosphor film, such as manganese-doped zinc sulfide ($ZnS:Mn$), which is sandwiched between two insulating films, e.g., Al_2O_3 or Al–Ti–O (ATO). These films are surrounded by two conducting films (one of them being transparent indium-tin-oxide (ITO) on glass and the other a good reflector), see Fig. 13.48. The light emission is generally induced by an alternating (pulsed) electrical potential[8] (about 120–200 V) applied between the two conducting electrodes. This generates an electric field amounting to about 10^6 V/cm across the phosphor layer, which causes an injection of electrons into the phosphor. Once the threshold voltage has been exceeded, the electrons become ballistic and excite the electrons of the activator atom in the phosphor (e.g., Mn) into a higher energy state. Upon reverting back into the ground state, photons of the respective wavelength are generated. As an example, a broad orange-yellow spectrum is emitted from the above-mentioned $ZnS:Mn$. Green color is observed for $ZnS:Tb$ or $SrS:Ce$, blue light is seen when utilizing $(Sr_{0.45}Ca_{0.55})_{1-x}$ $Ga_2S_4:Ce_x$ or $SrS:Cu$, Ag, and red light is emitted from $ZnS:Sm$, Cl, or $CaS:Eu$. To produce white light, a combination of $ZnS:Mn$ and $SrS:Ce$ has been used as phosphors. Green and red can also be obtained by filtering light from $ZnS:Mn$.

---

[8] A dc voltage is possible but electromigration of impurity ions (e.g. halides) which causes eventually a counter field shortens the lifetime of the device.

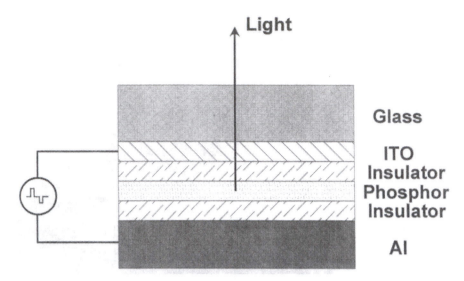

Figure 13.48. Schematic diagram of an electroluminescence device operated by alternating current pulses of about 200 V. The thin-film layers are about 300 nm thick except in the case of the phosphor, whose thickness is between 600 and 1,000 nm. The phosphor consists of the host matrix, such as a wide–band gap metal sulfide (ZnS, CuS, SrS), and an "activator", also called "luminescence center", such as Mn, Tb, Eu, Ce, Sm, Cu, Ag, etc.

The cost of electroluminescing devices is higher than for LCDs, but the viewing angle is wider and the usable temperature range is larger. They are rugged and have long lifetimes. Even though the luminescence efficiency is relatively good, readability in sunlight is still a problem. The response time is fast enough for video displays, but the power requirements (about 120–200 V) are unacceptable for small portable device applications. Electroluminescence devices comprise at present about 8% of the flat-panel market. They find applications in medical instruments, transportation, defense, and industrial equipment.

Plasma display devices operate quite similar to fluorescence light bulbs. A relatively high AC voltage (100 V) is applied across a discharge gas (such as a helium/neon mixture) to create a plasma. Recombination of electron-ion pairs in the plasma causes photons of high energy (e.g., in the UV range). They are absorbed by the phosphors which in turn emit visible light. Individually addressable compartments, which may contain different phosphors, yield the pixels needed for the three fundamental colors.

Flat-panel displays of this type are rugged, the viewing angles are wide, and the lifetimes are adequate. However, the pixel sizes are too large for small displays, and the colors tend to "wash out" in strong ambient lighting. Plasma displays currently share 7% of the flat-panel display market.

Field-emission displays are still in the experimental stage. They have much in common with cathode ray tubes, that is, with the bulky displays used for TVs and desk computers. The aim is to build them flat (about 10 mm thick), with wide viewing angles and fast response times. They consist of a large number of tip-shaped field-emitters (made of Mo, Si, Pt, etc.) that can be matrix-addressed and which emit, when hermetically encapsulated into a vacuum, substantial amounts of electrons under the influence of high electric fields. These electrons impinge on phosphors of various kinds to cause cathodoluminescence in different colors.

13.9. Integrated Optoelectronics

Integrated optoelectronics deals with a family of optical components, such as lasers, photodiodes, optical waveguides, optical modulators, optical storage devices, etc., which are integrated on a common substrate (if feasible) with the aim of fulfilling similar functions as electrical integrated circuits do. The main difference to electrical devices is that in optical integrated circuits (OICs) the signal is transmitted by light. Still, they need in most cases electrical energy to become functional, which explains the name *optoelectronic*. Among the advantages of optical devices are reduced weight, the capability of light of different wavelengths to travel independently and simultaneously in the same waveguide (*multiplexing*), the immunity against receiving extraneous signals from surrounding devices by stray electromagnetic coupling (*crosstalk*), the difficulty in performing wire taps (because of the lack of electromagnetic fields, which would extend beyond the optical fiber), high reliability, speeds greater than electrons in a metallic wire, larger bandwidth (10^{12} Hz compared to 10^5 Hz for telephones) and notably, the *low-loss transmission* (<2 dB/km) of signals in optical fibers. Most of all, however, telecomunication utilizing laser optics allows the simultaneous transmission of billions of telephone calls in one glass fiber, that is, as many simultaneous telephone calls as there are humans on earth! We have discussed in previous chapters two major optoelectronic components, the laser (Section 13.8) and the photodetector (Sections 8.7.6 and 8.7.7). A few more building blocks need to be added to complete the picture. This will be done now.

13.9.1. Passive Waveguides

The interconnecting medium between various optical devices is called a *waveguide*. It generally consists of a thin, transparent layer whose index of refraction, n_2, is *larger* than the refractive indices of the two surrounding media, n_1 and n_3. If this condition is fulfilled and if the light impinges on the boundary between n_2 and n_1 (or n_3) at an angle which is larger than the angle

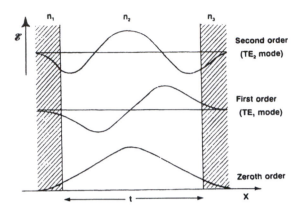

Figure 13.49. Electric field strength distribution (modes) in a waveguide assuming $n_1 = n_3$ (symmetric behavior). The zeroth order and higher-order modes are shown. (Compare with Fig. 4.8.)

of total reflection,[9] then the optical beam travels in zigzag paths between the internal boundaries of Region 2. In other words, by undergoing total reflection, the light wave is considered to remain in the center region. This statement needs, however, some refinement. As a rule, the light which travels in the center medium extends, to a certain degree, into the neighboring media. The spatial distribution of the optical energy within all three media is called a *mode*. This spatial distribution can be calculated by solving the wave equation (10.4) while taking the appropriate boundary conditions into consideration. We have done this twice in earlier parts of this book (Section 10.3 and Section 4.3). We learned there that the electric field strength or, equivalently, the intensity of a wave, decreases in the adjacent medium obeying an exponential function. If two boundaries need to be considered, as in the present case, and if the thickness, t, of the center region is comparable to the wavelength of light, then the solution of the wave equation yields an electric field strength distribution (as depicted in Fig. 13.49, lower curve). Now, we know from previous calculations (Section 4) that under certain conditions additional solutions, i.e., distribution functions, do exist (similarly, as a vibrating string can oscillate at higher harmonics). In the present case they are called first-order, second-order, etc., modes. They are likewise depicted in Fig. 13.49. The reader probably recognizes that this "optical tunnel effect" is equivalent to the quantum mechanical tunnel effect shown in Fig. 4.8.

We now consider the most common case, in which n_1 is considerably

---

[9] If light passes from an optically dense material (e.g., glass with $n_1 \approx 1.5$) into air ($n_2 \approx 1$), then the angle of the refracted beam, β, is larger than the angle of incidence, α. At a critical angle, α_T, β becomes $90°$ (grazing exit). Total reflection occurs when $\sin \alpha_T > n_2/n_1$; see also Section 10.2.

smaller than n_3, e.g., $n_1 = 1$ for air and $n_3 = 3.6$ for GaAs (whereas n_2 is still made larger than n_3!). For this *asymmetric* case the condition for containing the light in the waveguide is

$$n_2 - n_3 \geq \frac{(2\kappa + 1)^2 \lambda_0^2}{32 n_2 t^2},$$
(13.20)

where λ_0 is the wavelength of the light in vacuum, $\kappa = 0, 1, 2 \ldots$ is the mode number, and t is the thickness of the center layer. A calculation (see Problem 1) shows that the difference between n_2 and n_3 needs to be only about 1% in order to contain the light in the center medium.

13.9.2. Electro-Optical Waveguides (EOW)

So far we tacitly implied that the various layers of a waveguide structure have been permanently manufactured by some type of deposition process out of the gaseous or liquid phase on a semiconducting substrate. This is indeed quite often done by employing, for example, molecular beam epitaxy or liquid phase epitaxy processes. However, a rather ingenious alternative method can be utilized instead. This technique involves a Schottky-barrier contact which, when reverse biased, forms (as we know from Section 8.7.2) a wide depletion layer (Fig. 13.50). We shall show in a short calculation that a depletion of charge carriers increases the index of refraction of a solid.

   Recall that $n_2 > n_3$ (and n_1) is the prerequisite for a waveguide structure. We derived in Chapter 11 a relationship between the free carrier density (N_f) and the index of refraction,

$$\hat{n}^2 = 1 - \frac{e^2 N_f}{4\pi^2 \varepsilon_0 m^* v^2},$$
(11.7)

where m^* is the effective mass of the electrons in the medium, v is the frequency of light, e is the charge of the electrons, and $\hat{n} = n - ik$ is the complex

Figure 13.50. Electro-optical waveguide making use of a reverse-biased Schottky-barrier contact. (See also Fig. 8.15.) The light travels in Medium 2 (the depletion layer) when a high-enough voltage is applied to the device.

index of refraction. We rewrite (11.7) twice for the substrate (Medium 3) and for the depletion layer from which some free carriers have been removed by the applied electric field (Medium 2),

$$\hat{n}_2^2 = 1 - \frac{e^2 N_{f2}}{4\pi^2 \varepsilon_0 m^* v^2},$$

(13.21)

$$\hat{n}_3^2 = 1 - \frac{e^2 N_{f3}}{4\pi^2 \varepsilon_0 m^* v^2}.$$

(13.22)

The difference in the indices of refraction is then

$$\hat{n}_2^2 - \hat{n}_3^2 = \frac{e^2}{4\pi^2 \varepsilon_0 m^* v^2} (N_{f3} - N_{f2}),$$

(13.23)

which reduces with[10]

$$\hat{n}_2^2 - \hat{n}_3^2 = (\hat{n}_2 + \hat{n}_3)(\hat{n}_2 - \hat{n}_3) \approx 2\hat{n}_3(\hat{n}_2 - \hat{n}_3)$$

and $c = v \cdot \lambda$ to

$$\hat{n}_2 - \hat{n}_3 = \frac{e^2 \lambda^2}{2n_3 4\pi^2 \varepsilon_0 m^* c^2} (N_{f3} - N_{f2}).$$

(13.24)

In the present case (transparent media) we can assume that the damping constant, k, in $\hat{n} = n - ik$ is negligibly small, so that $\hat{n}$ in (13.24) becomes a real quantity:

$$n_2 - n_3 = \frac{e^2 \lambda^2}{2n_3 4\pi^2 \varepsilon_0 m^* c^2} (N_{f3} - N_{f2}).$$

(13.25)

Equation (13.25) demonstrates, as suggested above, that a reduction in the number of free carriers from N_{f3} to N_{f2} causes an increase in the index of refraction in Medium 2. Then, the device becomes an optical waveguide. For this to happen, the doping of the substrate needs to be reasonably high in order that an appreciable change in the index of refraction is achieved (see Problem 4).

13.9.3. Optical Modulators and Switches

When discussing electronic devices in Section 8.7.12, we encountered a digital *switch* that is capable of turning the electric current *on* or *off* by applying a voltage to the gate of a MOSFET. An equivalent optical device is obtained by making use of the electro-optical waveguide (Fig. 13.50). In the present case, this device is biased initially just below the threshold, i.e., at a voltage which barely prevents the lowest-mode optical wave from passing. Then, by

---

[10] $\hat{n}_2$ can be assumed to be approximately equal to $\hat{n}_3$ (see above and Problem 1) which yields $\hat{n}_2 + \hat{n}_3 \approx 2\hat{n}_3$.

Figure 13.51. Schematic representation of the Franz–Keldysh effect.

an additional voltage between metal and substrate, the EOW becomes transparent. In analogy to its electrical equivalent (Fig. 8.30), this device may be called an *enhancement-type* or *normally-off* electro-optical wave-guide. By varying the bias voltage periodically above the threshold, the EOW can serve as an effective modulator of light.

A *depletion-type* or *normally-on* EOW can also be built. This device exploits the Franz–Keldysh effect, i.e., the shift of the absorption edge to lower energies when an electric field is applied to a semiconductor (Fig. 13.51). The photon energy of the light is chosen to be slightly smaller than the band gap energy (dotted line in Fig. 13.51). Thus, the semiconductor is normally in the transparent mode. If, however, a large electric field (on the order of 10^5 V/cm) is applied to the device, then the band gap shifts to lower energies and the absorbance at that particular wavelength (photon energy) becomes several orders of magnitude larger, thus essentially blocking the light. (The **Franz–Keldysh shift** can be understood when inspecting Fig. 8.15, which shows a lowering of the conduction band and thus a reduction in the band gap energy when a reverse bias is applied to a semiconductor.)

Finally, if a piezoelectric transducer imparts some pressure on a wave-guide, the index of refraction changes. This **photoelastic effect** can also be utilized for modulation and switching.

Electro-optical modulators can be switched rapidly. The range of frequencies over which the devices can operate is quite wide.

13.9.4. Coupling and Device Integration

We now need to discuss some procedures for transferring optical waves (i.e., information) from one optical (or optoelectronic) device to the next. Of

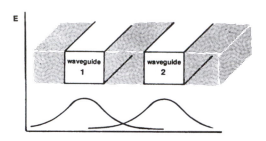

Figure 13.52. Schematic representation of energy transfer between two waveguides (or a waveguide and an optical fiber) by optical tunneling. Compare with Fig. 13.49. ($n_2 > n_1, n_3$.)

course, *butting*, i.e., the end-on attachment of two devices, is always an option, particularly if their cross-sectional areas are comparable in size. This technique is indeed frequently utilized for connecting optical fibers (used in long-distance transmission) to other components. A special fluid or layer which matches the indices of refraction is inserted between the two faces in order to reduce reflection losses. Optical alignment and permanent mechanical attachment are nontrivial tasks. They can be mastered, however. In those cases where no end faces are exposed for butting, a *prism coupler* may be used. This device transfers the light through a *longitudinal surface*. In order to achieve low-loss coupling, the index of refraction of the prism must be larger than that of the underlying materials. This is quite possible for glass fibers ($n \approx 1.5$) in conjunction with prisms made out of strontium titanate ($n = 2.3$) or rutile ($n = 2.5$), but is difficult for semiconductors ($n \approx 3.6$).

Phase coherent energy transfer between two *parallel* waveguides (or an optical fiber and a waveguide) can be achieved by optical tunneling (Fig. 13.52). For this to occur, the indices of refraction of the two waveguides must be larger than those of the adjacent substrates. Further, the width of the layer between the two waveguides must be small enough to allow the tails of the energy profiles to overlap.

The most elegant solution for efficient energy transfer is the monolithic integration of optical components on *one* chip. For example, a laser and a waveguide may be arranged in one building block, as schematically depicted in Fig. 13.53. Several points need to be observed however. First, the wavelength of the light emitted by the laser needs to be matched to a wavelength at which the absorption in the waveguide is minimal. Second, the end faces of the laser need to be properly coated (e.g., with SiO_2) to provide adequate feedback for stimulated emission.

Another useful integrated structure involves a transverse photodiode that is coupled to a waveguide, see Fig. 13.54. As explained in Section 8.7.6, this photodiode is reverse biased. The electron–hole pairs are created in or near a long and wide depletion layer by photon absorption. The losses are minimized owing to the fact that the light does not have to penetrate the (inac-

Figure 13.53. Schematic representation of a monolithic laser/waveguide structure. Compare with Fig. 13.43.

tive) p-region as in flat-plate photovoltaics. The quantum efficiency of the transverse photodiode can be considerably enhanced by increasing the *length* of the depletion layer.

All taken, the apparently difficult task of connecting optical fibers, waveguides, lasers, or photodetectors and their integration on one chip have progressed considerably in the last decade and have found wide application in a multitude of commercial devices.

13.9.5. Energy Losses

Optical devices lose energy through absorption, radiation, or light scattering, similarly as the electrical resistance causes energy losses in wires, etc. The optical loss is expressed by the *attenuation* (or absorbance), α, which was

Figure 13.54. Schematic representation of a monolithic transverse photodiode/ waveguide structure. A wide depletion layer (active region) is formed in the n-region by the reverse bias. For details, see Section 8.7.6.

defined in Section 10.4. It is measured in cm^{-1} or, when multiplied by 4.3, in decibels per centimeter.

Scattering losses take place when the direction of the light is changed by multiple reflections on the "rough" surfaces in waveguides or glass fibers, or to a lesser extent by impurity elements and lattice defects.

Absorption losses occur when photons excite electrons from the valence band into the conduction band (interband transitions), as discussed in Chapter 12. They can be avoided by using light whose photon energy is smaller than the band gap energy. Free carrier absorption losses take place when electrons *in the conduction band* (or in shallow donor states) are raised to higher energies by *intra*band transitions. These losses are therefore restricted to semiconductor waveguides, etc., and essentially do not occur in dielectric materials. We know from (10.21a) that the absorbance, α, is related to the imaginary part of the dielectric constant, ε_2, through

$$\alpha = \frac{2\pi}{\lambda n}\varepsilon_2. \tag{13.26}$$

On the other hand, the free electron theory provides us with an expression for ε_2 (11.27), which is, for $v^2 \gg v_2^2$,

$$\varepsilon_2 = \frac{v_2 v_1^2}{v^3}, \tag{13.27}$$

where

$$v_1^2 = \frac{e^2 N_f}{4\pi^2 \varepsilon_0 m^*} \tag{13.28}$$

(see (11.8)) and

$$v_2 = \frac{2\pi\varepsilon_0 v_1^2}{\sigma_0} \tag{13.29}$$

(see (11.23)) and

$$\sigma_0 = N_f e\mu \tag{13.30}$$

(see (8.13)). Combining equations (13.26) through (13.30) yields

$$\alpha = \frac{e^3 N_f \lambda^2}{4\pi^2 \varepsilon_0 n (m^*)^2 c^3 \mu}. \tag{13.31}$$

We note in (13.31) that the free carrier absorbance is a linear function of N_f and is inversely proportional to the mobility of the carriers. The absorbance is also a function of the square of the wavelength.

Radiation losses are, in essence, only significant for curved-channel waveguides, in which case photons are emitted into the surrounding media. A detailed calculation reveals that the radiation loss depends exponentially on the radius of the curvature. The minimal tolerable radius differs considerably in different materials and ranges between a few micrometers to a few centimeters. The energy loss is particularly large when the difference in the indices of refraction between the waveguide and the surrounding medium is small.

13.9.6. Photonics

A short note on the recently coined term "photonics" shall be added. *Electronics* deals with electrons and materials in which electrons propagate. Similarly *photonics* relates to photons and their interaction with *photonic crystals*. These crystals are materials that possess a periodicity of the dielectric constant so that they can affect the properties of photons in much the same way as electrons are affected by periodically arranged atoms, that is, by the lattice structure. However, photonic crystals need to be created artificially. The "lattice constant" of photonic crystals must be comparable to the wavelength of light, that is, the periodicity needs to be on the order of 500 nm. This requires high-resolution microlithography techniques, as known from semiconductor processing, involving X-rays or electron beams.

The solution of the Maxwell equations for this particular case (rather than the Schrödinger equation) leads to photonic band structures, Brillouin zones, and occasionally to band gaps quite similarly as known from electronics. Rather than displaying *s*- or *p*-bands, photonic band structures contain transverse magnetic (TM) or transverse electric (TE) modes. Doping can be accomplished by introducing point defects that affect the periodicity of the photonic crystal. This leads to localized photonic states within the gap similar to donor or acceptor states. Furthermore, a line defect acts like a waveguide and a planar defect behaves like a mirror. Photonic band structures are quite similar to *phononic* band structures (see Chapter 20.2) and, naturally, to electronic bands.

The research results of this field should be followed with considerable anticipation.

13.10. Optical Storage Devices

Optical techniques have been used for thousands of years to retrieve stored information. Examples are ancient papyrus scrolls or stone carvings. The book you are presently reading likewise belongs in this category. It is of the *random-access* type, because a particular page can be viewed immediately without first exposing all previous pages. Other examples of optical storage devices are the conventional photographic movie film (with or without optical sound track) or the microfilm used in libraries. The latter are *sequential* storage media because all previous material has to be scanned before the

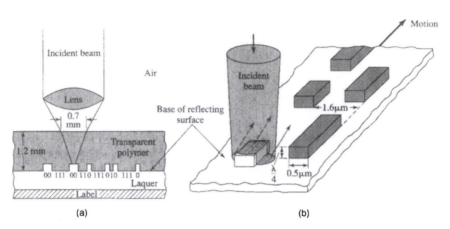

Figure 13.55. Schematic of a compact disk optical storage device. Readout mode. (Not drawn to scale.) The reflected beams in Fig. 13.55(b) are drawn under an angle for clarity. The land and bump areas covered by the probing light have to be of equal size in order that destructive interference can occur (see the hatched areas covered by the incident beam in Fig. 13.55(b)).

information of interest can be accessed. They are also called *read-only memories* (ROM) because the information content *cannot* be changed by the user. All examples given so far are *analog* storage devices.

Another form of storage utilizes the optical disk, which has recently gained wide-spread popularity. Here, the information is generally stored in *digital* form. The most common application, the compact disk (CD), is a *random-access, read-only memory* device. However, *"write-once, read-many"* (WORM) and erasable magneto-optical disks (Section 17.5) are also available. Recently, rewritable CD-RW disks came on the market which can be "burned", i.e. reused up to 40 times. The main advantage of optical techniques is that the readout involves a noncontact process (in contrast to magnetic tape or mechanical systems). Thus, no wear is encountered.

Let us now discuss the optical compact disk. Here, the information is stored below a transparent, polymeric medium in the form of bumps, as shown in Fig. 13.55. The height of these bumps is one-quarter of a wavelength ($\lambda/4$) of the probing light. Thus, the light which is reflected from the base of these bumps (called the "land") travels half a wavelength farther than the light reflected from the bumps. If a bump is encountered, the combined light reflected from bump and land is extinguished by destructive interference. *No light* may be interpreted as a *zero* in binary code, whereas *full intensity* of the reflected beam would then constitute a *one*. 14 *ones* and *zeros* represent one byte of data. For audio purposes, the initial analog signal is sampled at a frequency of 44.1 kHz (about twice the audible frequency) to digitize the information into a series of ones and zeros (similarly as known for computers, Section 8.7.12). Quantization of the signal into 16-digit binary numbers gives a scale of 2^{16} or 65,536 different values. This information is transferred to a disk (see below) in the form of bumps and absences of bumps. For readout from the disk, the probing light is pulsed with the same frequency so that it is synchronized with the digitized storage content.

The spiral path on the useful area of a 120 mm diameter CD is 5.7 km long and contains 22,188 tracks spaced 1.6 μm apart. (As a comparison, 30 tracks can be accommodated on a human hair.) The spot diameter of the readout beam near the bumps is about 1.2 μm. The information density on a CD is 800 kbits/mm^2, i.e., a standard CD can hold about 7×10^9 bits. This number will increase by a factor of four when blue lasers ($\lambda = 450$ nm) are used! The current playback time is about one hour. A disk of the same diameter can also be digitally encoded with 600–800 megabyte of computer data, which is equivalent to three times the text of a standard 24-volume encyclopedia.

The manufacturing process of CDs requires an optically flat glass plate which has been covered with a light-sensitive layer (photoresist) about $\lambda/4$ in thickness. Then, a helium–neon laser whose intensity is modulated (pulsed) by the digitized information is directed onto this surface while the disk is rotated. Developing of the photoresist causes a hardening of the unexposed areas. Subsequent etching removes the exposed areas and thus creates pits in the photoresist. The pitted surface is then coated with silver (to facilitate electrical conduction) and then electroplated with nickel. The nickel mold thus created (or a copy of it) is used to transfer the pit structure to a transparent polymeric material by injection molding. The disk is then coated with a reflective aluminum film and finally covered by a protective lacquer and a label.

The CD is read from the back side, i.e., the information is now contained in the form of bumps (see Fig. 13.55). In order to facilitate focusing onto a narrow spot, monochromatic light, as provided by a laser, is essential. At present, a GaAlAs heterojunction laser having a wavelength in air of 780 nm is utilized. This will change as soon as blue semiconductors, such as InGaN lasers ($\lambda = 450$ nm), are commercially available. The beam size at the surface of the disk is relatively large (0.7 mm in diameter) to minimize possible light obstruction by small dust particles. However, the beam converges as it traverses through the polymer disk to reach the reflecting surface that contains the information. Small scratches on the polymer surface are also tolerated quite well. The aligning of the laser beam on the extremely narrow tracks is a nontrivial task, but it can be managed. It involves, actually, three light beams, obtained by dividing into three parts the impinging laser beam shown in Fig. 13.55(b) utilizing a grating or a holographic element. One of these parts (the center one) is the above-described read beam. The other two are tracking beams which strike the inner and outer edges of the groove. The reflected signals from the tracking beams are subtracted from each other. A null signal indicates correct tracking while positive or negative signals cause the servo to move the read head to one or the other side. The tracking is accurate to about 0.1 μm.

An alternative to the above-described CD-ROM is the magneto-optical device, which employs a laser to read the data on the disk while the information is written by simultaneously exposing a small area on the disk to a strong laser pulse in addition to a magnetic field. This device will be further described in Section 17.5. As of this writing, 4.6 GB can be stored on a $5\frac{1}{4}$ inch (130 mm) magneto-optical disk (compared to 0.62 GB on a CD-ROM). The data can be erased and rewritten many times.

13.11. The Optical Computer

We have learned in Section 8.7 that transistors are used as switching devices. We know that a small voltage applied to the base terminal of a transistor triggers a large electron flow from emitter to collector. The question arises whether or not a purely optical switching device can be built for which a light beam, having a small intensity, is capable of triggering the emission of a light beam that has a large intensity. Such an *optical transistor* (called *transphasor*) has indeed been constructed which may switch as much as 1000 times faster (picoseconds) than an electronic switch (based on a transistor).

The main element of a transphasor is a small (a few millimeters long) piece of nonlinear optical material (see below) which has, similar to a laser (or a Fabry–Perot interferometer), two exactly parallel surfaces at its longitudinal ends. These surfaces are coated with a suitable thin film in order to render them semitransparent. Once monochromatic light, stemming from a laser, has entered this "cavity" through one of its semitransparent windows, some of the light is reflected back and forth between the interior windows, whereas another part of the light eventually escapes *through* the windows (Fig. 13.56). If the length between the two windows just happens to be an integer multiple of half a wavelength of the light, then *constructive interference* occurs and the amplitude (or the intensity) of the light in the "cavity" increases rapidly (Fig. 13.56(a)). As a consequence, the intensity of the transmitted light is also strong. In contrast to this, if the distance between the two windows is not an integer multiple of half a wavelength of the light, the many forward and reflected beams in the "cavity" weaken each other mutually, with the result that the intensity of the transmitted light is rather small (Fig. 13.56(b)). In other words, all conditions which do not lead to constructive (or near constructive) interference produce rather small transmitted intensities (particularly if the reflectivity of the windows is made large).

The key ingredient of a transphasor is a specific substance, namely, the above-mentioned nonlinear optical material which changes its index of refraction as a function of the intensity of light. As we know from Section 10.2 the index of refraction is

$$n_{\text{med}} = \frac{c_{\text{vac}}}{c_{\text{med}}} = \frac{\lambda_{\text{vac}}}{\lambda_{\text{med}}}. \qquad (13.32)$$

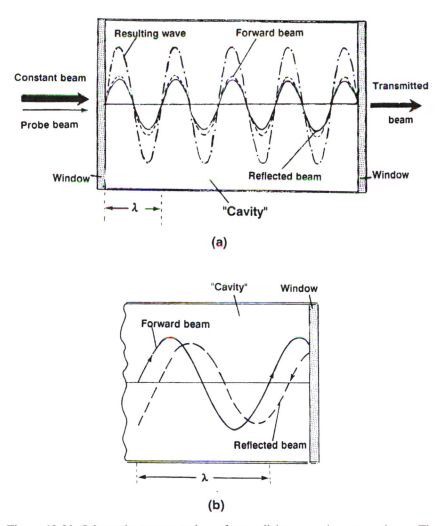

Figure 13.56. Schematic representation of some light waves in a *transphasor*. The reflectivity of the windows is about 90%. (a) Constructive interference. The length of the "cavity" equals an integer multiple of $\lambda/2$. (b) Condition (a) above is not fulfilled. The sum of many forward and reflected beams decreases the total intensity of the light. (*Note*: No phase shift occurs on the boundaries inside the "cavity", because $n_{\text{cavity}} > n_{\text{air}}$.)

Thus, we have at our disposal a material which, as a result of high light intensity, changes its index of refraction, which, in turn, changes λ_{med} until an integer multiple of $\lambda_{\text{med}}/2$ equals the cavity length and constructive interference may take place. Moreover, just shortly before this condition has been reached, a positive feedback mechanism mutually reinforces the parameters involved and brings the beams rapidly closer to the constructive interference state.

Figure 13.57. Schematic representation of an optical AND gate as obtained from an optical transistor (transphasor) constructed from a material with nonlinear refractive index. The low transmission state may represent a "zero" in binary logic, whereas the high transmission of light may stand for a "one."

An **optical switch** involves a "constant laser beam" whose intensity is not yet strong enough to trigger constructive interference (Fig. 13.56(a)). This light intensity is supplemented by a second laser beam, the *"probe beam,"* which is directed onto the same spot of the window of the transphasor and which provides the extra light energy to trigger a large change in n and thus constructive interference (Fig. 13.57). All taken, a small intensity change caused by the probe beam invokes a large intensity of the transmitted beam. This combination of two signals that interact with a switching device can be utilized as an "AND" logic circuit, as described in Section 8.7.12. Likewise, "OR" gates (either of the two beams is already strong enough to trigger critically a change in n) or "NOT" gates (which involve the reflected light) can be constructed.

One important question still remains to be answered. It pertains to the mechanisms involved in a **nonlinear optical material**. Such a material consists, for example, of indium antimonide, a narrow–band gap semiconductor having a gap energy of only 0.2 eV. (It therefore needs to be cooled to 77 K in order to suppress thermally-induced conduction band electrons.) Now, we know from Chapter 12 that when photons of sufficiently high energy interact with the valence electrons of semiconductors, some of these electrons are excited across the gap into the conduction band. The number of excited electrons is, of course, larger the smaller the gap energy (see Chapter 12) and the larger the number of impinging photons. On the other hand, the index of refraction, n, depends on the number of free electrons, N_f (in the conduction band, for example), as we know from (11.7),

$$\hat{n}^2 = 1 - \frac{e^2 N_f}{4\pi^2 \varepsilon_0 m v^2}. \tag{13.33}$$

Thus, a high light intensity substantially changes N_f and therefore n as stated above.

A crude photonic computer was introduced in 1990 by Bell Laboratories. However, new nonlinear materials need to be found before optical computers become competitive with their electronic counterparts.

13.12. X-Ray Emission

Electromagnetic radiation of energy higher than that characteristic for UV light is called **X-rays**. (Still higher-energy radiation are γ-rays). X-rays were discovered in 1895 by Wilhelm Conrad Röntgen, a German scientist. In 1901, he received the first Nobel Prize in physics for this discovery. The wavelength of X-rays is in the order of 10^{-10} m; see Figure 10.1. For its production, a beam of electrons emitted from a hot filament is accelerated in a high electric field towards a metallic (or other) electrode. On impact, the energy of the electrons is lost either by **white X-radiation**, that is, in the form of a continuous spectrum (within limits), or by essentially monochromatic X-rays (called characteristic X-rays) that are specific for the target material. The white X-rays are emitted as a consequence of the deceleration of the electrons in the electric field of a series of atoms, where each interaction with an atom may lead to photons of different energies. The maximal energy that can be emitted this way (assuming only *one* interaction with an atom) is proportional to the acceleration voltage, V, and the charge of the electron, e, that is

$$E_{\max} = eV = h\nu = \frac{hc}{\lambda} \qquad (13.34)$$

[see Eqs. (2.1) and (1.5)]. From this equation the minimum wavelength, λ (in nm), can be calculated using the values of the constants as listed in Appendix 4 and inserting V in volts, that is

$$\lambda = \frac{1240}{V}. \qquad (13.35)$$

Figure 13.58 depicts the voltage dependence of several white X-ray spectra. The cutoff wavelengths, as calculated by Eq. (13.35), are clearly detected. White X-radiation is mostly used for medical and industrial applications such as dentistry, bone fracture detection, chest X-rays, and so on. Different densities of the materials under investigation yield variations in the blackening of the exposed photographic film which has been placed behind the specimen.

The wavelength of **characteristic X-rays** depends on the material on which the accelerated electrons impinge. Let us assume that the impinging electrons possess a high enough energy to excite inner electrons, for example, electrons from the K-shell, to leave the atom. As a consequence, an L electron may immediately revert into the thus created vacancy while emitting a photon

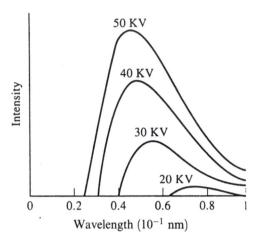

Figure 13.58. Schematic representation of the wavelength dependence of the intensity of white X-ray emission for selected acceleration voltages.

having a narrow and characteristic wavelength. This mechanism is said to produce K_α *X-rays*; see Figure 13.59. Alternately and/or simultaneously, an electron from the M shell may revert to the K shell. This is termed K_β-radiation.

For the case of copper, the respective wavelengths are 0.1542 nm and 0.1392 nm. (As a second example, aluminum yields K_α and K_β radiations having characteristic wavelengths of 0.8337 nm and 0.7981 nm.) Character-

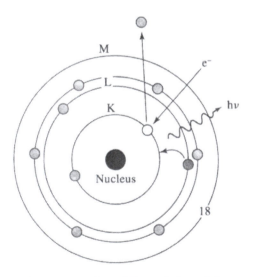

Figure 13.59. Schematic representation of the emission of characteristic X-radiation by exciting a K-electron and refilling the vacancy thus created with an L-electron.

istic (monochromatic) X-radiation is frequently used in materials science, for example, for investigating the crystal structure of materials. For this, only one of the possible wavelengths is used by eliminating the others utilizing appropriate filters, made, for example, of nickel foils, which strongly absorb the K_β-radiation of copper while the stronger K_α-radiation is only weakly absorbed. The characteristic X-radiation is superimposed on the often weaker, white X-ray spectrum.

Problems

1. Calculate the difference in the refractive indices which is necessary in order that an asymmetric waveguide operates in the zeroth mode. Take $\lambda_0 = 840$ nm, $t = 800$ nm, and $n_2 = 3.61$.

2. How thick is the depletion layer for an electro-optical waveguide when the index of refraction ($n_3 = 3.6$) increases in Medium 2 by 0.1%? Take $n_1 = 1$, $\lambda_0 = 1.3\ \mu$m, and zeroth-order mode.

3. Calculate the angle of total reflection in (a) a GaAs waveguide ($n = 3.6$), and (b) a glass waveguide ($n = 1.5$) against air.

4. Of which order of magnitude does the doping of an electro-optical waveguide need to be in order that the index of refraction changes by one-tenth of one percent? Take $n_3 = 3.6$, $m^* = 0.067\ m_0$, and $\lambda = 1.3\ \mu$m.

5. Calculate the free carrier absorption loss in a semiconductor assuming $n = 3.4$, $m^* = 0.08\ m_0$, $\lambda = 1.15\ \mu$m, $N_f = 10^{18}$ cm^{-3}, and $\mu = 2 \times 10^3$ cm^2/Vs.

6. Show that the energy loss in an optical device, expressed in decibels per centimeter, indeed equals 4.3α.

7. Calculate the necessary step height of a "bump" on a compact disk in order that destructive interference can occur. (Laser wavelength in air, 780 nm; index of refraction of transparent polymeric materials, 1.55.)

8. Calculate the gap energy and the emitting wavelength of a GaAs laser that is operated at 100°C. Take the necessary data from the tables in Appendix 4 and Table 19.2.

Suggestions for Further Reading (Part III)

F. Abelès, ed., Optical properties and electronic structure of metals and alloys, *Proceedings of the International Conference*, Paris, 13–16 Sept. 1965, North-Holland, Amsterdam (1966).
F. Abelès, ed., *Optical Properties of Solids*, North-Holland, Amsterdam (1972).
M. Born and E. Wolf, *Principles of Optics*, 3rd ed., Pergamon Press, Oxford (1965).
G. Bouwhuis, ed., *Principles of Optical Disc Systems*, Adam Hilger, Bristol (1985).

M. Cardona, *Modulation Spectroscopy, Solid State Physics*, Suppl. 11, Academic Press, New York, (1969).

W.W. Duley, *Laser Processing and Analysis of Materials*, Plenum Press, New York (1983).

K.J. Ebeling, *Integrierte Optoelektronik*, Springer-Verlag, Berlin (1989).

G. Fasol, S. Nakamura, I. Davies, *The Blue Laser Diode: GaN-Based Light emitters*, Springer, New York (1997).

M.P. Givens, Optical properties of metals, in *Solid State Physics*, Vol. 6, Academic Press, New York (1958).

O.S. Heavens, *Optical Properties of Thin Solid Films*, Academic Press, New York (1955).

L.L. Hench and J.K. West, *Principles of Electronic Ceramics*, Wiley, New York (1990).

P.H. Holloway, S. Jones, P. Rack, J. Sebastian, T. Trottier, *Flat Panel Displays: How Bright and Colorful is the Future?* Proceedings ISAF'96 Vol I IEEE page 127 (1996).

R.E. Hummel, *Optische Eigenschaften von Metallen und Legierungen*, Springer-Verlag, Berlin (1971).

R.G. Hunsperger, *Integrated Optics, Theory and Technology*, 3rd ed., Springer-Verlag, New York (1991).

T.S. Moss, *Optical Properties of Semiconductors*, Butterworth, London (1959).

P.O. Nilsson, Optical properties of metals and alloys, *Solid State Physics*, Vol. 29, Academic Press, New York (1974).

F.A. Ponce and D.P. Bour, *Nitrogen-Based Semiconductors for Blue and Green light-emitting devices*, Nature, *386*, 351 (1997).

B.O. Seraphin, ed., *Optical Properties of Solids—New Developments*, North-Holland/American Elsevier, Amsterdam, New York (1976).

J.H. Simmons and K.S. Potter, *Optical Materials*, Academic Press, San Diego (2000).

A.V. Sokolov, *Optical Properties of Metals*, American Elsevier, New York (1967).

O. Svelto, *Principles of Lasers*, 2nd ed., Plenum Press, New York (1982).

A. Vašíček, *Optics of Thin Films*, North-Holland, Amsterdam (1960).

F. Wooten, *Optical Properties of Solids*, Academic Press, New York (1972).

PART IV

MAGNETIC PROPERTIES
OF MATERIALS

CHAPTER 14

Foundations of Magnetism

14.1. Introduction

The phenomenon of magnetism, i.e., the mutual attraction of two pieces of iron or iron ore, was surely known to the antique world. The ancient Greeks have been reported to experiment with this "mysterious" force. The designation *magnetism* is said to be derived from a region in Turkey which was known by the name of *Magnesia* and which had plenty of iron ore.

Interestingly enough, a piece of magnetic material such as iron ore does not immediately attract other pieces of the same material. For this, at least one of the pieces has to be *magnetized*. Simply said, its internal "elementary magnets" need alignment in order for it to become a permanent magnet. Magnetizing causes no problem in modern days. One merely places iron into an electric coil through which a direct current passes for a short time. (This was discovered by Oersted at the beginning of the 19th century.) But how did the ancients do it? There may have been at least three possibilities. First, a bolt of lightning could have caused a magnetic field large enough to magnetize a piece of iron ore. Once one permanent magnet had been produced and identified, more magnets could have been obtained by rubbing virgin pieces of iron ore with the first magnet. There is another possibility. It is known that if a piece of iron is repeatedly hit very hard, the "elementary magnets" will be "shaken loose" and align in the direction of the earth's magnetic field. An iron hammer, for example, is *north-magnetic* on its face of impact in the northern hemisphere. Could it have been that a piece of iron ore was used as a hammer and thus it became a permanent magnet? A third possibility is that iron or nickel-containing meteorites responded with an alignment of their "elementary magnets" in an electromagnetic field during their immersion into the earth's atmosphere.

Magnetic materials made an important contribution to the development of the consciousness of mankind, because they paved the way to discoveries of new continents once the compass had been invented. (A compass needle is a pivoted bar magnet which positions itself approximately in the north–south direction. We call the tip that points to geographic north, the north-seeking pole, or simply the north pole, and the opposite end the south pole.) Around 1500, the British coined the word *lodestone* for the iron ore Fe_3O_4, which is derived from the old English word *lode* and which means to *lead* or to *guide*. Our modern technology would be unthinkable without magnetic materials and magnetic properties. Magnetic tapes or disks (computers), television, motors, generators, telephones, and transformers are only a few examples of their applications.

Thus far, we have used the word magnetism very loosely when implying the mutual magnetic attraction of pieces of iron. There are, however, several classes of magnetic materials that differ in kind and degree in their mutual interaction. We shall distinguish in the following between ferromagnetism (which term we restrict to the classical magnetism in iron and a few other metals and alloys) and para-, dia-, antiferro-, and ferrimagnetism. The oldest known magnetic ore, the magnetite, or lodestone, Fe_3O_4, is actually a *ferri*-magnet $(FeO) \cdot Fe_2O_3$ called *iron ferrite*.

In the sections to come, we will first define the magnetic constants and then remind the reader of some fundamental equations in magnetism before discussing magnetism by classical and quantum theory. Practical applications of magnetic materials are presented in the final chapter.

14.2. Basic Concepts in Magnetism

The goal of this chapter is to characterize the magnetic properties of materials. At least five different types of magnetic materials exist, as mentioned in the Introduction. A qualitative, as well as a quantitative, distinction between these different types can be achieved in a relatively simple way by utilizing a method proposed by Faraday. The magnetic material to be investigated is suspended from one of the arms of a sensitive balance and is allowed to reach into an inhomogeneous magnetic field (Fig. 14.1). Diamagnetic materials are expelled from this field, whereas para-, ferro-, antiferro-, and ferri-magnetic materials are attracted in different degrees. It has been found empirically that the apparent loss or gain in mass, i.e., the force, F, on the sample exerted by the magnetic field, H, is

$$F = V\chi\mu_0 H \frac{dH}{dx}, \qquad (14.1)$$

where V is the volume of the sample, μ_0 is a universal constant called the permeability of free space (1.257×10^{-6} H/m or Vs/Am), and χ is the **sus-**

Figure 14.1. Measurement of the magnetic susceptibility in an inhomogeneous magnetic field. The electromagnet is driven by an electric current, which flows through the helical windings of a long insulated wire called a solenoid. The magnetic flux lines (dashed) follow the iron core.

ceptibility, which expresses how responsive a material is to an applied magnetic field. Characteristic values for χ are given in Table 14.1. The term dH/dx in Eq. (14.1) is the change of the **magnetic field strength**, H, in the x-direction. The field strength, H, of an electromagnet (consisting of helical windings of a long, insulated wire as seen in the lower portion of Figure 14.1) is proportional to the current, I, which flows through this coil, and on the number, n, of the windings (called *turns*) that have been used to make the coil. Further, the magnetic field strength is inversely proportional to the length, L, of the *solenoid*. Thus, the magnetic field strength is expressed by

$$H = \frac{In}{L}. \tag{14.2}$$

The field strength is measured (in SI units) in "Amp-turns per meter", or shortly, in A/m.

The magnetic field can be enhanced by inserting, say, iron, into a solenoid, as shown in Figure 14.1. The parameter that expresses the amount of enhancement of the magnetic field is called the **permeability**, μ. The magnetic field strength within a material is known by the names **magnetic induction**[1]

---

[1] Calling B "magnetic induction" is common practice but should be discouraged because it may be confused with electromagnetic induction. Thus, some authors call B "magnetic field" and H "applied field".

Table 14.1. Magnetic constants of some materials at room temperature

Material	χ (SI) unitless	χ (cgs) unitless	μ unitless	Type of magnetism
Bi	-165×10^{-6}	-13.13×10^{-6}	0.99983	
Ge	-71.1×10^{-6}	-5.66×10^{-6}	0.99992	
Au	-34.4×10^{-6}	-2.74×10^{-6}	0.99996	Diamagnetic
Ag	-23.8×10^{-6}	-1.90×10^{-6}	0.99997	
Be	-23.2×10^{-6}	-1.85×10^{-6}	0.99998	
Cu	-9.7×10^{-6}	-0.77×10^{-6}	0.99999	
Water	-9.14×10^{-6}	-0.73×10^{-6}	0.99999	
Si	-4.1×10^{-6}	-0.32×10^{-6}	0.99999	
Superconductors[a]	-1.0	$\sim -8 \times 10^{-2}$	0	
β-Sn	$+2.4 \times 10^{-6}$	$+0.19 \times 10^{-6}$	1	
Al	$+20.7 \times 10^{-6}$	$+1.65 \times 10^{-6}$	1.00002	Paramagnetic
W	$+77.7 \times 10^{-6}$	$+6.18 \times 10^{-6}$	1.00008	
Pt	$+264.4 \times 10^{-6}$	$+21.04 \times 10^{-6}$	1.00026	
Low carbon steel	$\approx 5 \times 10^3$	3.98×10^2	5×10^3	
Fe–3%Si (grain-oriented)	4×10^4	3.18×10^3	4×10^4	Ferromagnetic
Ni–Fe–Mo (supermalloy)	10^6	7.96×10^4	10^6	

[a] See Section 7.6.
Note: The table lists the unitless susceptibility, χ, in SI and cgs units. (The difference is a factor of 4π, see Appendix 4.) Other sources may provide mass, atomic, molar, volume, or gram equivalent susceptibilities in cgs or SI units. μ has the same value in both unit systems, see Section 14.3.
Source: Landolt-Börnstein, *Zahlenwerte der Physik*, Vol. 11/9, 6th Edition, Springer-Verlag, Berlin (1962).

(or *magnetic flux density*) and is denoted by *B*. Magnetic field strength and magnetic induction are related by the equation

$$B = \mu\mu_0 H. \tag{14.3}$$

The SI unit for B is the tesla (T); see Appendix 4. The permeability (sometimes called relative permeability, μ_r) in Eq. (14.3) is unitless and is listed in Table 14.1 for some materials. The relationship between the susceptibility and the permeability is

$$\mu = 1 + \chi. \tag{14.4}$$

For empty space and, for all practical purposes, also for air, one defines $\chi = 0$ and thus $\mu = 1$ [See Eq. (14.4)]. The susceptibility is small and negative for diamagnetic materials. As a consequence, μ is slightly less than 1 (see Table 14.1). For para- and antiferromagnetic materials, χ is again small, but positive. Thus, μ is slightly larger than 1. Finally, χ and μ are large and positive for ferro- and ferrimagnetic materials.

The magnetic constants are temperature-dependent, except for diamagnetic materials, as we will see later. Further, the susceptibility for ferromagnetic materials depends on the field strength, *H*.

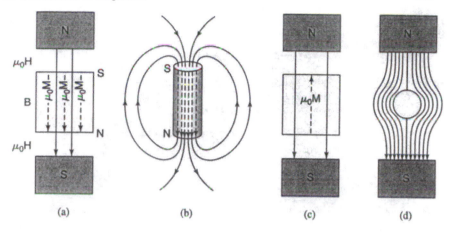

Figure 14.2. Schematic representation of magnetic field lines in and around different types of materials. (a) *Para- or ferromagnetics*. The magnetic induction (**B**) inside the material consists of the free-space component ($\mu_0 H$) plus a contribution by the material ($\mu_0 M$); see Eq. (14.5). (b) The magnetic field lines outside a material point from the north to the south poles, whereas inside of para- or ferromagnetics, **B** and $\mu_0 M$ point from south to north in order to maintain continuity. (c) In *diamagnetics*, the response of the material counteracts (weakens) the external magnetic field. (d) In a thin surface layer of a *superconductor*, a supercurrent is created (below its transition temperature) which causes a magnetic field that opposes the external field. As a consequence, the magnetic flux lines are expelled from the interior of the material. Compare to Figure 9.18.

The magnetic field parameters at a given point in space are, as explained above, the magnetic field strength, H, and the magnetic induction, B. In free (empty) space, B and $\mu_0 H$ are identical, as seen in Eq. (14.3). Inside a magnetic material the induction, B, consists of the free-space component ($\mu_0 H$) plus a contribution to the magnetic field ($\mu_0 M$) which is due to the presence of matter [Figure 14.2(a)], that is,

$$B = \mu_0 H + \mu_0 M, \tag{14.5}$$

where M is called the **magnetization** of the material. Combining Eqs. (14.3) through (14.5) yields

$$M = \chi H. \tag{14.6}$$

H, B, and M are actually vectors. Specifically, outside a material, **H** (and **B**) point from the north to the south pole. Inside of a ferro- or paramagnetic material, **B** and **M** point from the south to the north; see Figures 14.2(a) and (b). However, we will mostly utilize their moduli in the following sections and thus use lightface italic letters.

B was said above to be the magnetic flux density in a material, that is, the magnetic flux per unit area. The magnetic flux, ϕ, is then defined as the product of B and area, A, that is, by

$$\phi = BA. \tag{14.7}$$

In free space, for which $M = 0$, we obtain instead, by using (14.5),

$$\phi = \mu_0 HA. \tag{14.7a}$$

Finally, we need to define the **magnetic moment**, μ_m, (also a vector) through the following equation:

$$M = \frac{\mu_m}{V}, \tag{14.8}$$

which means that the magnetization is the magnetic moment per unit volume.

*14.3. Units

It needs to be noted that in magnetic theory several unit systems are commonly in use. The scientific and technical literature on magnetism, particularly in the USA, is still widely written in electromagnetic cgs (emu) units. In some European countries, and in many international scientific journals, the SI units are mandatory. Conversion factors from emu into SI units are given in Appendix 4. The magnetic field strength in cgs units is measured in oersted and the magnetic induction in gauss. In SI units H is measured in A/m and B is given in tesla (T). Equation (14.5) reads in cgs units

$$B = H + 4\pi M. \tag{14.9}$$

Writing (14.3) in cgs (emu) units yields

$$B = \mu H, \tag{14.10}$$

The permeability in cgs units is

$$\mu = 1 + 4\pi\chi. \tag{14.11}$$

Comparison of (14.4) with (14.11) indicates a difference by a factor 4π between the susceptibilities in the two unit systems. As a result, μ has the same value in both unit systems, see Appendix 4 and Table 14.1. It needs to be further stressed that the electric charge (e.g., of an electron) in electromagnetic cgs units is written in "abcoulombs" or ($g^{1/2}$ $cm^{1/2}$), see Appendix 4.

Problems

1. Show that the unit for 1 Oe is equivalent to $[g^{1/2}/cm^{1/2} \cdot s]$ by making use of (14.1).

2. An electromagnet is a helical winding of wire through which an electric current flows. Such a "solenoid" of 1000 turns is 10 cm long and is passed through by a current of 2A. What is the field strength in Oe and A/m?

3. Familiarize yourself with the units of **H**, **B**, and **M** in the different unit systems. Convert (14.9) into (14.5) by making use of the conversion table in Appendix 4.

4. Calculate the (relative) permeability of Bi and Al from their susceptibilities ($\chi_{Bi} = -165 \times 10^{-6}$ and $\chi_{Al} = 20.7 \times 10^{-6}$) and compare your values with those in Table 14.1. Perform the same calculation for Ni–Fe–Mo ($\chi = 10^6$). What do you observe?

CHAPTER 15

Magnetic Phenomena and Their Interpretation—Classical Approach

15.1. Overview

We stated in the last chapter that different types of magnetism exist, and that they are characterized by the magnitude and the sign of the susceptibility (see Table 14.1).

Since various materials respond so differently in a magnetic field, we suspect that several fundamentally different mechanisms must be responsible for the magnetic properties. In the first part of this chapter we shall attempt to unfold the multiplicity of the magnetic behavior of materials by describing some pertinent experimental findings and giving some brief interpretations. In the sections to follow, we shall treat the atomistic theory of magnetism in more detail.

15.1.1. Diamagnetism

Ampère postulated more than one hundred years ago that *molecular currents* are responsible for the magnetism in a solid. He compared the molecular currents to an electric current in a loop-shaped piece of wire, which is known to cause a magnetic moment. Today, we replace Ampère's molecular currents by *orbiting valence electrons*.

For the understanding of diamagnetism, a second aspect needs to be considered. It was found by Lenz that a current is induced in a wire loop whenever a bar magnet is moved toward (or from) this loop. The current thus induced causes, in turn, a magnetic moment which is opposite to the one of the bar magnet (Fig. 15.1(a)). (This has to be so in order for mechanical work to be expended in producing the current; otherwise, a perpet-

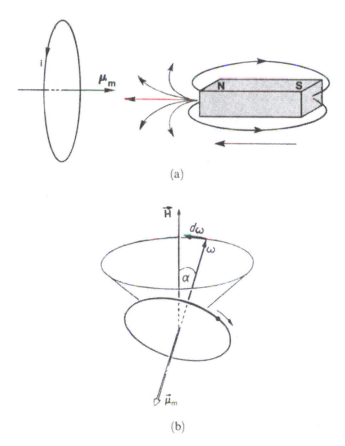

(a)

(b)

Figure 15.1. Explanation of diamagnetism. (a) Induction of a current in a loop-shaped piece of wire by moving a bar magnet toward the wire loop. The current in the loop causes a magnetic field that is directed opposite to the magnetic field of the bar magnet (Lenz's law). (b) Precession of an orbiting electron in an external magnetic field. Precession is the motion which arises as a result of external torque acting on a spinning body (such as a spinning top) or, as here, on an orbiting electron.

ual motion would be created!) Diamagnetism may then be explained by postulating that the external magnetic field induces a change in the magnitude of inner-atomic currents, i.e., *the external field accelerates or decelerates the orbiting electrons,* in order that their magnetic moment is in the opposite direction from the external magnetic field. In other words, the responses of the orbiting electrons counteract the external field (Fig. 14.2(c)) whereas the outermost electrons provide the largest contribution. A more accurate and quantitative explanation of diamagnetism replaces the induced currents by precessions of the electron orbits about the magnetic field direction (Larmor precession, see Fig. 15.1(b)).

So far, we implicitly considered only electrons that are *bound* to their respective nuclei. Now, metals are known also to have *free electrons*. They are forced to move in a magnetic field in a circular path. This leads to a second contribution to the diamagnetic moment; specifically, the circulating *free* electrons cause a magnetic moment, similarly as described above.

It has been observed that **superconducting materials** (Section 7.6) expel the magnetic flux lines when in the superconducting state (**Meissner effect**). In other words, a superconductor behaves in an external magnetic field as if **B** is zero inside the superconductor (Fig. 14.2(d)). Thus, with (14.5), we obtain

$$\mathbf{H} = -\mathbf{M},$$

which means that the magnetization is equal and opposite to the external magnetic field strength. The result is a perfect diamagnet. The susceptibility (14.6)

$$\chi = \frac{\mathbf{M}}{\mathbf{H}}$$

in superconductors is -1 compared to -10^{-6} in the normal state (see Table 14.1). This strong diamagnetism can be used for frictionless bearings, i.e., for support of loads by a repelling magnetic field. The levitation effect in which a magnet hovers above a superconducting material, and the suspension effect where a chip of superconducting material hangs some distance beneath a magnet can be explained with the strong diamagnetic properties of super- conductors. (See also Problem 12.)

15.1.2. Paramagnetism

Paramagnetism in *solids* is attributed, to a large extent, to a magnetic mo- ment that results from electrons which spin around their own axes, Fig.

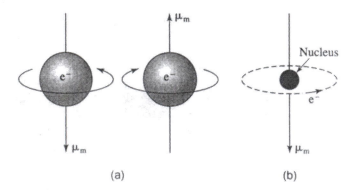

(a)   (b)

Figure 15.2. (a) Schematic representation of electrons which spin around their own axes. A (para)magnetic moment μ_m results; its direction depends on the mode of rotation. Only two spin directions are shown (called "spin up" and "spin down"). (b) An orbiting electron is the source of *electron-orbit paramagnetism*.

15.2(a). We have already introduced the *electron spin* in Section 6.4 and mentioned there that, because of the Pauli principle, no two electrons having the same energy can have the same value and sign for the spin moment. In other words, each electron state can be occupied by two electrons only; one with positive spin and one with negative spin, or, as is often said, one with spin up and one with spin down. An external magnetic field tries to turn the unfavorably oriented spin moments in the direction of the external field. We will talk about the quantum mechanical aspect of spin paramagnetism in more detail in Chapter 16. Spin paramagnetism is slightly temperature-dependent. It is in general very weak and is observed in some metals and in salts of the transition elements.

Free atoms (dilute gases) as well as rare earth elements and their salts and oxides possess an additional source of paramagnetism. It stems from the magnetic moment of the orbiting electrons Fig. 15.2(b). Without an external magnetic field, these magnetic moments are randomly oriented and thus they mutually cancel one another. As a result, the net magnetization is zero. However, when an external field is applied, the individual magnetic vectors tend to turn into the field direction. Thermal agitation counteracts the alignment. Thus, electron-orbit paramagnetism is temperature-dependent.

The temperature dependence of many paramagnetic materials is governed by the experimentally found Curie law, which states that the susceptibility, χ, is inversely proportional to the absolute temperature T,

$$\chi = \frac{C}{T}, \tag{15.1}$$

where C is called the Curie constant. For many other substances, a more general relationship is observed, which is known as the Curie–Weiss law,

$$\chi = \frac{C}{T - \theta}, \tag{15.2}$$

where θ is another constant that has the same unit as the temperature and may have positive as well as negative values (see Fig. 15.3). We will explain the meaning of the constants C and θ in Section 15.3.

Metals, with a few exceptions, do *not* obey the Curie–Weiss law, as we shall see in Chapter 16. However, Ni (above the Curie temperature, see Section 15.1.3) and, in a limited temperature interval, also Fe and β-Co, the rare earth elements, and salts of the transition elements (e.g., the carbonates, chlorides, and sulfates of Fe, Co, Cr, Mn) obey the Curie–Weiss law quite well.

We have just mentioned that in most solids only spin paramagnetism is observed. This is believed to be due to the fact that in crystals the electron orbits are essentially coupled to the lattice, which prevents the orbital magnetic moments from turning into the field direction. One says in this case that the orbital moments are "quenched." Exceptions are the rare earth elements

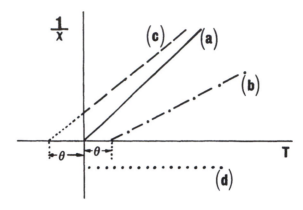

Figure 15.3. Schematic representation of (a) the Curie law and (b) and (c) the Curie–Weiss law. (d) The diamagnetic behavior is also shown for comparison.

and their derivatives, which have "deep-lying" 4f-electrons.[2] The latter ones are shielded by the outer electrons from the crystalline field of the neighboring ions. Thus, the orbital magnetic moments of the f-electrons may turn into the external field direction and contribute to electron-orbit paramagnetism. The fraction of the total magnetic moment contributed by orbital motion versus by spin is defined as the "**g-factor**."

It is now possible to make some general statements about whether para- or diamagnetism might be expected in certain materials. For paramagnetic materials, the magnetic moment of the electrons is thought to point *in* the direction of the external field, i.e., the magnetic moment enhances the external field. Diamagnetism counteracts an external field, as we have seen in Section 15.1.1. Thus, para- and diamagnetism oppose each other. Solids that have both orbital as well as spin paramagnetism are clearly paramagnetic since the sum of both paramagnetic components is commonly larger than the diamagnetism. Rare earth metals with unfilled 4f-electron bands are an example of this. In most other solids, however, the orbital paramagnetism is "quenched," as we said above. Yet, they still might have spin paramagnetism. The possible presence of a net spin-paramagnetic moment depends upon whether or not the magnetic moments of the individual spins cancel each other. More specifically, if a solid has *completely filled electron bands*, we anticipate (because of the Pauli principle) the same number of electrons with spins up as well as with spins down. For example, a completely filled d-band contains 5N electrons with spins up and 5N electrons with spins down. This results in a cancellation of the spin moments and no net spin paramagnetism is expected. These materials are thus diamagnetic (no orbital and no spin paramagnetic moment). We mention as examples for filled bands

---

[2] See Appendix 3.

Figure 15.4. Schematic representation of the spin alignment in a *d*-band which is partially filled with eight electrons (Hund's rule).

intrinsic semiconductors, insulators, and ionic crystals such as NaCl. (In the latter case, an electron transfer occurs between cations and anions, which causes closed electron shells, i.e., filled bands.)

In materials with partially filled bands, the electron spins are arranged, according to "**Hund's rule**," in such a manner that the total spin moment is maximized. This condition is energetically more favorable, as quantum mechanics shows. For example, in an atom with eight valence *d*-electrons, five of the spins align, say, up, and three spins point down, which results in a net total of two spins up (Fig. 15.4). The atom is then expected to have two units of (para-)magnetism.

The smallest unit (or quantum) of the magnetic moment is called one Bohr magneton

$$\mu_B = \frac{eh}{4\pi m} = 9.274 \times 10^{-24} \left(\frac{J}{T}\right) \equiv (A \cdot m^2) \qquad (15.3)$$

(The symbols have the usual meaning.) We shall derive equation (15.3) in Chapter 16. In the above example, the metal is said to have two Bohr magnetons per atom.

One word of caution should be added about applying the above general principles too rigidly. Some important exceptions do exist. They must be explained by considering additional information (see Chapter 16). For example, copper, which has one *s*-electron in its valence band, should be paramagnetic according to our considerations brought forward so far. In reality, copper is diamagnetic. Other examples are superconductors, which are perfect diamagnetics below a transition temperature; they repel the magnetic flux lines from their interior, as we explained in Section 15.1.1.

15.1.3. Ferromagnetism

We turn now to ferromagnetics and commence with the experimentally found magnetization curve for these materials. A newly cast piece of iron (often called *virgin iron*) is inserted into a ring-shaped solenoid (Fig. 15.5). (The ring shape is used to contain the magnetic field within the coil.) If the external field strength is increased (by increasing the current in the primary winding), then the magnetization (measured in a secondary winding with a flux meter) rises at first slowly and then more rapidly (Fig. 15.6). Finally, *M* levels off and reaches a constant value, called the saturation magnetization, M_s. When *H* is reduced to zero, the magnetization retains a positive value,

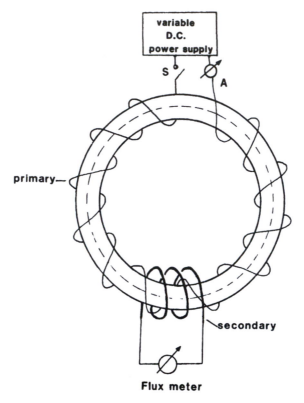

Flux meter

Figure 15.5. A ring-shaped solenoid with primary and secondary windings. The magnetic flux lines are indicated by a dashed circle. Note, that a current can flow in the secondary circuit only if the current (and therefore the magnetic flux) in the primary winding changes with time. An on–off switch in the primary circuit may serve for this purpose.

called the remanent magnetization, or **remanence**, M_r. It is this retained magnetization that is utilized in permanent magnets. The remanent magnetization can be removed by reversing the magnetic field strength to a value H_c, called the **coercive field**. Solids having a large combination of M_r and H_c are called **hard magnetic materials** (in contrast to *soft* magnetic materials for which the area inside the loop of Fig. 15.6 is very small and the slope dM/dH about the origin is quite steep). A complete cycle through positive and negative H-values, as shown in Fig. 15.6, is called a **hysteresis loop**. It should be noted that a second type of hysteresis curve is often used, in which B (instead of M) is plotted versus H. No saturation value for B is observed. (The residual induction B_r at $H = 0$ is called the **retentivity**. Removal of B_r requires a field which is called **coercivity**. However, remanence and retentivity, as well as coercive field, coercive force, and coercivity are often used interchangeably.)

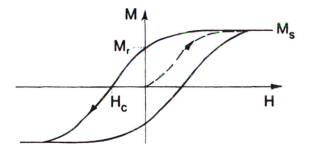

Figure 15.6. Schematic representation of a hysteresis loop of a ferromagnetic material. The dashed curve is for virgin material.

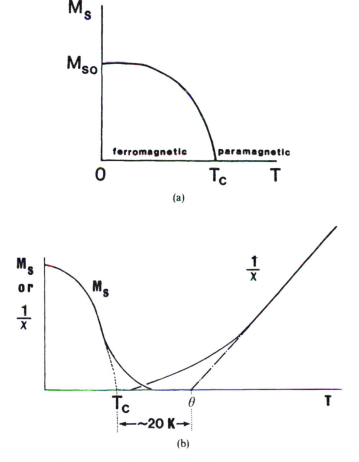

Figure 15.7. (a) Temperature dependence of the saturation magnetization of ferromagnetic materials. (b) Enlarged area near the Curie temperature showing the paramagnetic Curie point θ (see Fig. 15.3) and the ferromagnetic Curie temperature T_C.

Table 15.1. Saturation Magnetization at 0 K and Curie Temperature (T_C) for Some Ferromagnetic Materials.

Metal	M_{S0} (A/m)	M_{S0} (Maxwells/cm^2)	T_C (K)
Fe	1.75×10^6	2.20×10^4	1043
Co	1.45×10^6	1.82×10^4	1404
Ni	0.51×10^6	0.64×10^4	631
Gd	5.66×10^6	7.11×10^4	289

The saturation magnetization is temperature-dependent (Fig. 15.7(a)). Above the **Curie temperature**, T_C, ferromagnetics become paramagnetic. Table 15.1 lists saturation magnetizations and Curie temperatures of some elements. For ferromagnetics the Curie temperature, T_C, and the constant θ in the Curie–Weiss law are nearly identical. A small difference exists, however, because the transition from ferromagnetism to paramagnetism is gradual, as can be seen in Fig. 15.7(b).

Piezomagnetism. The magnetization of ferromagnetics is also stress-dependent (Fig. 15.8). As an example, a compressive stress increases the magnetization for nickel, while a tensile stress reduces M and therefore μ. This effect is just the opposite in certain nickel–iron alloys (permalloy, see Table 17.1) where a tensile stress *increases* M or μ. In polycrystalline iron the situation is more complex. At low fields, iron behaves like permalloy, whereas at high fields it behaves similar to nickel.

The inverse of piezomagnetism is called **magnetostriction**, an effect which

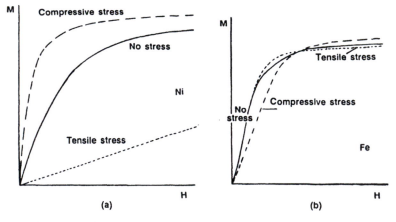

Figure 15.8. Schematic representation of the effect of tensile and compressive stresses on the magnetization behavior of (a) nickel and (b) iron. (Piezomagnetism.)

describes a change in dimensions when a ferromagnetic substance is exposed
to a magnetic field. (Incidentally, the periodic dimensional change caused by
an alternating magnetic field produces the humming noise in transformers
and "ballasts" for fluorescence lights; see Section 17.2.) Magnetostriction is
also observed in ferrimagnetic and antiferromagnetic materials. Moreover,
terbium-disprosium-iron displays magnetostriction which is about 3 orders
of magnitude larger than in iron and iron–nickel alloys.

A few preliminary words should be said to explain the above-mentioned
observations. In ferromagnetic materials, the spins of unfilled d-bands spontaneously align parallel to each other below T_C, i.e., they align within small
domains without the presence of an external magnetic field (Fig. 15.9). The
individual domains are magnetized to saturation. The spin direction in each
domain is, however, different, so that the individual magnetic moments for
the material as a whole cancel each other and the net magnetization is zero.
An external magnetic field causes those domains whose spins are parallel or
nearly parallel to the external field to grow at the expense of the unfavorably
aligned domains. (See the transition from Fig. 15.9(c) to Fig. 15.9(d).) When
the entire crystal finally contains *one* single domain, having all spins aligned
parallel to the external field direction, the material is said to have reached
technical saturation magnetization, M_s. Nevertheless, if the external magnetic
field is further increased a small, additional rise in M is observed. This is
caused by the forced alignment of those spins which precess about the field
direction due to thermal activation. The largest magnetization (M_{S0}) is obtained at 0 K. An increase in temperature progressively destroys the spontaneous alignment. The gradual transition from ferromagnetism to paramagnetism (Fig. 15.7(b)) is believed to be due to the fact that, slightly above
T_C, small clusters of spins are still aligned parallel to each other, a phenomenon which is called **magnetic short-range order**.

There are a number of fundamental questions which come immediately to
mind; e.g., in the virgin state, why is the spontaneous division into many
individual domains apparently preferred to one single domain? To answer
this, let us assume for a moment that all electron spins in a *crystal* are indeed
aligned in parallel, Fig. 15.9(a). As a consequence, north and south poles
would be created on opposite ends of the solid. This would be energetically
unfavorable because it would be the source of a large external magnetic field.
The **magnetostatic energy** of this field can be approximately halved if the
crystal contains *two* domains that are magnetized in opposite directions. This
way, north and south poles are closer together and the external magnetic
field is confined to a smaller area (Fig. 15.9(b)). Further divisions into still
smaller and smaller domains with concomitant reductions in magnetostatic
energies lead, however, eventually to an optimal domain size. Apparently,
an opposing mechanism must be active. The energy involved for the latter
has been found to be the **quantum mechanical exchange energy**. As we will
learn in Section 16.2, this exchange energy causes adjacent spins to align

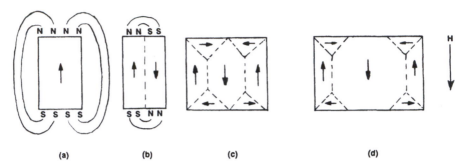

Figure 15.9. (a) Spontaneous alignment of all spins in a single direction. (b) Division into two magnetic domains having opposite spin directions. (c) Closure domains in a cubic crystal. (d) Growth of a domain whose spins are parallel to an external magnetic field. (The domain walls are *not* identical with the grain boundaries.)

parallel to each other. It is this interplay between exchange energy, which demands parallel spin alignment, and magnetostatic energy, which supports antiparallel spins, that leads eventually to an energetically most favorable domain size (which is about 1–100 μm).

A further reduction in magnetostatic energy can be obtained if the magnetic flux follows a completely closed path within a crystal so that no exterior poles are formed. Indeed, **"closure" domain structures**, as shown in Fig. 15.9(c), are observed in cubic crystals.

Another question which needs to be answered pertains to whether the flip from one spin direction into the other occurs in one step, i.e., between two adjacent atoms, or instead over an extended range of atoms. Again, the above-mentioned exchange energy, which supports a parallel spin alignment, hinders a spontaneous flip-over. Instead, a gradual rotation over several hundred atomic distances is energetically more favorable. The region between individual domains in which the spins rotate from one direction into the next is called a domain wall or a **Bloch wall**.

Finally, we may ask the question whether and how those domain walls can be made visible. The most common method, devised by Bitter in 1931, utilizes an aqueous suspension of very finely dispersed Fe_3O_4 particles which is applied to the polished surface of a test material. These particles are attracted to the domain wall endings and can then be observed as fine lines under an optical microscope. Another method exploits the rotation of the plane of polarization of reflected light from differently magnetized areas (**Kerr effect**).

We mentioned above that an external magnetic field causes a movement of the domain walls. The movement is, as a rule, not continuous, but occurs most of the time in distinct jumps. This is known as the **Barkhausen effect**, which utilizes an induction coil wound around a ferromagnetic rod. The former is connected to an amplifier and a loudspeaker. Audible clicks are heard when a permanent magnet approaches the iron rod. The wall motions may be impeded by imperfections in the crystal, such as by particles of a second phase, oxides, holes, or cracks. A second type of impediment to free domain wall motion stems from dislocations, i.e., from residual stresses in the crystal caused by grinding, polishing, or plastic deformation.

Cold work enlarges the coercivity and the area within the hysteresis loop. Further, cold work decreases the permeability and causes a clockwise rotation of the hysteresis curve. In short, mechanical hardness and magnetic hardness parallel each other in many cases. (There exist exceptions, however, such as in the case of silicon additions to iron, which makes the material magnetically softer and mechanically harder, see Section 17.2.3.) Recrystallization and grain growth by annealing at suitable temperatures relieve the stresses and restore the soft-magnetic properties.

We shall return to ferromagnetism in Section 15.4 and Chapter 16.

15.1.4. Antiferromagnetism

Antiferromagnetic materials exhibit, just as ferromagnetics, a spontaneous alignment of moments below a critical temperature. However, the responsible neighboring atoms in antiferromagnetics are aligned in an antiparallel fashion (Fig. 15.10). Actually, one may consider an antiferromagnetic crystal to be divided into two interpenetrating sublattices, A and B, each of which has a spontaneous parallel alignment of spins. Figure 15.10 depicts the spin alignments for two manganese compounds. (Only the spins of the manganese ions contribute to the antiferromagnetic behavior.) Figure 15.10(a) im-

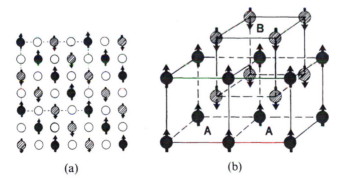

(a)   (b)

Figure 15.10. Schematic representation of spin alignments for antiferromagnetics at 0 K. (a) Display of a (100) plane of MnO. The gray (spin down) and black (spin up) circles represent the Mn ions. The oxygen ions (open circles) do not contribute to the antiferromagnetic behavior. MnO has a NaCl structure. (b) Three-dimensional representation of the spin alignment of manganese ions in MnF$_2$. (The fluorine ions are not shown.) This figure demonstrates the interpenetration of two manganese sublattices, A and B, having antiparallel aligned moments.

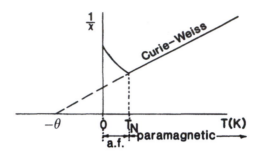

Figure 15.11. Schematic representation of the temperature dependence of a poly-crystalline antiferromagnetic (a.f.) material.

plies that the ions in a given {110} plane possess parallel spin alignment, whereas ions in the adjacent plane have antiparallel spins with respect to the first plane. Thus, the magnetic moments of the solid cancel each other and the material as a whole has no net magnetic moment.

Antiferromagnetic materials are paramagnetic above the **Néel temperature** T_N, i.e., they obey there a linear $T = f(1/\chi)$ law (see Fig. 15.11). Below T_N, however, the inverse susceptibility may rise with decreasing temperature. The extrapolation of the paramagnetic line to $1/\chi = 0$ yields a negative θ. Thus, the Curie–Weiss law (15.2) needs to be modified for anti-ferromagnetics to read

$$\chi = \frac{C}{T - (-\theta)} = \frac{C}{T + \theta}. \tag{15.4}$$

The Néel temperature is often below room temperature (Table 15.2). Most antiferromagnetics are found among ionic compounds. They are insulators or semiconductors. Essentially no practical application for antiferromagnetism is known at this time. (See, however, the use of "canted" anti-ferromagnetics, described in Section 17.5, which are materials in which the magnetic moments of the two sublattices are not completely antiparallel. This results in a small net magnetization.)

Table 15.2. Characteristic Data for Some
Antiferromagnetic Materials.

Substance	T_N (K)	$-\theta$ (K)
MnO	116	610
MnF_2	67	82
α-Mn	100	?
FeO	198	570
NiO	523	~2000
CoO	293	330
Cr	310	?

15.1.5. Ferrimagnetism

Ferrimagnetic materials are of great technical importance. They exhibit a spontaneous magnetic moment (Fig. 15.9) and hysteresis (Fig. 15.6) below a Curie temperature, just as iron, cobalt, or nickel. In other words, ferrimagnetic materials possess, similarly as ferromagnetics, small domains in which the electron spins are spontaneously aligned in parallel. The main difference from ferromagnetics is, however, that ferrimagnetics are ceramic materials (oxides) and that they are poor electrical conductors. A large resistivity is often desired for high-frequency applications (e.g., to prevent eddy currents in cores of coils, see Chapter 17).

To explain the spontaneous magnetization in ferrimagnetics, Néel proposed that two sublattices should exist in these materials (just as in antiferromagnetics) each of which contains ions whose spins are aligned parallel to each other. The crucial point is that each of the sublattices contain *different* numbers of magnetic ions. This causes some of the magnetic moments to remain uncancelled. As a consequence, a net magnetic moment results. Ferrimagnetic materials can thus be described as *imperfect antiferromagnetics*. The crystallography of ferrites is rather complex. We defer its discussion until later. For the time being, it suffices to know that there are two types of lattice sites which are available to be occupied by the metal ions. They are called A sites and B sites. (As before, oxygen ions do not contribute to the magnetic moments).

We will now discuss as an example nickel ferrite, $NiO \cdot Fe_2O_3$. The Fe^{3+} ions are equally distributed between A and B sites (Fig. 15.12), and since ions on A and B sites exhibit spontaneous magnetization in opposite directions, we expect overall cancellation of spins for these ions. Specifically, atomic iron possesses six $3d$-electrons and two $4s$-electrons ($3d^64s^2$, see Appendix 3). The Fe^{3+} ions are deprived of three electrons, so that five d-electrons, or five spin moments per atom, remain in its outermost shell. This is indicated in Fig. 15.12.

A sites	B sites
8 Fe^{3+}	8 Fe^{3+} 8 Ni^{2+}
↑↑↑↑↑	↓↓↓↓↓ ↓↓

Figure 15.12. Distribution of spins upon A and B sites for the inverse spinel $NiO \cdot Fe_2O_3$. The spins within one site are arranged considering Hund's rule (Fig. 15.4). The iron ions are equally distributed among the A and B sites. The nickel ions are only situated on B sites. The relevance of the number of ions per unit cell is explained later on in the text.

Table 15.3. Calculated and Measured Number of Bohr Magnetons for Some Ferrites.

Ferrite	Mn	Fe	Co	Ni	Cu
Calculated μ_B	5	4	3	2	1
Measured μ_B	4.6	4.1	3.7	2.3	1.3

The electron configuration of nickel in its atomic state is $3d^8 4s^2$. Two electrons are stripped in the Ni^{2+} ion so that eight d-electrons per atom remain. They are arranged, according to Hund's rule (Fig. 15.4), to yield two net magnetic moments (Fig. 15.12). All nickel ions are accommodated on the B sites. Nickel ferrite is thus expected to have two uncancelled spins, i.e., two Bohr magnetons (per formula unit), which is essentially observed (see Table 15.3).

The small discrepancy between experiment and calculation is believed to be caused by some contributions of orbital effects to the overall magnetic moment, and by a slight deviation of the distribution of metal ions on the A and B sites from that shown in Fig. 15.12.

The unit cell of cubic ferrites contains a total of 56 ions. Some of the metal ions are situated inside a *tetrahedron formed by the oxygen ions*. These are the above-mentioned A sites (Fig. 15.13(a)). Other metal ions are arranged in the center of an *octahedron* and are said to be on the B sites (Fig. 15.13(b)). The A and B sites are nestled inside a unit cell (Fig. 15.13(c)).

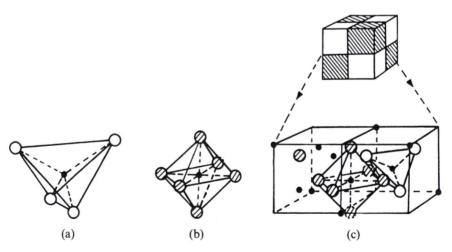

(a)   (b)   (c)

Figure 15.13. Crystal structure of cubic ferrites. The small filled circles represent metal ions, the large open or shaded circles represent oxygen ions: (a) tetrahedral or A sites; (b) octahedral or B sites; and (c) one-fourth of the unit cell of a cubic ferrite. A tetrahedron and an octahedron are marked. Adapted from J. Smit, and H.P.J. Wijn, *Ferrites*, Wiley, New York (1959).

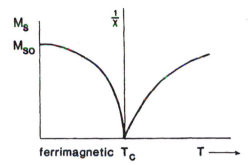

Figure 15.14. Schematic representation of the temperature dependence of the saturation magnetization, M_s, and the reciprocal susceptibility for ferrites.

Now, only 8 tetrahedral sites and 16 octahedral sites are occupied by metal ions. In $NiO \cdot Fe_2O_3$ twice as many iron ions as nickel ions are present. Eight of the Fe^{3+} ions per unit cell occupy the A sites, eight of them occupy some of the B sites and the eight Ni^{2+} ions fill the remaining B sites (Fig. 15.12). This distribution is called an *inverse spinel structure* (in contrast to a *normal spinel*, such as for $ZnO \cdot Fe_2O_3$, in which *all* Fe^{3+} ions occupy the B sites).

The temperature dependence of most ferrimagnetics is very similar to ferromagnetics (Fig. 15.14): The saturation magnetization decreases with increasing temperature until it vanishes at a Curie temperature, T_C. Above T_C, ferrimagnetics behave paramagnetically, having a nonlinear $1/\chi$ versus T relationship.

In conclusion, this section described, in a mostly qualitative way, the difference between dia-, para-, ferro-, antiferro-, and ferrimagnetism. In the sections to come, we shall again pick up the different forms of magnetism and deepen our understanding of these phenomena by following essentially the train of thought brought forward by Langevin, Weiss, and Néel.

15.2. Langevin Theory of Diamagnetism

We shall now develop the classical theory of diamagnetism in a quantitative way as put forward by Langevin at the turn of the 20th century.

We stated before that the orbital motion of an electron about its nucleus induces a magnetic moment, μ_m. We compared the latter with a magnetic moment which is created by a current passing through a loop-shaped wire. This magnetic moment is, naturally, larger, the larger the current, I, and the larger the area, A, of the orbit or loop:

$$\mu_m = I \cdot A = \frac{e}{t}A = \frac{e}{s/v}A = \frac{ev\pi r^2}{2\pi r} = \frac{evr}{2} \qquad (15.5)$$

(e is the electron charge, r is the radius of the orbit, $s = 2\pi r =$ length of the orbit, $v =$ velocity of the orbiting electrons, and $t =$ orbiting time).

We know that an external magnetic field accelerates (or decelerates) the orbiting electrons, which results in a change in magnetic moment. We shall now calculate this change in μ_m.

The external magnetic field induces an electric field (Section 15.1.1), which, in turn, exerts an electrostatic force $|\mathbf{F}|$ on the orbiting electron, which is

$$F = ma = \mathscr{E}e, \tag{15.6}$$

where $|\mathscr{E}|$ is the electric field strength and m is the mass of the electron. From this equation we obtain the acceleration of the electron,

$$a = \frac{dv}{dt} = \frac{\mathscr{E}e}{m}. \tag{15.7}$$

To calculate the acceleration we need to know the electric field strength, $\mathscr{E}$. It is defined as the ratio of the induced voltage (or emf), V_e, per orbit length, L, (see Section 7.1), i.e.,

$$\mathscr{E} = \frac{V_\mathrm{e}}{L}. \tag{15.8}$$

As we said earlier, a change in an external magnetic flux, ϕ, induces in a loop-shaped wire an emf which opposes, according to Lenz's law, the change in flux:

$$V_\mathrm{e} = -\frac{d\phi}{dt} = -\frac{d(\mu_0 H A)}{dt} \tag{15.9}$$

(see (14.7a)). Thus, the acceleration of the electron becomes, by combining (15.7)–(15.9),

$$\frac{dv}{dt} = \frac{\mathscr{E}e}{m} = \frac{V_\mathrm{e}e}{Lm} = -\frac{eA\mu_0}{Lm}\frac{dH}{dt} = -\frac{e\pi r^2 \mu_0}{2\pi r m}\frac{dH}{dt} = -\frac{er\mu_0}{2m}\frac{dH}{dt}. \tag{15.10}$$

A change in the magnetic field strength from 0 to H yields a change in the velocity of the electrons:

$$\int_{v_1}^{v_2} dv = -\frac{er\mu_0}{2m}\int_0^H dH \tag{15.11}$$

or

$$\Delta v = -\frac{er\mu_0 H}{2m}. \tag{15.12}$$

This change in electron velocity yields in turn a change in magnetic moment, as we see by combining (15.5) with (15.12):

$$\Delta\mu_\mathrm{m} = \frac{e\Delta v r}{2} = -\frac{e^2 r^2 \mu_0 H}{4m}. \tag{15.13}$$

So far we tacitly assumed that the magnetic field is perpendicular to the plane of the orbiting electron. In reality, however, the orbit plane varies constantly in direction with respect to the external field. Thus, we have to find an average value for $\overline{\Delta\mu_m}$ which we expect to be slightly smaller than that given in (15.13) since $\Delta\mu_m$ approaches zero when the field direction and the orbit plane become parallel. A simple calculation (see Problem 2) yields

$$\overline{\Delta\mu_m} = -\frac{e^2 r^2 \mu_0 H}{6m}. \tag{15.14}$$

One further consideration needs to be made. Up to now, we treated only *one* electron. If we take all Z electrons into account (Z = atomic number), then the *average change in magnetic moment per atom* is

$$\overline{\Delta\mu_m} = -\frac{e^2 Z \bar{r}^2 \mu_0 H}{6m}, \tag{15.15}$$

where $\bar{r}$ is the average radius of all electronic orbits ($\bar{r} \approx 1$ Å). The *magnetization caused by this change of magnetic moment* is, according to (14.8),

$$M = \frac{\mu_m}{V} \equiv -\frac{e^2 Z \bar{r}^2 \mu_0 H}{6mV}. \tag{15.16}$$

This finally yields, together with (14.6), the diamagnetic susceptibility,

$$\boxed{\chi_{\text{dia}} = \frac{M}{H} = -\frac{e^2 Z \bar{r}^2 \mu_0}{6mV} = -\frac{e^2 Z \bar{r}^2 \mu_0}{6m} \frac{N_0 \delta}{W},} \tag{15.17}$$

where $N_0\delta/W$ is the number of atoms per unit volume (with N_0 = Avogadro constant, δ = density, and W = atomic mass. Inserting specific numbers into (15.17) yields susceptibilities between -10^{-5} and -10^{-7}, quite in agreement with the experimental values listed in Table 14.1 (see Problem 1).

The quantities in (15.17) are essentially temperature-independent, which is in agreement with the experimental observation that χ does not vary much with temperature for diamagnetic materials.

*15.3. Langevin Theory of (Electron Orbit) Paramagnetism

We turn now to the atomistic theory of paramagnetism as brought forward by Langevin. This theory should explain the observations made by Curie and Weiss, i.e., it should explain the temperature dependence of the susceptibility, as shown in Fig. 15.3. The Langevin theory does *not* treat spin paramagnetism, which is, as we said before, responsible for the paramagnetic behavior of many metals and which is only *slightly* temperature-dependent.

Langevin postulated that the magnetic moments of the orbiting electrons

Figure 15.15. Schematic representation of the magnetic moment of an electron that has been partially aligned by an external magnetic field.

are responsible for paramagnetism. The magnetic moments of these electrons are thought to point in random directions. An external magnetic field tries to align the individual magnetic moments, μ_m, parallel to the field direction. Once aligned, the magnetic moments have a potential energy, E_p, that is naturally greater the larger the field strength, H, and the larger μ_m. As a matter of fact, the maximum potential energy is reached when the magnetic moments are completely aligned, i.e., when μ_m is parallel to H. In general, the potential energy is

$$E_p = -\mu_m \mu_0 H \cos \alpha, \tag{15.18}$$

where α is the angle between field direction and μ_m (see Fig. 15.15). The sign in (15.18) defines the direction in which μ_m points with respect to H.

As we explained earlier, thermal agitation tends to counteract the alignment caused by the external magnetic field. The randomizing effect obeys, as usual, the laws of Boltzmann statistics. The probability of an electron to have the energy E_p is thus proportional to $\exp(-E_p/k_B T)$, where k_B is the Boltzmann constant and T is the absolute temperature.

Let us assume the electrons to be situated at the center of a sphere. The vectors, representing their magnetic moments, may point in all possible directions. Let us consider at present a small number, dn, of these vectors per unit volume only. They are thought to point in the direction interval $d\alpha$ and thus penetrate an area, dA, situated at the surface of the unit sphere; see Fig. 15.16. This infinitesimal number dn of magnetic moments per unit volume which have the energy E_p is

$$dn = \text{const.}\, dA \exp(-E_p/k_B T). \tag{15.19}$$

We relate the area dA to the angle interval $d\alpha$, which yields, because of trigonometric considerations (see Problem 2),

$$dA = 2\pi R^2 \sin \alpha \, d\alpha, \tag{15.20}$$

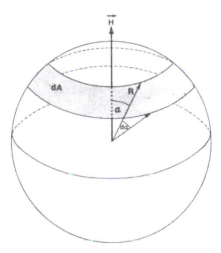

Figure 15.16. Schematic representation of a unit sphere in whose center the electrons are thought to be located.

where $R = 1$ is the radius of the unit sphere. Combining (15.18)–(15.20) gives

$$dn = \text{const.}\, 2\pi \sin \alpha \, d\alpha \exp\left(\frac{\mu_m \mu_0 H}{k_B T} \cos \alpha\right). \qquad (15.21)$$

We use for abbreviation

$$\xi = \frac{\mu_m \mu_0 H}{k_B T}. \qquad (15.22)$$

Integrating (15.21) provides

$$n = 2\pi \,\text{const.} \int_0^\pi \sin \alpha \exp(\xi \cos \alpha)\, d\alpha, \qquad (15.23)$$

which yields

$$\text{const.} = \frac{n}{2\pi \displaystyle\int_0^\pi \sin \alpha \exp(\xi \cos \alpha)\, d\alpha}. \qquad (15.24)$$

Now, the magnetization M is, according to (14.8), the magnetic moment μ_m per unit volume. In our case, the total magnetization must be the sum of all individual magnetic moments. And, if we consider the magnetic moments in the field direction, then the magnetization is

$$M = \int_0^n \mu_m \cos \alpha \, dn, \qquad (15.25)$$

which yields, with (15.21),

$$M = \text{const.} \; 2\pi\mu_m \int_0^\pi \cos\alpha \sin\alpha \exp(\xi\cos\alpha)\, d\alpha, \tag{15.26}$$

and, with (15.24),

$$M = \frac{n\mu_m \displaystyle\int_0^\pi \cos\alpha \sin\alpha \exp(\xi\cos\alpha)\, d\alpha}{\displaystyle\int_0^\pi \sin\alpha \exp(\xi\cos\alpha)\, d\alpha}. \tag{15.27}$$

This function can be brought into a standard form by setting $x = \cos\alpha$ and $dx = -\sin\alpha\, d\alpha$ (see Problem 5), which yields

$$M = n\mu_m\left(\coth\zeta - \frac{1}{\zeta}\right) = n\mu_m\left(\frac{\zeta}{3} - \frac{\zeta^3}{45} + \frac{2\zeta^5}{945} - \cdots\right) \tag{15.28}$$

where the expression in parenthesis is called the Langevin function $L(\zeta)$. The term $\zeta = \mu_m\mu_0 H/k_B T$ is usually much smaller than one (Problem 6), so that (15.28) reduces to

$$M = n\mu_m \frac{\zeta}{3} = \frac{n\mu_m^2\mu_0 H}{3k_B T}, \tag{15.29}$$

which yields, for the susceptibility (14.6) at not-too-high field strengths,

$$\boxed{\chi_{\text{para}}^{\text{orbit}} = \frac{M}{H} = \frac{n\mu_m^2\mu_0}{3k_B}\frac{1}{T} \equiv C \cdot \frac{1}{T}.} \tag{15.30}$$

This is Curie's law (15.1), which expresses that the susceptibility is inversely proportional to the temperature. The Curie constant is thus

$$C = \frac{n\mu_m^2\mu_0}{3k_B}. \tag{15.31}$$

Let us now discuss the results of the Langevin theory for electron-orbit paramagnetism. If we insert actual values in (15.30), we obtain susceptibilities that are small and positive, which is quite in agreement with experimental findings (see Table 14.1 and Problem 7).

The Langevin theory for paramagnetism yields that for a given temperature and for small values of the field strength the magnetization is a linear function of H (Fig. 15.17 and Equation (15.29)). For large field strengths the magnetization eventually reaches a saturation value, M_s. (This behavior is quite similar to the one observed for virgin iron or other ferromagnetics.) It indicates that eventually a limit is reached at which all magnetic moments are aligned to their maximum value. The Langevin model yields a temperature dependence of the susceptibility as found experimentally by Curie for

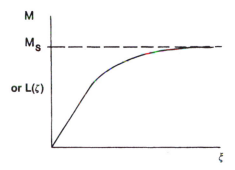

Figure 15.17. Schematic representation of the Langevin function $L(\zeta) = \coth \zeta - 1/\zeta$, where $\xi = \mu_m \mu_0 H / k_B T$.

many substances. The $1/\chi$ dependence of T is characteristic for electron-orbital paramagnetism.

One can refine the Langevin result by applying quantum theory. This was done by Brillouin, who took into account that not all values for the magnetic moment (or the angular moment) are allowed, i.e., that the angular moments are quantized in an external magnetic field (Appendix 3). This restriction is termed *space quantization*. The calculation leads to the *Brillouin function*, which improves the quantitative agreement between theory and experiment.

Finally, we know from Section 15.1.2 that the temperature dependence of the susceptibility for many solids does not always obey the Curie (or the Curie–Weiss) law. Actually, the susceptibility for most metals and alloys varies only very little with temperature. We have learned that in these solids the spin paramagnetism is predominant, which is not considered in the atomistic Langevin model. Quantum theory can explain the relative temperature insensitivity of spin paramagnetism, as we shall see in Section 16.1.

*15.4. Molecular Field Theory

So far, we implied that the magnetic field, which tries to align the magnetic moments, stems from an external source only. This assumption seems to be not always correct. Weiss observed that some materials obey a somewhat modified Curie law, as shown in Fig. 15.3(b) and (c). He postulated, therefore, that the magnetic moments of the individual electrons (or atoms) interact with each other. In this case, the total magnetic field, H_t, acting on a magnetic moment, is thought to be composed of two parts, namely, the external field, H_e, and the *molecular field*, H_m,

$$H_t = H_e + H_m, \tag{15.32}$$

where

$$H_{\mathrm{m}} = \gamma M \tag{15.33}$$

contains the molecular field constant, γ. The susceptibility is calculated by using (15.30), (15.32), and (15.33)

$$\chi = \frac{M}{H_{\mathrm{t}}} = \frac{M}{H_{\mathrm{e}} + \gamma M} = \frac{C}{T}. \tag{15.34}$$

Solving (15.34) for M yields

$$M = \frac{H_{\mathrm{e}} C}{T - \gamma C}. \tag{15.35}$$

Finally, we obtain

$$\chi = \frac{M}{H_{\mathrm{e}}} = \frac{C}{T - \gamma C} = \frac{C}{T - \theta}, \tag{15.36}$$

which is the experimentally observed Curie–Weiss law (15.2). If θ is found to be positive, then the interactions of the individual magnetic moments reinforce each other, i.e., the magnetic moments align parallel. In this case the susceptibility becomes larger, as can be deduced from (15.36).

We now attempt to interpret **ferromagnetism** by making use of the molecular field theory. We already know from Section 15.1.3 that, in ferromagnetic materials, the neighboring magnetic moments interact with each other, which leads to a spontaneous magnetization in small domains below T_{C}. Weiss postulated that the above-introduced internal or molecular field is responsible for this parallel alignment of spins, and considered ferromagnetics to be essentially paramagnetics having a very large molecular field. In essence, he applied the Langevin theory to ferromagnetics. In the light of quantum theory, the molecular field is essentially the *exchange force*, as we shall see in Section 16.2.

We follow the train of thought put forward by Weiss. Let us consider the case for no external magnetic field. Then the spins are only subjected to the molecular field H_{m}. This yields for the Langevin variable ζ (see (15.22)), with (15.33),

$$\zeta = \frac{\mu_{\mathrm{m}} \mu_0 H_{\mathrm{m}}}{k_{\mathrm{B}} T} = \frac{\mu_{\mathrm{m}} \mu_0 \gamma M}{k_{\mathrm{B}} T}, \tag{15.37}$$

and provides for the magnetization by rearranging (15.37):

$$M = \frac{k_{\mathrm{B}} T}{\mu_{\mathrm{m}} \mu_0 \gamma} \zeta. \tag{15.38}$$

We note from (15.38) that for the present case the magnetization is a linear function of ζ with the temperature as a proportionality factor (see Fig. 15.18). The intersection I of a given temperature line with the Langevin function $L(\zeta)$ represents the finite spontaneous magnetization, M_I, at this

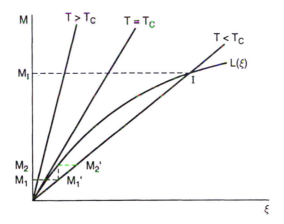

Figure 15.18. Langevin function $L(\zeta)$, i.e., (15.28) and plot of (15.38) for three temperatures.

temperature.[3] With increasing temperature, the straight lines in Fig. 15.18 increase in slope, thus decreasing the point of intercept, I, and therefore the value for the spontaneous magnetization. Finally, at the Curie temperature, T_C, no intercept, i.e., no spontaneous magnetization, is present anymore. The slope $k_B T/\mu_m \mu_0 \gamma$ in (15.38) is then identical to the slope of the Langevin function near the origin, which is $n\mu_m/3 = M/3$ according to (15.29) and (14.8). This yields, for T_C,

$$\frac{k_B T_C}{\mu_m \mu_0 \gamma} = \frac{M}{3}. \qquad (15.39)$$

A value for the molecular field constant, γ, can then be calculated by measuring the Curie temperature and inserting T_C into the rearranged equation (15.39):

$$\gamma = \frac{3k_B T_C}{\mu_m \mu_0 M}. \qquad (15.40)$$

This yields, for the molecular magnetic field strength (15.33),

$$H_m = \gamma M = \frac{3k_B T_C}{\mu_m \mu_0}. \qquad (15.41)$$

Numerical values for the molecular field are around 10^9 A/m (10^7 Oe) (see

---

[3] The intersection at the origin is an unstable state, as can easily be seen: If the ferromagnetic material is exposed to, say, the magnetic field of the earth, its magnetization will be, say, M_1. This causes a molecular field of the same value (M_1') which in turn magnetizes the material to the value M_2, and so on until the point I is reached.

Problem 10). This hypothetical field is several orders of magnitude larger than any steady magnetic field that can be produced in a laboratory. We should note that even though the molecular field theory gives some explanation of ferromagnetism, it cannot predict which solids are ferromagnetic. Quantum theory extends considerably our understanding of this matter.

We mention in closing that the molecular field theory can also be applied to antiferromagnetics and to ferrimagnetic materials. As we know from Section 15.1.4, we need to consider in this case two interpenetrating sublattices, A and B, each having mutually antiparallel aligned spins. This means that we now have to consider a molecular field, H_{mA}, acting on the A ions which stems from the magnetization, M_B, of the B ions. Since the magnetization of A and B ions point in opposite directions, the molecular field from an adjacent ion is now negative. The calculations, which follow similar lines as shown above, yield equation (15.4), i.e., the Curie–Weiss law for antiferromagnetics.

Problems

1. Calculate the diamagnetic susceptibility of germanium. Take $\bar{r} = 0.92\,\text{Å}$. (*Note*: Check your units! Does χ come out unitless? Compare your result with that listed in Table 14.1.)

2. In the text, we introduced an average value for the magnetic moment, which we said is somewhat smaller than the maximal value for $\mathbf{\mu}_m \| \mathbf{H}$. Calculate this $\Delta\mu_m$. (*Hint*: Consider all orbits projected on a plane perpendicular to the field direction and calculate thus an average value for the square of the orbit radius. Refer to the figure below. Show at first that $dA = 2\pi R^2 \sin\alpha\, d\alpha$.)

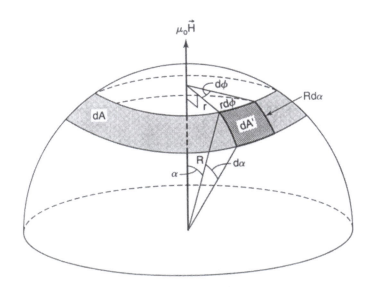

3. Convince yourself that the units in (15.15), (15.16), (15.17), and (15.5) are consistent with the SI system.

4. Confirm the numerical value of the Bohr magneton listed in (15.3) and confirm the unit given there.

5. Evaluate the function

$$\frac{\displaystyle\int_0^\pi \cos\alpha \sin\alpha \exp(\xi\cos\alpha)\,d\alpha}{\displaystyle\int_0^\pi \sin\alpha \exp(\xi\cos\alpha)\,d\alpha}$$

by substituting $x = \cos\alpha$ and $dx = -\sin\alpha\,d\alpha$. Compare your result with (15.28).

6. Calculate a value for ξ in the Langevin function assuming $\mu_m = 3\mu_B$, $H = 8 \times 10^5$ A/m, and room temperature.

7. Calculate the susceptibility for a paramagnetic substance at room temperature, assuming $\mu_m = \mu_B$ and 10^{23} magnetic moments per cubic centimeter. Compare your result with Table 14.1. What is the implication of $n = 10^{23}$ magnetic moments per cubic centimeter?

8. Estimate the number of Bohr magnetons for iron and cobalt ferrite from their electron configuration, as done in the text. Compare your results with those listed in Table 15.3. Explain the discrepancy between experiment and calculation. Give the chemical formula for these ferrites.

9. Explain the term 'mixed ferrites." Explain also why the *lodestone*, Fe_3O_4, is a ferrimagnetic material. Give its chemical formula.

10. Calculate the molecular field for iron ($\mu_m = 2.22\ \mu_B$, $T_C = 1043$ K).

11. You are given two identical rectangular iron rods. One of the rods is a permanent magnet, the other is a plain piece of iron. The rods are now placed on a wooden table. Using only the two rods and nothing else, you are asked to determine which is which. Can this be done?

12. Explain the "suspension effect" of superconductors mentioned in Section 15.1.1. (*Hint:* Refer to Fig. 15.1 and keep in mind that if the bar magnet is moved in the opposite direction from that shown, the current direction in the loop is reversed.)

13. *Computer problem.* Plot the Langevin function using various parameters. For which values of H does one obtain saturation magnetization?

CHAPTER 16

Quantum Mechanical Considerations

We have seen in the previous chapter that the classical electromagnetic theory is quite capable of explaining the essentials of the magnetic properties of materials. Some discrepancies between theory and experiment have come to light, however, which need to be explained. Therefore, we now refine and deepen our understanding by considering the contributions which quantum mechanics provides to magnetism. We will see in the following that quantum mechanics yields answers to some basic questions. We will discuss why certain metals that we expect to be paramagnetic are in reality diamagnetic; why the paramagnetic susceptibility is relatively small for most metals; and why most metals do not obey the Curie–Weiss law. We will also see that ferromagnetism can be better understood by applying elements of quantum mechanics.

16.1. Paramagnetism and Diamagnetism

We mentioned at the beginning of the previous chapter that, for most solids, the dominant contribution to paramagnetism stems from the magnetic moment of the *spinning electrons*. We recall from Chapter 6 that each electron state may be occupied by a maximum of two electrons, one having positive spin and the other having negative spin (called spin up and spin down). To visualize the distribution of spins, we consider an electron band to be divided into two halves, each of which is thought to be occupied under normal conditions by an identical amount of electrons of opposite spin, as shown in Fig. 16.1(a). Now, if we apply an external magnetic field to a free electron solid, some of the electrons having unfavorably oriented spins tend to change their

(a)　　　　　　　　　　(b)　　　　　　　　　　(c)

Figure 16.1. Schematic representation of the effect of an external magnetic field on the electron distribution in a partially filled electron band, (a) without magnetic field, (b) and (c) with magnetic field.

field direction. This can only be achieved, however, when the affected electrons assume an energy which is higher than the Fermi energy, E_F, since all lower electron states of opposite spin direction are already occupied (Fig. 16.1(b)). Thus, theoretically, the transfer of electrons from one half-band into the other would cause two individual Fermi energies (E_F' and E_F'') to occur. Of course, this is not possible. In reality the two band halves shift relative to each other until equilibrium, i.e., a common Fermi energy, is reached (Fig. 16.1(c)).

Now, we recall from Chapter 6 that the electron distribution within a band is not uniform. We rather observe a parabolic distribution of energy states, as shown in Fig. 6.4. Thus, we refine our treatment by replacing Fig. 16.1(c) with Fig. 16.2, which depicts the density of states of the two half-bands. We observe a relatively large $Z(E)$ near E_F. Thus, a small change in energy (provided by the external magnetic field) may cause a large number of electrons to switch to the opposite spin direction.

We calculate now the susceptibility from this change in energy, ΔE. It is evident that ΔE is larger, the larger the external magnetic field strength $|\mathbf{H}|$,

Figure 16.2. Schematic representation of the density of states $Z(E)$ in two half-bands. The shift of the two half-bands occurs as a result of an external magnetic field. Free electron case. (See also Fig. 16.1(c).) The area ΔN equals $\Delta E \cdot Z(E)$.

and the larger the magnetic moment of the spinning electrons $|\mathbf{\mu}_{ms}|$, i.e.,

$$\Delta E = \mu_0 H \mu_{ms}. \tag{16.1}$$

As mentioned already, the number of electrons, ΔN, transferred from the spin down into the spin up direction depends on the density of states at the Fermi energy, $Z(E_F)$, and the energy difference, ΔE (Fig. 16.2), i.e.,

$$\Delta N = \Delta E Z(E_F) = \mu_0 H \mu_{ms} Z(E_F). \tag{16.2}$$

The magnetization $|\mathbf{M}|$ of a solid, caused by an external magnetic field is, according to (14.8),

$$M = \frac{\mu_m}{V}. \tag{16.3}$$

The magnetization is, of course, larger, the more electrons are transferred from spin down into spin up states. We thus obtain, for the present case,

$$M = \frac{\mu_{ms}}{V} \Delta N = \frac{\mu_{ms}^2 \mu_0 H Z(E_F)}{V}, \tag{16.4}$$

which yields for the susceptibility

$$\chi = \frac{M}{H} = \frac{\mu_{ms}^2 \mu_0 Z(E_F)}{V}. \tag{16.5}$$

The spin magnetic moment of one electron equals one Bohr magneton, μ_B (see below). Thus, (16.5) finally becomes

$$\boxed{\chi_{para}^{spin} = \frac{\mu_B^2 \mu_0 Z(E_F)}{V}.} \tag{16.6}$$

The susceptibilities for paramagnetic metals calculated with this equation agree fairly well with those listed in Table 14.1 (see Problem 1). Thus, (16.6) substantiates, in essence, that only the electrons close to the Fermi energy are capable of realigning in the magnetic field direction. If we postulate instead that *all* valence electrons contribute to χ_{para} we would wrongfully calculate a susceptibility which is two or even three orders of magnitude larger than that obtained by (16.6).

It is important to realize that the ever-present diamagnetism makes a sizable contribution to the overall susceptibility, so that χ for metals might be positive or negative depending on which of the two components predominates. This will be elucidated now in a few examples.

To begin with, we discuss beryllium, which is a bivalent metal having a filled 2s-shell in its atomic state (see Appendix 3). However, in the crystalline state, we observe band overlapping (see Chapter 6), which causes some of the 2s-electrons to spill over into the 2p-band. They populate the very bottom

Figure 16.3. Overlapping of 2s- and 2p-bands in Be and the density of states curve for the 2p-band.

of this band (see Fig. 16.3). Thus, the density of states at the Fermi level, and consequently, χ_{para}, is very small. In effect, the diamagnetic susceptibility predominates, which makes Be diamagnetic.

In order to understand why copper is diamagnetic, we need to remember that for this metal the Fermi energy is close to the band edge (Fig. 5.22). Thus, the density of states near E_F and the paramagnetic susceptibility (16.6) are relatively small. Furthermore, we have to recall that the diamagnetic susceptibility (15.17),

$$\chi_{\text{dia}} = -\frac{e^2 Z r^2 \mu_0}{6mV}, \qquad (16.7)$$

is proportional to the square of an electron orbit radius, r, and proportional to the total number of electrons, Z, in that orbit. Copper has about ten $3d$-electrons, which makes $Z \approx 10$. Further, the radius of d-shells is fairly large. Thus, for copper, χ_{dia} is large because of two contributions. The diamagnetic contribution predominates over the paramagnetic one. As a result, copper is diamagnetic. The same is true for silver and gold and the elements which follow copper in the Periodic Table, such as zinc and gallium.

Intrinsic semiconductors, which have filled valence bands and whose density of states at the top of the valence band is zero (Fig. 6.6) have, according to (16.6), no paramagnetic susceptibility and are therefore diamagnetic. However, a small paramagnetic contribution might be expected for highly doped *extrinsic* semiconductors, which have, at high enough temperatures, a considerable number of electrons in the conduction band (see Chapter 8).

We turn now to the temperature dependence of the susceptibility of metals. The relevant terms in both (16.6) as well as (16.7) do not vary much with temperature. Thus, it is conceivable that the susceptibility of diamagnetic metals is not temperature-dependent, and that the susceptibility of paramagnetic metals often does *not* obey the Curie–Weiss law. In fact, the temperature dependence of the susceptibility for different paramagnetic *metals* has been observed to decrease, to increase, or to remain essentially constant

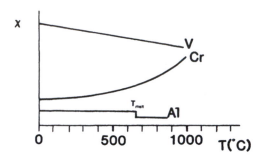

Figure 16.4 Temperature dependence of the paramagnetic susceptibility for vanadium, chromium, and aluminum in arbitrary units. From Landolt–Börnstein, *Zahlenwerte der Physik*, 6th ed., Vol. II/9, Springer-Verlag, Berlin (1962).

(Fig. 16.4). However, nickel (above T_C) and rare earth metals obey the Curie–Weiss law reasonably well.

At the end of this section we remind the reader that in dilute gases (and also in rare earth metals and their salts) a second component contributes to paramagnetism. It stems from a magnetic moment which is caused by the angular momentum of the orbiting electrons (Section 15.3). We mentioned already in Section 15.1 that this contribution is said to be "quenched" (nonexistent) in most solids.

Finally, we want to find a numerical value for the magnetic moment of an orbiting electron from a quantum-mechanical point of view. We recall from (15.5):

$$\mu_m = \frac{evr}{2}. \tag{16.8}$$

Now, quantum theory postulates that the angular momentum, mvr, of an electron is not continuously variable but that it rather changes in discrete amounts of integer multiples of $\hbar$ only, i.e.,

$$mvr = n\hbar = \frac{nh}{2\pi}. \tag{16.9}$$

If one combines (16.8) with (16.9) one obtains

$$\mu_m = \frac{enh}{4\pi m}. \tag{16.10}$$

Using $n = 1$ for the first electron orbit (ground state) yields, for the magnetic moment of an orbiting electron,

$$\mu_m = \frac{eh}{4\pi m}. \tag{16.11}$$

It was found experimentally and theoretically that the magnetic moment of an electron due to orbital motion as well as the magnetic moment of the spinning electron are identical. This smallest unit of the magnetic moment is given by (16.11) and is called the Bohr magneton,

$$\mu_B = \frac{eh}{4\pi m} = 9.274 \times 10^{-24} \left(\frac{J}{T}\right),$$ (16.12)

which we already introduced without further explanation in (15.3).

16.2. Ferromagnetism and Antiferromagnetism

The ferromagnetic metals iron, cobalt, and nickel are characterized by unfilled d-bands (see Appendix 3). These d-bands overlap the next higher s-band in a similar manner as shown in the band structure of Fig. 5.22. The density of states for a d-band is relatively large because of its potential to accommodate up to ten electrons. This is schematically shown in Fig. 16.5, along with the Fermi energies for iron, cobalt, nickel, and copper. Since the density of states for, say, nickel is comparatively large at the Fermi energy, one needs only a relatively small amount of energy to transfer a considerable number of electrons from spin down into spin up configurations, i.e., from one half-band into the other. We have already discussed in the previous section this transfer of electrons under the influence of an external magnetic field (Fig. 16.1). Now, there is an important difference between paramagnetics and ferromagnetics. In the former case, an external energy (i.e., the magnetic field) is needed to accomplish the flip in spin alignment, whereas for ferromagnetic materials the parallel alignment of spins occurs spontaneously in small domains of about $1-100$ μm diameter. Any theory of

Figure 16.5. Schematic representation of the density of states for $4s$- and $3d$-bands and the Fermi energies for iron, cobalt, nickel, and copper. The population of the bands by the ten nickel $(3d + 4s)$-electrons is indicated by the shaded area.

Figure 16.6. Amplitude modulation resulting from the coupling of two pendula. The vibrational pattern shows *beats*, similarly as known for two oscillators that have almost identical pitch. Compare with Fig. 2.1.

ferromagnetism must be capable of satisfactorily explaining the origin of this energy which transfers electrons into a higher energy state.

The energy in question was found to be the **exchange energy**. It is "set free" when equal atomic systems are closely coupled, and in this way exchange their energy. This needs some further explanation.

We digress for a moment and compare two ferromagnetic atoms with two identical pendula that are interconnected by a spring. (The spring represents the interactions of the electrical and magnetic fields.) If one of the pendula is deflected, its amplitude slowly decreases until all energy has been transferred to the second pendulum, which then in turn transfers its energy back to the first one and so on. Thus, the amplitudes decrease and increase periodically with time, as shown in Fig. 16.6. The resulting vibrational pattern is similar to that of two violin strings tuned at almost equal pitch. A mathematical expression for this pattern is obtained by adding the equations for two oscillators having similar frequencies, ω_1 and ω_2,

$$X_1 = b \sin \omega_1 t, \tag{16.13}$$

$$X_2 = b \sin \omega_2 t, \tag{16.14}$$

which yields

$$X_1 + X_2 = X = 2b \cos \frac{\omega_1 - \omega_2}{2} t \cdot \sin \frac{\omega_1 + \omega_2}{2} t. \tag{16.15}$$

Equation (16.15) provides two frequencies, $(\omega_1 - \omega_2)/2$ and $(\omega_1 + \omega_2)/2$, which can be identified in Fig. 16.6. The difference between the resulting frequencies is larger, the stronger the coupling. If the two pendula vibrate in a parallel fashion, the "pull" on the spring, i.e., the restoring force, κx, is small. As a consequence, the frequency

$$\nu_0 = \frac{1}{2\pi} \sqrt{\frac{\kappa}{m}} \tag{16.16}$$

(see Appendix 1) is likewise small and is smaller than for independent vibrations. (On the other hand, antisymmetric vibrations cause large values of κ and ν_0.) This classical example demonstrates that two coupled and symmetrically vibrating systems may have a lower energy than two individually vibrating systems would have.

Figure 16.7. Exchange integral, I_{ex}, versus the ratio of interatomic distance, r_{ab}, and the radius of an unfilled d-shell. The position of the rare earth elements (which have unfilled f-shells) are also shown for completeness.

Quantum mechanics treats ferromagnetism in a similar way. The exact calculation involving many atoms is, however, not a trivial task. Thus, one simplifies the problem by solving the appropriate Schrödinger equation for two atoms only. The potential energy in the Schrödinger equation then contains the exchange forces between the nuclei a and b, the forces between two electrons 1 and 2, and the interactions between the nuclei and their neighboring electrons. This simplification seems to be justified, because the exchange forces decrease rapidly with distance.

The calculation, first performed by Slater and Bethe, leads to an *exchange integral*,

$$I_{ex} = \int \psi_a(1)\psi_b(2)\psi_a(2)\psi_b(1)\left[\frac{1}{r_{ab}} - \frac{1}{r_{a2}} - \frac{1}{r_{b1}} + \frac{1}{r_{12}}\right] d\tau. \qquad (16.17)$$

A positive value for I_{ex} means that parallel spins are energetically more favorable than antiparallel spins (and vice versa). We see immediately from (16.17) that I_{ex} becomes positive for a small distance r_{12} between the electrons, i.e., a small radius of the d-orbit, r_d. Similarly, I_{ex} becomes positive for a large distance between the nuclei and neighboring electrons r_{a2} and r_{b1}.

I_{ex} is plotted in Fig. 16.7 versus the ratio r_{ab}/r_d. The curve correctly separates the ferromagnetics from manganese, which is not ferromagnetic. Figure 16.7 suggests that if the interatomic distance r_{ab} in manganese is increased (e.g., by inserting nitrogen atoms into the manganese lattice), the crystal thus obtained should become ferromagnetic. This is indeed observed. The ferromagnetic alloys named after Heusler, such as Cu_2MnAl or Cu_2MnSn, are particularly interesting in this context because they contain constituents which are not ferromagnetic, but all contain manganese.

The Bethe–Slater curve (Fig. 16.7) suggests that cobalt should have the highest, and nickel (and the rare earth elements) the lowest, Curie temperature among the ferromagnetics because of the magnitude of their I_{ex} values. This is indeed observed (Table 15.1). Overall, quantum theory is capable of explaining some ferromagnetic properties that cannot be understood with classical electromagnetic theory.

Table 16.1. Magnetic
Moment, μ_m, at 0 K for
Ferromagnetic Metals.

Metal	μ_m
Fe	2.22 μ_B
Co	1.72 μ_B
Ni	0.60 μ_B
Gd	7.12 μ_B

We turn now to a discussion on the number of Bohr magnetons in ferro-magnetic metals as listed in Table 16.1. Let us consider nickel as an example and reinspect, in this context, Fig. 16.5. We notice that because of band overlapping the combined ten $(3d + 4s)$-electrons occupy the lower s-band and fill, almost completely, the $3d$-band. It thus comes as no surprise that nickel behaves experimentally as if the $3d$-band is filled by 9.4 electrons. To estimate μ_B we need to apply Hund's rule (Fig. 15.4), which states that the electrons in a solid occupy the available electron states in a manner which maximizes the imbalance of spin moments. For the present case, this rule would suggest five electrons with, say, spin up, and an average of 4.4 elec-trons with spin down, i.e., we obtain a spin imbalance of 0.6 spin moments or 0.6 Bohr magnetons per atom (see Table 16.1). The average number of Bohr magnetons may also be calculated from experimental values of the saturation magnetization, M_{s0} (see Table 15.1). Similar considerations can be made for the remaining ferromagnetics as listed in Table 16.1.

We now proceed one step further and discuss the magnetic behavior of certain nickel-based alloys. We use nickel–copper alloys as an example. Copper has one valence electron more than nickel. If copper is alloyed to nickel, the extra copper electrons progressively fill the d-band and therefore compensate some of the unsaturated spins of nickel. Thus, the magnetic moment per atom of this alloy (and also its Curie temperature) is reduced. Nickel lacks about 0.6 electrons per atom for complete spin saturation, be-cause the $3d$-band of nickel is filled by only 9.4 electrons (see above). Thus, about 60% copper atoms are needed until the magnetic moment (and μ_B) of nickel has reached a zero value (Fig. 16.8). Nickel–copper alloys, having a copper concentration of more than about 60% are consequently no longer ferromagnetic; one would expect them to be diamagnetic. (In reality, how-ever, they are strongly paramagnetic, probably owing to small traces of un-dissolved nickel.)

Zinc contributes about two extra valence electrons to the electron gas when alloyed to nickel. Thus, we expect a zero magnetic moment at about 30 at.% Zn, etc. Palladium, on the other hand, has the same number of valence electrons as nickel and thus does not change the magnetic moment of the nickel atoms when alloyed to nickel. The total magnetization (14.8) of the

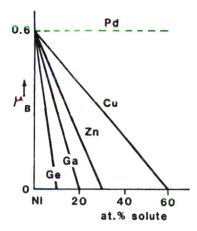

Figure 16.8. Magnetic moment per nickel atom as a function of solute concentration.

alloy is, of course, diluted by the nonferromagnetic palladium. The same is also true for the other alloys.

We conclude our discussion by adding a few interesting details. The rare earth elements are weakly ferromagnetic. They are characterized by unfilled f-shells. Thus, their electronic structure and their density of states have several features in common with iron, cobalt, and nickel. They have a positive I_{ex} (see Fig. 16.7).

Copper has one more valence electron than nickel, which locates its Fermi energy slightly above the d-band (Fig. 16.5). Thus, the condition for ferromagnetism, i.e., an unfilled d- or f-band is not fulfilled for copper. The same is true for the following elements such as zinc or gallium.

We noted already that manganese is characterized by a negative value of the exchange integral. The distance between the manganese atoms is so small that their electron spins assume an antiparallel alignment. Thus, manganese and many manganese compounds are antiferromagnetic (see Fig. 15.10). Chromium has also a negative I_{ex} and thus is likewise antiferromagnetic (see Table 15.2).

Problems

1. The density of states near the Fermi surface of 1 cm^3 of a paramagnetic metal at $T = 0$ K is approximately 5×10^{41} energy states per Joule. Calculate the volume susceptibility. Compare your value with those of Table 14.1. What metal could this value represent? Explain possible discrepancies between experiment and calculation.

2. Derive (16.15) by adding (16.13) and (16.14).

3. Compare the experimental saturation magnetization, M_{s0} (Table 15.1 third column), with the magnetic moment, μ_m, at 0 K for ferromagnetic metals (Table 16.1). What do you notice? Estimate the degree of d-band filling for iron and cobalt.

4. From the results obtained in Problem 3 above, calculate the number of Bohr magnetons for crystalline (solid) iron and cobalt and compare your results with those listed in Table 16.1. What is the number of Bohr magnetons for an iron *atom* and a cobalt atom? What is the number of Bohr magnetons for iron and cobalt ferrite?

5. Refer to Figure 16.1(b). Why are two different Fermi energies not possible within the same metal?

CHAPTER 17

Applications

17.1. Introduction

The production of ferro- and ferrimagnetic materials is a large-scale operation, measured in quantity as well as in currency. (This is in contrast to the products of the computer industry, where the price of the *material* that goes into a chip is a minute fraction of the device fabrication cost.) As an example, the annual sales of so-called electrical steel, used for electromotors and similar devices, reach the millions of tons and their market values are in the hundreds of millions of dollars. Other large-scale production items are permanent magnets for loudspeakers, etc., and magnetic recording materials. The following sections will give some impression about the technology (i.e., mostly materials science) which has been developed to improve the properties of magnetic materials.

17.2. Electrical Steels (Soft Magnetic Materials)

Electrical steel is used to multiply the magnetic flux in the cores of electromagnetic coils. These materials are therefore widely incorporated in many electrical machines in daily use. Among their applications are cores of transformers, electromotors, generators, or electromagnets.

In order to make these devices most energy efficient and economical, one needs to find magnetic materials which have the highest possible permeability (at the lowest possible price). Furthermore, magnetic core materials should be capable of being easily magnetized or demagnetized. In other words, the area within the hysteresis loop (or the coercive force, H_c) should be as small as possible (Fig. 15.6). We remember that materials whose hysteresis loops are narrow are called *soft* magnetic materials.

Electrical steels are classified by some of their properties, for example, by the amount of their core losses, by their composition, by their permeability, and whether or not they are *grain-oriented*. We shall discuss these different properties in detail.

The energy losses which are encountered in electromotors (efficiency between 50% and 90%) or transformers (efficiency 95–99.5%) are estimated to be, in the United States, as high as 3×10^{10} kWh per year, which is equivalent to the energy consumption by about 3 million households, and which wastes about 2×10^9 per year. If by means of improved design of the magnetic cores, the energy losses would be reduced by only 5%, one could save about 10^8 per year and several electric power stations. Thus, there is a clear incentive for improving the properties of magnetic materials.

17.2.1. Core Losses

The core loss is the energy that is dissipated in the form of heat within the core of electromagnetic devices when the core is subjected to an alternating magnetic field. Several types of losses are known, among which the eddy current loss and the hysteresis loss contribute the most. Typical core losses are between 0.3 and 3 watts per kilogram of core material (Table 17.1).

Let us first discuss the **eddy current**. Consider a transformer whose primary and secondary coils are wound around the legs of a rectangular iron yoke (Fig. 17.1(a)). An alternating electric current in the primary coil causes an alternating magnetic flux in the core, which, in turn, induces in the secondary coil an alternating electromotive force, V_e, proportional to $d\phi/dt$, see (14.7) and (15.9),

$$V_e \propto -\frac{d\phi}{dt} = -A\frac{dB}{dt}. \tag{17.1}$$

Concurrently, an alternating emf is induced within the core itself, as shown in Fig. 17.1(a). This emf gives rise to the eddy current, I_e. The eddy current is larger, the larger the permeability, μ (because $B = \mu_0\mu \cdot H$), the larger the conductivity, σ, of the core material, the higher the applied frequency, and the larger the cross-sectional area, A, of the core. (A is perpendicular to the magnetic flux, ϕ, see Fig. 17.1(a).) We note in passing that, particularly at high frequencies, the eddy current shields the interior of the core from the magnetic field, so that only a thin exterior layer of the core contributes to the flux multiplication (skin effect).

In order to decrease the eddy current, several remedies are possible. First, the core can be made of an insulator in order to decrease σ. Ferrites are thus effective but also expensive materials to build magnetic cores (see Section 15.1.5). They are indeed used for high–frequency applications. Second, the

Table 17.1. Properties of Some Soft Magnetic Materials.

Name	Composition (mass %)	Permeability, μ_{max} (unitless)	Coercivity, H_c		Saturation induction[a], B_s		Resistivity, $\rho\,(\mu\Omega \cdot cm)$	Core loss at 1.5 T and 60 Hz (W/kg)
			(Oe)	(A/m)	(kG)	(T)		
Low carbon steel	Fe–0.05% C	5×10^3	1.0	80	21.5	2.1	10	2.8
Nonoriented silicon iron	Fe–3% Si, 0.005% C, 0.15% Mn	7×10^3	0.5	40	19.7	2	60	0.9
Grain-oriented silicon iron	Fe–3% Si, 0.003% C, 0.07% Mn	4×10^4	0.1	8	20	2	47	0.3
78 Permalloy	Ni–22% Fe	10^5	0.05	4	10.8	1.1	16	≈ 2
Mumetal	77% Ni; 16% Fe, 5% Cu, 2% Cr	10^5	0.05	4	6.5	0.6	62	
Supermalloy	79% Ni; 16% Fe, 5% Mo	10^6	0.002	0.1	7.9	0.8	60	
Supermendur	49% Fe, 49% Co, 2% V	6×10^4	0.2	16	24	2.4	27	
Metglas #2605 annealed	$Fe_{80}B_{20}$	3×10^5	0.04	3.2	15	1.5	≈ 200	0.3

[a] Above B_s the magnetization is constant and $dB/d(\mu_0 H)$ is unity.

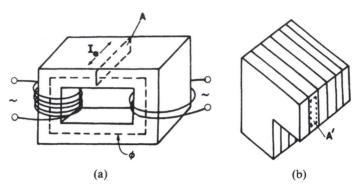

Figure 17.1. (a) Solid transformer core with eddy current, I_e, in a cross-sectional area A. Note the magnetic flux lines ϕ. (b) Cross section of a laminated transformer core. The area A' is smaller than area A in (a).

core can be manufactured from pressed iron powder whereby each particle (which is about 50–100 μm in diameter) is covered by an insulating coating. However, the decrease in σ, in this case, is at the expense of a large decrease in μ. Third, the most widely applied method to reduce eddy currents is the utilization of cores made out of thin sheets which are electrically insulated from each other (Fig. 17.1(b)). This way, the cross-sectional area, A, is reduced, which in turn decreases V_e (17.1), and additionally reduces losses due to the skin effect. Despite the lamination, a residual eddy current loss still exists, which is caused by current losses within the individual laminations and interlaminar losses that may arise if laminations are not sufficiently insulated from each other. These losses are, however, less than 1% of the total energy transferred.

Hysteresis losses are encountered when the magnetic core is subjected to a complete hysteresis cycle (Fig. 15.6). The work thus dissipated into heat is proportional to the area enclosed by a B/H loop. Proper materials selection and rolling of the materials with subsequent heat treatment greatly reduces the area of a hysteresis loop (see below).

17.2.2. Grain Orientation

The permeability of electrical steel can be substantially increased and the hysteresis losses can be decreased by making use of favorable grain orientations in the material. This needs some explanation. The magnetic properties of crystalline ferromagnetic materials depend on the crystallographic direction in which an external field is applied, an effect which is called **magnetic anisotropy**. Let us use iron as an example. Figure 17.2(a) shows magnetization curves of single crystals for three crystallographic directions. We observe that if the external field is applied in the $\langle 100 \rangle$ direction, saturation is achieved with the smallest possible field strength. The $\langle 100 \rangle$ direction is

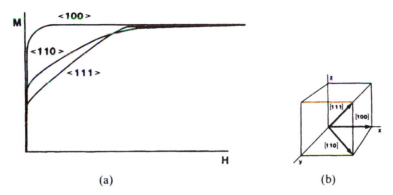

Figure 17.2. (a) Schematic magnetization curves for rod-shaped *iron single crystals* having different orientations (virgin curves). The magnetic field was applied in three different crystallographic directions. (Compare with Fig. 15.6, which refers to poly-crystalline material). (b) Reminder of the indices which identify directions in space. (See also Footnote 14 in Section 5.6).

thus called the "**easy direction.**" (In nickel, on the other hand, the $\langle 111 \rangle$ direction is the easy direction and $\langle 100 \rangle$ is the hard direction.)

This experimental finding gives us, incidentally, some clues about the spontaneous orientation of the spin magnetic moments in the demagnetized state. They are aligned in the easy direction. As an example, in virgin iron the spins are aligned along the $\langle 100 \rangle$ directions. Now, suppose that an external field is applied parallel to an easy direction. Then, the domains already having favorable alignment grow without effort at the expense of other domains until eventually the crystal contains one single domain (Fig. 15.9). The energy consumed during this process (which is proportional to the area between the magnetization curve and the horizontal line through M_s) is used to move the domain walls through the crystal.

A second piece of information needs to be considered, too. Metal sheets, which have been manufactured by rolling and heating, often possess a "**texture,**" i.e., they have a preferred orientation of the grains. It just happens that in α-iron and α-iron alloys the $\langle 100 \rangle$ direction is parallel to the rolling direction. This property is exploited when utilizing electrical steel.

Grain-oriented electrical steel is produced by initially hot-rolling the alloy followed by two stages of cold reduction with intervening anneals. During the rolling, the grains are elongated and their orientation is altered. Finally, the sheets are recrystallized, whereby some crystals grow in size at the expense of others (occupying the entire sheet thickness).

In summary, the magnetic properties of grain-oriented steels are best in the direction parallel to the direction of rolling. Electrical machines having core material of grain-oriented steel need less iron and are therefore smaller; the price increase due to the more elaborate fabrication procedure is often compensated by the savings in material. For details, see Table 17.1.

17.2.3. Composition of Core Materials

The least expensive core material is commercial **low carbon steel** (0.05% C). It possesses a relatively small permeability and has about ten times higher core losses than grain-oriented silicon iron (Table 17.1). Low carbon steel is used where low cost is more important than the efficient operation of a device. Purification of iron increases the permeability but also increases conductivity (eddy current!) and price.

Iron–silicon alloys containing between 1.4 and 3.5% Si and very little carbon have a higher permeability and a lower conductivity than low carbon steel (see Table 17.1). Furthermore, because of special features in the phase diagram ("γ-loop"), heat treatments of these alloys can be performed at much higher temperatures without interference from phase changes during cooling. The core losses decrease with increasing silicon content. However, for silicon concentrations above 4 or 5 weight %, the material becomes too brittle to allow rolling. Grain orientation in iron–silicon alloys (see above) further increases the permeability and decreases the hysteresis losses. Other constituents in iron–silicon alloys are aluminum and manganese in amounts less than 1%. They are added mainly for metallurgical reasons, because of their favorable influence on the grain structure and their tendency to reduce hysteresis losses. Grain-oriented silicon "steel" is the favored commercial product for highly efficient–high flux multiplying core applications.

The highest permeability is achieved for certain multicomponent nickel-based alloys such as **Permalloy**, **Supermalloy**, or **Mumetal** (Table 17.1). The latter can be rolled into thin sheets and is used to shield electronic equipment from stray magnetic fields.

17.2.4. Amorphous Ferromagnets

The electrical properties of amorphous metals (metallic glasses) and their methods of production have already been discussed in Section 9.4. In the present context, we are interested only in their magnetic properties, in particular, as flux multipliers in transformers, motors, etc. Some amorphous metals (consisting of iron, nickel, or cobalt with boron, silicon, or phosphorus) have, when properly annealed below the crystallization temperature (for strain relaxation), a considerably higher permeability and a lower coercivity than the commonly used grain-oriented silicon–iron, see Table 17.1. Further, the electrical resistivity of amorphous alloys is generally larger than their crystalline counterparts, which results in smaller eddy current losses. However, amorphous ferromagnets possess a somewhat lower saturation induction (Table 17.1) (which sharply decreases even further at elevated temperatures) and their core losses increase rapidly at higher flux densities (e.g., above 1.4 T). Thus, the application of metallic glasses for flux multiplication purposes is, at the present, limited to devices with small flux densities, i.e., low currents, such as for transformers (e.g., for communication equipment), magnetic sensors, or magnetostrictive transducers.

17.3. Permanent Magnets (Hard Magnetic Materials)

Permanent magnets are devices that retain their magnetic field indefinitely. They are characterized by a large **remanence** B_r (or M_r), a relatively large **coercivity** H_c, and a large area within the hysteresis loop. They are called *hard magnetic materials* (see Section 15.1.3).

The best means to visualize the properties of permanent magnets is to inspect their **demagnetization curve** (Fig. 17.3), which is a part of a hysteresis loop, as shown in Fig. 15.6. Another parameter which is used to characterize hard magnetic materials is the **maximum energy product**, $(BH)_{max}$, which is related to the area within the hysteresis loop. We see immediately from Fig. 17.3 that B times H is zero at the intercepts of the demagnetization curve with the coordinate axes, and that the energy product peaks somewhere between these extreme values, depending on the shape and size of the hysteresis curve. The values of B_r, H_c, and $(BH)_{max}$ for some materials which are used as permanent magnets are listed in Table 17.2.

The remanence, B_r, shown in Fig. 17.3 or listed in Table 17.2 is the maximal residual induction which can be obtained in a circular, close-loop magnet inserted in a coil. However, all permanent magnets need to have exposed poles in order to be useful. The necessary *air gap* between the north and south poles reduces the remanence, because the exposed poles create a **demagnetizing field**, H_d, which acts in the opposite direction to the B lines. We understand intuitively that the demagnetizing field depends on the shape, size, and gap length of a magnet. Thus, a reduced value for the residual in-

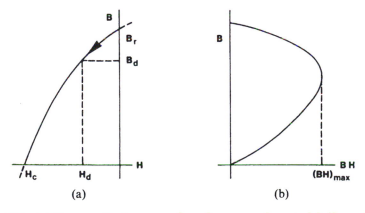

Figure 17.3. (a) Demagnetization curve for a ferromagnetic material. (Second quadrant in a B–H diagram.) (b) Energy product, BH, as a function of induction, B.

Table 17.2. Properties of Materials Used for Permanent Magnets.

Material	Composition (mass %)	Remanence B_r		Coercivity H_c		Maximum energy product $(BH)_{max}$ per Volume	
		(kG)	(T)	(Oe)	(A/m)	(MGOe)	(kJ/m^3)
Steel	Fe–1% C	9	0.9	51	4×10^3	0.2	1.6
36 Co steel	36 Co, 3.75 W, 5.75 Cr, 0.8 C	9.6	0.96	228	1.8×10^4	0.93	7.4
Alnico 2	12 Al, 26 Ni, 3 Cu, 63 Fe	7	0.7	650	5.2×10^4	1.7	13
Alnico 5	8 Al, 15 Ni, 24 Co, 3 Cu, 50 Fe	12	1.2	720	5.7×10^4	5.0	40
Alnico 5 DG	same as above	13.1	1.3	700	5.6×10^4	6.5	52
Ba-ferrite (Ceramic 5)	BaO $\cdot$ 6 Fe$_2$O$_3$	3.95	0.4	2,400	1.9×10^5	3.5	28
PtCo	77 Pt, 24 Co	6.45	0.6	4,300	3.4×10^5	9.5	76
Remalloy	12 Co, 17 Mo, 71 Fe	10	1	230	1.8×10^4	1.1	8.7
Vicalloy 2	13 V, 52 Co, 35 Fe	10	1	450	3.6×10^4	3.0	24
Cobalt–Samarium	Co$_5$Sm	9	0.9	8,700	6.9×10^5	20	159
Iron–Neodymium–Boron	Fe$_{14}$Nd$_2$B$_1$	13	1.3	14,000	1.1×10^6	40	318

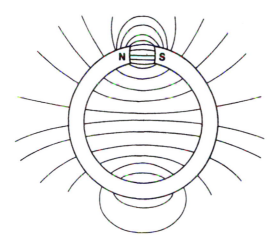

Figure 17.4. Fringing and leakage of a permanent magnet.

duction, termed B_d, is obtained as shown in Fig. 17.3. Another effect which reduces the useful magnetic field is **fringing** near the air gap and *leakage* from the sides of a magnet (Fig. 17.4).

We now turn to the properties of some common hard magnetic materials. Today, many permanent magnets are made of **Alnico** alloys, which contain various amounts of aluminum, nickel, cobalt, and iron, along with some minor constituents such as copper and titanium (Table 17.2). Their properties are improved by heat treatments (homogenization at 1250°C, fast cooling, and tempering at 600°C, **Alnico 2**). Further improvement is accomplished by cooling the alloys in a magnetic field (**Alnico 5**). The best properties are achieved when the grains are made to have a preferred orientation. This is obtained by cooling the bottom of the crucible after melting, thus forming long columnar grains with a preferred ⟨100⟩ axis in the direction of heat flow. A magnetic field parallel to the ⟨100⟩ axis yields Alnico 5-DG (directional grain).

The superior properties of heat-treated Alnico stem from the fact that during cooling and tempering of these alloys, rod-shaped iron and cobalt-rich α-precipitates are formed which are parallel to the ⟨100⟩ directions (**shape anisotropy**). These strongly magnetic precipitates are single-domain particles and are imbedded in a weakly magnetic nickel and aluminum matrix (α). Alnico alloys possess, just as iron, a ⟨100⟩ easy direction (see Fig. 17.2) and have also a cubic crystal structure. Alnico alloys are mechanically hard and brittle and can, therefore, only be shaped by casting or by pressing and sintering of metal powders.

The newest hard magnetic materials are made of neodymium–boron–iron, see Table 17.2. They possess a superior coercivity and thus a larger $(BH)_{max}$. The disadvantage is a relatively low Curie temperature of about 300°C.

Ceramic ferrite magnets, such as barium or strontium ferrite ($BaO \cdot 6Fe_2O_3$ or $SrO \cdot 6Fe_2O_3$), are brittle and relatively inexpensive. They crystallize in the form of plates with the hexagonal c-axis (which is the easy axis) perpendicular to the plates. Some preferred orientation is observed, because the flat plates arrange parallel to each other during pressing and sintering. Ferrite powder is often imbedded in plastic materials, which yields flexible magnets. They are used, for example, in the gaskets of refrigerator doors.

High carbon steel magnets with or without cobalt, tungsten, or chromium are only of historic interest. Their properties are inferior to other magnets. It is believed that the permanent magnetization of quenched steel stems from the martensite-induced internal stress, which impedes the domain walls from moving through the crystal.

Research on permanent magnetic materials still proceeds with unbroken intensity. The goal is to improve corrosion resistance, price, remanence, coercivity, magnetic ordering temperature, and processing procedures. A few examples are given here. Carbon and nitrogen are increasingly used as the metalloid in iron/rare earth magnets such as in Fe–Nd–C or in $Fe_{17}Sm_2N_x$. Nitrogen treatment of sintered $Fe_{14}Nd_2B$ raises the Curie temperature by more than 100 K. Nitriding of $F_{17}Sm_2$ (at 400° to 500°C) yields a room temperature coercivity as high as 2.4×10^6 A/m (30 kOe), a remanence of 1.5 T (15.4 kG) and a T_c of 470°C. Corrosion of the Fe–Nd–B sintered magnets is a serious problem. The principal corrosion product is $Nd(OH)_3$. The corrosion resistance can be improved by utilizing intermetallic compounds such as Fe–Nd–Al or Fe–Nd–Ga, or by applying a moisture-impervious coating. Other approaches are the rare-earth-free Co–Zr–B alloys (with or without silicon) which have a Curie temperature around 500°C and a coercivity of 5.3×10^5 A/m (6.7 kOe).

17.4. Magnetic Recording and Magnetic Memories

Magnetic recording **tapes**, disks, drums, or magnetic strips on credit cards consist of small, needlelike oxide particles about 0.1×0.5 μm in size which are imbedded in a nonmagnetic binder. The particles are too small to sustain a domain wall. They consist therefore of a single magnetic domain which is magnetized to saturation along the major axis (shape anisotropy). The elongated particles are aligned by a field during manufacturing so that their long axes are parallel with the length of the tape or the track. The most popular magnetic material has been ferrimagnetic γ-Fe_2O_3. Its coercivity is 20–28 kA/m (250–350 Oe). More recently, ferromagnetic chromium dioxide has been used having a coercivity between 40–80 kA/m (500 and 1000 Oe) and a particle size of 0.05 μm by 0.4 μm. High coercivity and high remanence prevent self-demagnetization and accidental erasure, they provide strong signals, and permit thinner coatings. A high H_c also allows tape du-

Figure 17.5. Schematic arrangement of a recording (playback) head and a magnetic tape. (Recording mode.) The gap width is exaggerated. The plastic substrate is about 25 μm thick.

plication by "contact printing." However, CrO_2 has a relatively low Curie temperature (128°C compared to 600°C for γ-Fe_2O_3). Thus, chromium dioxide tapes which are exposed to excessive heat (glove compartment!) may lose their stored information. Lately, most video tapes use cobalt-doped γ-Fe_2O_3, which has a somewhat higher Curie temperature than chromium dioxide and a coercivity of 48 kA/m (600 Oe). Most recently, iron particles have been utilized ($H_c = 120$ kA/m, i.e., 1500 Oe). This technology requires, however, a surface coating of tin to prevent coalescence of the individual particles and corrosion.

The **recording head** of a tape machine consists of a laminated electromagnet made of permalloy or soft ferrite (Table 17.1) which has an air gap about 0.3 μm wide (Fig. 17.5). The tape is passed along this electromagnet, whose fringing field redirects the spin moments of the particles in a certain pattern proportional to the current which is applied to the recording head coil. This leaves a permanent record of the signal. In the playback mode, the moving tape induces an alternating emf in the coil of the same head. The emf is amplified, filtered, and fed to a loudspeaker.

Some modern recording heads utilize conventional ferrites whose gap surfaces are coated with a micrometer-thick metal layer composed of aluminum, iron, and silicon (*Sendust*). This *metal-in-gap* (M-I-G) technology combines the superior high-frequency behavior and good wear properties of ferrites with the higher coercivity of ferromagnetic metals. Thus, fields two or three times as intense as for pure ferrites can be supported. Such high fields are necessary to record efficiently on high density (i.e., on high coercivity) media, in which tiny regions of alternating magnetization are closely spaced and should not mutually demagnetize each other.

For ultrahigh recording densities (extremely small bit sizes) the signal strength produced in the reading heads diminishes considerably. Thus, the latest head technology utilizes a thin **magnetoresistive** *element*, made out of permalloy, which senses the slight variation in resistance (about 2%) that occurs as the angle of magnetization is changed when the magnetized data bits pass beneath the head (see below). 1.8 Mbits/mm^2 have been achieved in this way. In contrast to an inductive head (see above), whose output voltage is directly proportional to the tape speed, magnetoresistivity is governed by the flux density. This is an advantage for low-speed applications (credit cards).

A note of explanation about magnetoresistance should be added: If a conductor is exposed to a magnetic field that is perpendicular to an electric field, the Lorentz force causes the paths of the drifting electrons to bend in near circular form, as explained in Section 8.5 (Hall effect). This bending leads to a decrease of the electron mobility, μ_e. Thus, because of (8.13),

$$\sigma_0 = N_e \cdot \mu_e \cdot e = \frac{1}{\rho_0}, \tag{17.2}$$

the conductivity, σ_0, decreases and the resistivity, ρ_0, increases. (N_e is the free electron concentration and e is the charge of an electron). The relative change in resistivity,

$$\frac{\Delta\rho}{\rho_0} = (\mu_e \Delta B)^2, \tag{17.3}$$

is proportional to the square of the variation in magnetic field strength, ΔB. The magnetoresistive head senses this change in magnetic field strength and, thus, yields a resistance change.

The materials for magnetoresistive read-heads have undergone a stormy development. First, "giant magnetoresistive materials" (MnFe, MnNi, NiO) having a resistance response of about 20% were discovered. Later, "**colossal magnetoresistive materials**" (lanthanum manganate, etc.) showed 50% resistance changes, allowing a further increase in areal densities.

Historically, **ferrite-core memories** used to be the dominant devices for random-access storage in computers. The principle is simple: a donut-shaped piece of ferrimagnetic material, having a nearly square-shaped hysteresis loop and a low coercivity, is threaded with a wire (Fig. 17.6(a)). If a sufficiently high current pulse is sent through this wire, then the core becomes magnetically saturated. Now, suppose the flux lines point clockwise. An opposite-directed current pulse of sufficient strength magnetizes the ferrite core counterclockwise. These two magnetization directions constitute the two possible values (0 and 1) in a binary system (see Section 8.7.12). A toroid-shaped memory core is used because a close-flux structure reacts efficiently to currents from a center wire but is not disturbed by external stray fields.

The actual configuration of a complete memory system consists of a stack of identical memory planes, each of which contains a set of wires in the x- as well as in the y-directions. The toroids are placed at the intersections (Fig. 17.6(c)). In order to switch the, say, X_3/Y_2 core from *zero* to *one*, a current proportional to half the saturation field ($H_s/2$) is sent through each of the X_3

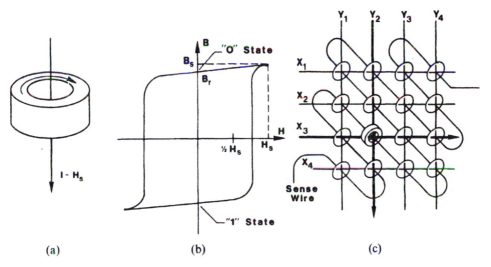

Figure 17.6. (a) Single ferrite core which is magnetized by a current-induced magnetic field; (b) square-shaped hysteresis loop of a soft ferrite memory core; and (c) one plane of a "coincident-current core memory device."

and the Y_2 wires (Fig. 17.6(b)). This provides only the X_3/Y_2 core with the necessary field for switching—the other cores stay at their present state. The information thus permanently stored can be read by again sending a current pulse proportional to $H_s/2$ through the X_3/Y_2 wires. A third wire, the sensing wire, which passes through all the cores of a given plane, senses whether or not the core was switched during the reading process. Since the reading process destroys the stored information, a special circuit is needed to rewrite the information back into the core. Ferrite-core memories (like other magnetic storage devices) do not need an electrical current to maintain their stored information. The weight/bit ratio for ferrite-core memories is, however, considerably larger than for electrical or optical storage devices. Thus, their usage is now limited to a few specialized applications.

Another magnetic storage device that has been heavily researched in the past, but is presently not much in use, is the **bubble domain memory**. Here, tiny cylindrical regions (as small as 1 μm in diameter), having a reversed magnetization compared to the matrix, are formed in thin crystals of "canted" anti-ferromagnetic oxides[4] ($BaFe_{12}O_{19}$, $YFeO_3$), or in amorphous alloyed films (Gd–Co, Gd–Fe), or in ferrimagnetic materials such as yttrium–iron–garnet ($Y_3Fe_5O_{12}$). These bubbles, whose easy axis is perpendicular to the plane of the film, can be generated, moved, replicated, or erased by electric currents. The crystals are transparent to red light. Thus, the domains can be visibly observed and optically read by the way in which

---

[4] See Section 15.1.4.

they rotate the plane of polarization of polarized light (Faraday effect in transmission, or Kerr effect in reflection). Each such domain constitutes one bit of stored information.

Thin magnetic films consisting of Co–Ni–Pt or Co–Cr–Ta or Co75–Cr13–Pt12 are frequently used in hard-disk devices. They are laid down on an aluminum substrate and are covered by a 40 nm thick carbon layer for lubrication and corrosion resistance. The coercivities range between 60–120 kA/m (750 and 1500 Oe). Thin-film magnetic memories can be easily fabricated (vapor deposition, sputtering, or electroplating), they can be switched rapidly, and they have a small unit size. Thin-film recording media are not used for tapes, however, because of their rapid wear. They have a density of 1.8 Mbits/mm^2 with a track separation of 3 μm and a bit length of 150 nm.

Magneto-optical memories possess the advantage of having no mechanical contact between medium and beam. Thus, no wear is encountered. A polycarbonate disk is covered by a certain magnetic material, such as MnBi, EuO, amorphous Gd–Co, or GdFe-garnet, that can sustain small (1 μm wide) magnetic domains which are stable against stray fields. Their spins are initially vertically aligned, see Fig 17.7(a). A strong focused laser beam heats a given domain for about a microsecond above the Curie temperature (typically 150°C to 200°C). Once the heat is turned off, the domain is made to cool in a magnetic field that is created by an electromagnet placed on the opposite side of the laser and which delivers the information to be stored. This causes the spins in the magnetic domain to re-orient according to the strength and direction of the magnetic field. For read-out, the probing laser beam, which is plane polarized, (Section 13.1.2) senses that the plane of polarization of the newly oriented magnetic domain has been rotated (Kerr effect). The degree of rotation is converted into an intensity change of the light by passing the reflected beam through a second polarizer (called an analyzer), which is rotated 90° with respect to the first polarizer, Fig. 17.7(b). In other words, the content of the stored magnetic data is equivalent to a change in the polarization direction of the reflected light and thus equivalent to a change in light intensity. Each magnetic domain represents one bit of information, for example, spin up is a "one" and spin down is a "zero". (See, in this context, Section 13.10.) Magneto-optical disks have a one thousand times larger storage density than common floppy disks and a ten times faster access time.

Despite this relatively large number of possible magnetic storage devices, semiconductor technology (Section 8.7.12) is presently preferred for short-term information storage, mainly because of price, easy handling capability, fast access time, and size. On the other hand, magnetic disks (for random access) or tapes (mainly for music recordings, etc.) are the choices for long-term, large-scale information storage, particularly since no electric energy is needed to retain the information. It should be noted in closing that tapes and floppy disks make direct contact with the recording (and playback) head, and are therefore subject to wear, whereas hard drive systems utilize a "flying

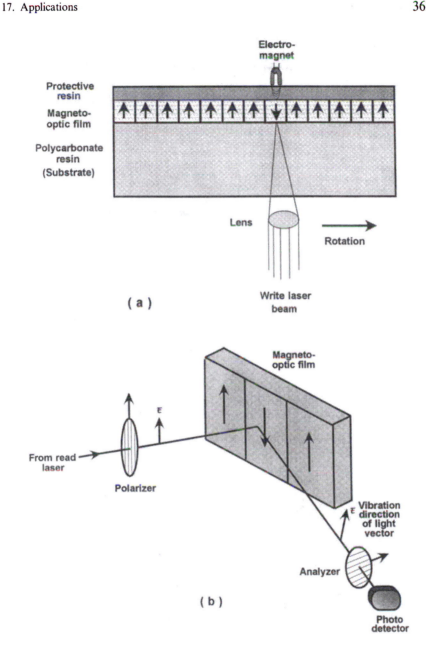

Figure 17.7. (a) Schematic representation of a magneto-optical disk in the writing mode (simplified). (b) Read-out mode of a magneto-optical device. (Polarizer and analyzer are identical devices.)

head" that hovers a few micrometers or less above the recording medium on an air cushion, caused by the high speed of the disk. On the other hand, the signal to noise ratio for contact recording is 90 dB, whereas for noncontact devices the signal to noise ratio is only 40 dB or lower. Magnetic recording is a \$140 billion annual business worldwide with a 12–14% growth rate and is said to be the biggest consumer of high-purity materials.

Problems

1. Calculate the energy expended during one full hysteresis cycle of a magnetic material having a rectangular hysteresis loop. Assume $H_c = 500$ A/m, $B_s = 2$ T and $V = 0.25$ cm^3. What units can be used?

2. Which core material should be utilized to supply a large-scale and constant magnetic field in a synchrotron? Justify your choice.

3. Pick an actual motor of your choice and find out (by analysis, by means of a data sheet, or by writing to the manufacturer) which type of electrical steel was used as the core material. Also, find out the core loss (often given in watts/1b).

4. Find from a manufacturer's data sheet the price of several qualities of electrical steel.

5. Inspect the magnets contained in a gasket of a modern refrigerator. How does the magnet work? Where are the south and north poles?

6. It was said in the text that transformers suffer eddy current as well as hysteresis losses. What other types of losses can be expected in a transformer? How can those losses be reduced?

Suggestions for Further Reading (Part IV)

R.M. Bozorth, *Ferromagnetism*, Van Nostrand, New York (1951).

S. Chikazumi, *Physics of Magnetism*, Wiley, New York (1964).

B.D. Cullity, *Introduction to Magnetic Materials*, Addison-Wesley, Reading, MA (1972).

J.D. Jackson, *Classical Electrodynamics*, Wiley, New York (1962).

D. Jiles, *Magnetism and Magnetic Materials*, Chapman and Hall, London, (1991).

E. Kneller, *Ferromagnetismus*, Springer-Verlag, Berlin (1962).

J.C. Mallionson, *The Foundation of Magnetic Recording*, Academic Press, San Diego (1987).

F.W. Sears, *Electricity and Magnetism*, Addison-Wesley, Reading, MA (1953).

T. Smit and H.P.T. Wijn, *Ferrites*, Wiley, New York (1959).

H.H. Stadelmaier and E.Th. Henig, Permanent magnetic materials—Developments during the last 18 months, *Journal of Metals*, **43** (1991), 32.

R.S. Tebble and D.J. Craik, *Magnetic Materials*, Wiley–Interscience, London (1969).

J.K. Watson, *Applications of Magnetism*, Wiley–Interscience, New York (1980).

H.P.J. Wijn (Editor), Magnetic properties of metals, *d*-elements, alloys, and compounds, in *Data in Science and Technology*, Springer-Verlag, Berlin (1991).

PART V

THERMAL PROPERTIES
OF MATERIALS

CHAPTER 18

Introduction

Heat was considered to be an invisible fluid, called *caloric*, until late into the eighteenth century. It was believed that a hot piece of material contained more caloric than a cold one and that an object would become warmer by transferring caloric into it. In the mid-1800s, Mayer, Helmholtz, and Joule discovered independently that heat is simply a form of energy. They realized that when two bodies have different temperatures, thermal energy is transferred from the hotter to the colder one when brought into contact. Count Rumford discovered, by observing the boring of cannons, that mechanical work expended in the boring process was responsible for the increase in temperature. He concluded that mechanical energy could be transformed into thermal energy. This observation lead eventually to the concept of a **mechanical heat equivalent**. Today, these results are treated in a different, more rigorous, scientific language (see next chapter).

The thermal properties of materials are important whenever heating and cooling devices are designed. Thermally induced expansion of materials has to be taken into account in the construction industry as well as in the design of precision instruments. Heat conduction plays a large role in thermal insulation, e.g., in homes, industry, and spacecraft. Some materials such as copper or silver conduct heat very well; other materials, like wood or rubber, are poor heat conductors. Good electrical conductors are generally also good heat conductors. This was discovered in 1853 by Wiedemann and Franz, who found that the ratio between heat conductivity and electrical conductivity (divided by the temperature) is essentially constant for all metals.

The thermal conductivity of materials only varies within four orders of magnitude (Fig. 18.1). This is in sharp contrast to the variation in electrical conductivity, which spans about twenty-five orders of magnitude (Fig. 7.1).

367

Figure 18.1. Room-temperature thermal conductivities for some materials.

The **thermal conductivity** of metals and alloys can be readily interpreted by making use of the electron theory that was developed in Part I of this book. The electron theory postulates that free electrons in the hot part of a metal bar pick up energy by interactions with the vibrating lattice atoms. This thermal energy is eventually transmitted to the cold end of the bar by a mechanism which we will treat in Chapter 21.

In electrical insulators, in which no *free* electrons exist, the conduction of thermal energy must occur by a different mechanism. This new mechanism was found by Einstein at the beginning of the century. He postulated the existence of **phonons**, or lattice vibration quanta, which are thought to be created in large numbers in the hot part of a solid and partially eliminated in the cold part. Transferral of heat in dielectric solids is thus linked to a flow of phonons from hot to cold.

Figure 18.1 indicates that in a transition region both electrons as well as phonons may contribute to thermal conduction. Actually, phonon-induced thermal conduction occurs even in metals, but its contribution is negligible to that of the electrons.

Another thermal property that will receive considerable attention in the following chapters is the **specific heat capacity**, as well as a related property, the molar heat capacity. Their importance can best be appreciated by the following experimental observations: Two substances with the same mass but different values for the specific heat capacity require different amounts of thermal energy to reach the same temperature. Water, for example, which has a relatively high specific heat capacity, needs more thermal energy to reach a given temperature than, say, copper or lead of the same mass.

The **molar heat capacity** is the product of the specific heat capacity and the molar mass. Its experimentally observed temperature dependence, as shown in Fig. 18.2, has stimulated various theories, among them the phonon model. Figure 18.2 shows schematically how the various theories for the interpretation of the heat capacity compare with the experimental findings. We will discuss these models in the chapters to come.

Figure 18.2. Schematic representation of the temperature dependence of the molar heat capacity—experimental, and according to four models.

CHAPTER 19

Fundamentals of Thermal Properties

Before we discuss the atomistic and quantum mechanical theories of the thermal properties of materials, we need to remind the reader on some relevant fundamental concepts and definitions which you might have been exposed to before in courses of physics and thermodynamics.

19.1. Heat, Work, and Energy

When two bodies of different temperatures are brought in contact with each other, heat, Q, flows from the hotter to the colder substance. Actually, an increase in temperature can be achieved in a number of ways, such as by mechanical work (friction), electrical work (resistive heating), radiation, or by the just-mentioned direct contact with a hotter medium. The change in energy, ΔE, of a "system" can be expressed by the **first law of thermodynamics**,

$$\Delta E = W + Q, \tag{19.1}$$

where W is the work done on the system and Q is the heat received by the system from the environment. The focus of this and subsequent chapters is on the thermal properties of materials. Thus, we limit our considerations to processes for which W can be considered to be zero, so that

$$\Delta E = Q. \tag{19.1a}$$

Energy, work, and **heat** have the same unit. The SI unit is the **joule** (J), which is related to the now obsolete thermomechanical calorie (cal) by

$$1 \text{ cal} = 4.184 \text{ J}, \tag{19.2}$$

i.e.,

$$1\,\text{J} = 0.239\,\text{cal.} \tag{19.2a}$$

A technique which links thermal energy and mechanical energy was proposed by Joule in 1850. The experiment involves rotating paddles which raise the temperature of a given amount of water by means of friction. The paddles are driven by the mechanical work provided by descending weights.

19.2. Heat Capacity, C'

Different substances need different amounts of heat to raise their temperatures by a given temperature interval. For example, it takes 4.18 J to raise 1 g of water by 1 K. But the same heat raises the temperature of 1 g of copper by about 11 K. In other words, water has a large heat capacity compared to copper. (The large heat capacity of water is, incidentally, the reason for the balanced climate in coastal regions and the heating of north European countries by the Gulf Stream.)

The heat capacity, C', is the amount of heat, dQ, which needs to be transferred to a substance in order to raise its temperature by a certain temperature interval. Units for the heat capacity are J/K.

The heat capacity is not defined uniquely, i.e., one needs to specify the conditions under which the heat is added to the system. Even though several choices for the heat capacities are possible, one is generally interested in only two: the **heat capacity at constant volume**, C'_v, and the **heat capacity at constant pressure**, C'_p. The former is the most useful quantity, because C'_v is obtained immediately from the energy of the system. The heat capacity at constant volume is defined as

$$C'_v = \left(\frac{\partial E}{\partial T}\right)_v . \tag{19.3}$$

On the other hand, it is much easier to measure the heat capacity of a solid at constant pressure than at constant volume. Fortunately, the difference between C'_p and C'_v *for solids* vanishes at low temperatures and is only about 5% at room temperature. C'_v can be calculated from C'^1_p if the volume expansion coefficient, α, and the compressibility, κ, of a material are known, by applying

$$C'_v = C'_p - \frac{\alpha^2 TV}{\kappa}, \tag{19.4}$$

---

[1] The heat capacity at constant pressure is defined as $C'_p = (\partial H/\partial T)_p$, where $H = U + P \cdot V$ is the enthalpy, and U is the internal energy.

where V is the volume of the solid. Equation (19.4) is derived in textbooks on thermodynamics.

19.3. Specific Heat Capacity, c

The **specific heat capacity** is the heat capacity *per unit mass*

$$c = \frac{C'}{m} \tag{19.5}$$

where m is the mass of the system. It is a materials constant and it is temperature-dependent. Characteristic values for the specific heat capacity (c_v and c_p) are given in Table 19.1. The unit of the specific heat capacity is $J/g \cdot K$. We note from Table 19.1 that values for the specific heat capacities of solids are considerably smaller than the specific heat capacity of water.

Combining (19.1a), (19.3), and (19.5) yields

$$\Delta E = Q = m\Delta T c_v, \tag{19.6}$$

which expresses that the thermal energy (or heat) which is transferred to a system equals the product of mass, increase in temperature, and specific heat capacity.

19.4. Molar Heat Capacity, C_v

A further useful materials constant is the **heat capacity per mole** (i.e., per amount of substance of a phase, n). It compares materials that contain the

Table 19.1. Experimental Thermal Parameters of Various Substances at Room Temperature and Ambient Pressure.

Substance	Specific heat capacity (c_p) $\left(\dfrac{J}{g \cdot K}\right)$	Molar (atomic) mass $\left(\dfrac{g}{mol}\right)$	Molar heat capacity (C_p) $\left(\dfrac{J}{mol \cdot K}\right)$	Molar heat capacity (C_v) $\left(\dfrac{J}{mol \cdot K}\right)$
Al	0.897	27.0	24.25	23.01
Fe	0.449	55.8	25.15	24.68
Ni	0.456	58.7	26.8	24.68
Cu	0.385	63.5	24.48	23.43
Pb	0.129	207.2	26.85	24.68
Ag	0.235	107.9	25.36	24.27
C (graphite)	0.904	12.0	10.9	9.20
Water	4.184	18.0	75.3	

same number of molecules or atoms. The molar heat capacity is obtained by multiplying the specific heat capacity, c_v (or c_p), by the molar mass, M (see Table 19.1):

$$C_v = \frac{C_v'}{n} = c_v \cdot M. \tag{19.7}$$

The units are J/mol $\cdot$ K. The amount of substance (in mol) is

$$n = N/N_0, \tag{19.7a}$$

where N is the number of particles (atoms, molecules, etc.), and N_0 is the Avogadro constant $(N_0 = 6.022 \times 10^{23} \text{ mol}^{-1})$.

We see from Table 19.1 that the room-temperature molar heat capacity at constant volume is approximately 25 J/mol $\cdot$ K (6 cal/mol $\cdot$ K) for most solids. This was experimentally discovered in 1819 by **Dulong and Petit**. We shall attempt to interpret this interesting result in a later section.

The experimental molar heat capacities for some materials are depicted in Fig. 19.1 as a function of temperature. We notice that some materials, such as carbon, reach the Dulong–Petit value of 25 J/mol $\cdot$ K only at high temperatures. Some other materials, such as lead, reach 25 J/mol $\cdot$ K at relatively low temperatures.

All heat capacities are zero at $T = 0$ K. The C_v values near $T = 0$ K climb in proportion to T^3 and reach 96% of their final value at a temperature θ_D, which is defined to be the Debye temperature. We shall see later that θ_D is an approximate dividing point between a high-temperature region, where clas-

Figure 19.1. Temperature dependence of the molar heat capacity, C_v, for some materials.

Table 19.2. Debye
Temperatures of Some
Materials.

Substance	θ_D (K)
Pb	95
Au	170
Ag	230
W	270
Cu	340
Fe	360
Al	375
Si	650
C	1850
GaAs	204
InP	162
InAs	280

sical models can be used for the interpretation of C_v, and a low-temperature region, where quantum theory needs to be applied. Selected Debye temperatures are listed in Table 19.2.

19.5. Thermal Conductivity, K

Heat conduction (or thermal conduction) is the transfer of thermal energy from a hot body to a cold body when both bodies are brought into contact. For best visualization we consider a bar of a material of length x whose ends are held at different temperatures. The heat that flows through a cross section of the bar divided by time and area, (i.e., the **heat flux**, J_Q) is proportional to the temperature gradient, dT/dx. The proportionality constant is called the **thermal conductivity**, K (or λ). We thus write

$$J_Q = -K\frac{dT}{dx}.$$

(19.8)

The negative sign indicates that the heat flows from the hot to the cold end (Fourier Law, 1822). Units for the heat conductivity are $(J/m \cdot s \cdot K)$ or $(W/m \cdot K)$. The heat flux, J_Q, is measured in $(J/m^2 \cdot s)$. Table 19.3 gives some characteristic values for K. The thermal conductivity decreases slightly with increasing temperature. For example, K for copper decreases by 20% within a temperature span of 1000°C. In the same temperature region, K for iron decreases by 10%.

Table 19.3. Thermal Conductivities at
Room Temperature.[a]

Substance	$K\left(\dfrac{W}{m \cdot K}\right) \equiv \left(\dfrac{J}{s \cdot m \cdot K}\right)$
Diamond, type IIa	2.3×10^3
SiC	4.9×10^2
Silver	4.29×10^2
Copper	4.01×10^2
Aluminum	2.37×10^2
Silicon	1.48×10^2
Brass (leaded)	1.2×10^2
Iron	8.02×10^1
GaAs	5×10^1
Ni-Silver[b]	2.3×10^1
Al_2O_3 (sintered)	3.5×10^1
SiO_2 (fused silica)	1.4
Concrete	9.3×10^{-1}
Soda-lime glass	9.5×10^{-1}
Water	6.3×10^{-1}
Polyethylene	3.8×10^{-1}
Teflon	2.25×10^{-1}
Snow (0°C)	1.6×10^{-1}
Wood (oak)	1.6×10^{-1}
Engine Oil	1.45×10^{-1}
Sulfur	2.0×10^{-2}
Cork	3×10^{-2}
Glass wool	5×10^{-3}
Air	2.3×10^{-4}

[a] See also Figure 18.1. Source: Handbook of Chemistry and Physics, CRC Press. Boca Raton, FL (1994).
[b] 62% Cu, 15% Ni, 22% Zn

19.6. The Ideal Gas Equation

Free electrons in metals and alloys can often be considered to behave like an ideal gas. An ideal gas is an abstraction which is frequently used in thermodynamics. It is usually defined to be a gas whose density is low enough in order for it to obey the equation

$$PV = nRT, \tag{19.9}$$

where P is the pressure of the gas, V is its volume, n is the amount of substance, T is the thermodynamic (absolute) temperature, R is the universal gas constant, and k_B is the Boltzmann constant. The gas constant is

$$R = k_B N_0 = 8.314 \ (\text{J/mol} \cdot \text{K})$$

$$= 1.986 \ (\text{cal/mol} \cdot \text{K}). \tag{19.10}$$

Equation (19.9) is a combination of two experimentally obtained thermo-dynamic laws: One, discovered by Boyle and Mariotte ($PV = \text{const.}$ at constant T), and the other, discovered by Gay-Lussac ($V \sim T$, at constant P). The reader who has taken classes in physics or thermodynamics is undoubtedly familiar with these equations.

19.7. Kinetic Energy of Gases

In the chapters to come, we need to know the kinetic energy of atoms, molecules, or electrons at a given temperature from a classical point of view. The calculation that is summarized below is usually contained in textbooks on thermodynamics.

We commence by quoting the number of molecules in a gas that interact in the unit time, t, with the end face of unit area of a bar which has the length dx. We assume that, because of thermal agitation, one-third of the particles move in the x-directions, i.e., one-sixth in the positive x-direction. The volume element, shown in Fig. 19.2, is

$$dV = A\,dx = Av\,dt, \tag{19.11}$$

where A is the unit area and v is the velocity of the particles that fly in the x-direction. The number of particles reaching the end face is naturally proportional to the number of particles, n_v, in the given volume, i.e.,

$$z' = \tfrac{1}{6} n_v \cdot dV = \tfrac{1}{6} n_v A v\,dt.$$

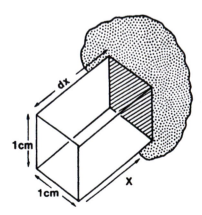

Figure 19.2. Diagram for the derivation of the kinetic energy of gases.

The number of particles per unit time and unit area that hit the end face is, consequently,

$$z = \tfrac{1}{6} n_v v, \tag{19.12}$$

where

$$n_v = \frac{N}{V} \tag{19.13}$$

is the *number of particles per unit volume*. Each particle transfers the momentum $2mv$ during its collision with the wall and subsequent reflection. The *momentum per unit time and unit area* is then

$$p^* = z2mv = \tfrac{1}{6} n_v v 2mv = \frac{1}{3}\frac{N}{V}mv^2. \tag{19.14}$$

This yields, for the pressure,

$$P = \frac{F}{A} = \frac{ma}{A} = \frac{d(mv)/dt}{A} = \frac{dp/dt}{A} = p^* = \frac{1}{3}\frac{N}{V}mv^2. \tag{19.15}$$

With

$$PV = nRT = nk_B N_0 T = Nk_B T \tag{19.16}$$

(see (19.9), (19.7a), and (19.10)), we obtain from (19.15)

$$PV = \tfrac{1}{3} Nmv^2 = k_B NT. \tag{19.17}$$

Inserting

$$E_{\text{kin}} = \tfrac{1}{2} mv^2 \tag{19.18}$$

into (19.17) yields

$$k_B NT = \tfrac{1}{3} N 2 \tfrac{1}{2} mv^2 = \tfrac{2}{3} N E_{\text{kin}}, \tag{19.19}$$

which finally yields the **kinetic energy of a particle**,

$$E_{\text{kin}} = \tfrac{3}{2} k_B T. \tag{19.20}$$

A more precise calculation, which considers the mutual collisions of the particles, and thus a velocity distribution, replaces the kinetic energy in (19.20) with an average kinetic energy.

Problems

Note: The problems in this chapter contain engineering applications in order to make the student aware of the importance of thermal properties in daily life.

1. Calculate the number of gas molecules that are left in an ultrahigh vacuum of 10^{-9} Pa ($\sim 7.5 \times 10^{-12}$ Torr) at room temperature.

2. Calculate the rate of heat loss per unit area in a 5 mm thick window glass when the exterior temperature is 0°C and the room temperature is 20°C. Compare your result with the heat loss in an aluminum and a wood frame of 10 mm thickness. How can you decrease the heat loss through the window?

3. A block of copper, whose mass is 100 g, is quenched directly from an annealing furnace into a 200 g glass container that holds 500 g of water. What is the temperature of the furnace when the water temperature rises from 0° to 15°C? ($c_{glass} = 0.5$ J/g · K.)

4. Explain in simple terms why wood has a smaller heat conductivity than copper.

5. What are the implications for the semiconductor industry that silicon has a relatively good heat conductivity?

6. Why is the fiberglass insulation used for buildings, etc., loose rather than compact? (*Hint*: Compare K for glass and air. Discuss also heat convection.)

7. Find in a handbook the relationship between J and BTU.

CHAPTER 20

Heat Capacity

20.1. Classical (Atomistic) Theory of Heat Capacity

This section attempts to interpret the thermal properties of materials using atomistic concepts. In particular, an interpretation of the experimentally observed molar heat capacity at high temperatures, $C_v = 25$ (J/mol · K) that is, 6 (cal/mol · K), is of interest.

We postulate that each atom in a crystal is bound to its site by a harmonic force. A given atom is thought to be capable of absorbing thermal energy, and in doing so it starts to vibrate about its point of rest. The amplitude of the oscillation is restricted by electrostatic repulsion forces of the nearest neighbors. The extent of this thermal vibration is therefore not more than 5 or 10% of the interatomic spacing, depending on the temperature. In short, we compare an atom with a sphere which is held at its site by two springs (Fig. 20.1(a)). The thermal energy that a harmonic oscillator of this kind can absorb is proportional to the absolute temperature of the environment. The proportionality factor has been found to be the Boltzmann constant, k_B (see below). The **average energy of the oscillator** is then

$$E = k_B T. \tag{20.1}$$

Now, solids are three-dimensional. Thus, a given atom in a cubic crystal also responds to the harmonic forces of lattice atoms in the other two directions. In other words, it is postulated that each atom in a cubic crystal represents three oscillators (Fig. 20.1(b)), each of which absorbs the thermal energy $k_B T$. Therefore, the **average energy per atom** is

$$E = 3k_B T. \tag{20.2}$$

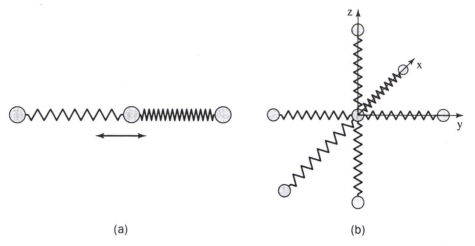

(a)   (b)

Figure 20.1. (a) A one-dimensional harmonic oscillator and (b) a three-dimensional harmonic oscillator.

We note in passing that the same result is obtained by using the kinetic theory of gases. It was shown in (19.20) that the average kinetic energy of a particle (or in the present case, an atom) is

$$E_{\text{kin}} = \tfrac{3}{2} k_{\text{B}} T. \qquad (20.3)$$

Now, each elastic vibration in a solid involves not only kinetic energy but also potential energy, which has the same average magnitude as the kinetic energy. The total energy of a vibrating lattice atom is thus

$$E = 2 \cdot \tfrac{3}{2} k_{\text{B}} T, \qquad (20.4)$$

which is the result of (20.2).

We consider now all N_0 atoms per mole. Then, the *total internal energy per mole* is

$$E = 3 N_0 k_{\text{B}} T. \qquad (20.5)$$

Finally, the **molar heat capacity** is given by combining (19.3), (19.7), and (20.5), which yields

$$\boxed{C_{\text{v}} = \left(\frac{\partial E}{\partial T}\right)_{\text{v}} = 3 N_0 k_{\text{B}}.} \qquad (20.6)$$

Inserting the numerical values for N_0 and k_{B} into (20.6) yields

$$C_{\text{v}} = 25 \ \text{J/mol} \cdot \text{K} \quad \text{or} \quad 5.98 \ \text{cal/mol} \cdot \text{K},$$

quite in agreement with the experimental findings at high temperatures (Figs. 18.2 and 19.1).

It is satisfying to see that a simple model involving three harmonic oscillators per atom can readily explain the experimentally observed heat capacity. However, one shortcoming is immediately evident: the calculated molar heat capacity turned out to be temperature-independent, according to (20.6), and also independent of the material. This discrepancy with the observed behavior (see Fig. 18.2) was puzzling to scientists in the 19th century and had to await quantum theory to be properly explained.

20.2. Quantum Mechanical Considerations— the Phonon

20.2.1. Einstein Model

Einstein postulated, in 1907, that the energies of the above-mentioned classical oscillators should be quantized, i.e., he postulated that only certain vibrational modes should be allowed, quite in analogy to the allowed energy states of electrons. These lattice vibration quanta were called **phonons**.

The term *phonon* stresses an analogy with *electrons* or *photons*. As we know from Chapter 2, photons are quanta of electromagnetic radiation, i.e., photons describe (in the appropriate frequency range) classical light. Phonons, on the other hand, are quanta of the ionic displacement field, which (in the appropriate frequency range) describe classical sound.

The phonon describes the *particle* nature of an oscillator. A phonon has, in analogy to the de Broglie relation (2.3), the momentum $p = h/\lambda$.

Furthermore, Einstein postulated a particle–wave duality. This suggests **phonon waves** which propagate through the crystal with the speed of sound. Phonon waves are *not* electromagnetic waves: they are *elastic waves*, vibrating in a longitudinal and/or in a transversal mode.

In analogy to the electron case shown in Part I of this book, one can describe the properties of phonons in terms of band diagrams, Brillouin zones, or density of states curves. Small differences exist, however. For example, the energy in the band diagram of an electron is replaced in a phonon band diagram by the vibrational frequency, ω, of the phonon. The branches in the phonon band diagram are sinusoidal in nature (compared to parabolic in the free electron case). The individual phonon bands are no longer called valence or conduction bands, but more appropriately **acoustic bands** and **optical bands**, mainly because the frequencies in which the branches are situated are in the acoustical and optical ranges, respectively. The density of states or, better, the *density of vibrational modes*, $D(\omega)$, for the phonon case is defined so that $D(\omega) \cdot d\omega$ is the number of modes whose frequencies lie in the interval ω and $\omega + d\omega$. For a continuous medium the **density of modes** is

$$D(\omega) = \frac{3V}{2\pi^2} \frac{\omega^2}{v_s^3}, \tag{20.7}$$

Figure 20.2. Allowed energy levels of a phonon: (a) average thermal energy at low temperatures and (b) average thermal energy at high temperatures.

where v_s is the sound velocity. This equation can be derived quite similarly as demonstrated in Section 6.3.

The allowed energies of a single oscillator are

$$E_n = n\hbar\omega, \tag{20.8}$$

similarly as in Section 4.2, where n is an integer.[2] A schematic energy level diagram for the allowed phonon energies is shown in Fig. 20.2.

One important difference between phonons and electrons needs to be emphasized. Phonons are created by *raising* the temperature, and eliminated by lowering it, i.e., the number of phonons is *not* conserved, as we shall show momentarily. (In contrast to this, the number of electrons is constant.) Einstein postulated that with increasing temperature more and more phonons are created, each of which has the same energy, $\hbar\omega$, or the same frequency of vibration, ω. The average **number of phonons**, $\bar{N}_{ph}$, at a given temperature was found by **Bose** and **Einstein** to obey a special type of **statistics**:

$$\bar{N}_{ph} = \frac{1}{\exp\left(\dfrac{\hbar\omega}{k_B T}\right) - 1}. \tag{20.10}$$

This equation is similar in form to the Fermi distribution function (6.1).

We note in passing that for high phonon energies, $\hbar\omega$, the exponential term in (20.10) becomes large when compared to unity so that the number of phonons can be approximated by Boltzmann statistics, i.e., by the laws of classical thermodynamics,

$$\bar{N}_{ph} \approx e^{-\hbar\omega/k_B T}. \tag{20.11}$$

---

[2] Since the ground state ($n = 0$) still has a zero-point energy of $\frac{1}{2}\hbar\omega$, we should, more appropriately, write

$$E_n = n\hbar\omega + \tfrac{1}{2}\hbar\omega = (n + \tfrac{1}{2})\hbar\omega. \tag{20.9}$$

The zero-point energy is, however, of no importance for the present considerations.

We already made a similar statement in Section 6.2. We also see from (20.10) that the number of phonons decreases rapidly when the temperature approaches 0 K.

The average energy of an isolated oscillator is then the average number of phonons times the energy of a phonon:

$$E_{osc} = \hbar\omega \bar{N}_{ph} = \frac{\hbar\omega}{\exp\left(\dfrac{\hbar\omega}{k_B T}\right) - 1}.$$

(20.12)

The *thermal energy of a solid* can now be calculated by taking into account (as in Section 20.1) that a mole of a substance contains $3N_0$ oscillators. This yields for the thermal energy per mole

$$E = 3N_0 \frac{\hbar\omega}{\exp\left(\dfrac{\hbar\omega}{k_B T}\right) - 1}.$$

(20.13)

The **molar heat capacity** is, finally,

$$C_v = \left(\frac{\partial E}{\partial T}\right)_v = 3N_0 k_B \left(\frac{\hbar\omega}{k_B T}\right)^2 \frac{\exp\left(\dfrac{\hbar\omega}{k_B T}\right)}{\left(\exp\left(\dfrac{\hbar\omega}{k_B T}\right) - 1\right)^2}.$$

(20.14)

We discuss C_v for two special temperature regions. For large temperatures the approximation $e^x \simeq 1 + x$ can be applied, which yields $C_v \simeq 3N_0 k_B$ (see Problem 5) in agreement with (20.6), i.e., we obtain the classical Dulong–Petit value. For $T \to 0$, C_v approaches zero, again in agreement with experimental observations. Thus, the temperature dependence of C_v is now in qualitative accord with the experimental findings. One minor discrepancy has to be noted however: At very small temperatures the experimental C_v decreases by T^3, as stated in Section 19.4. The Einstein theory predicts, instead, an exponential reduction. The Debye theory, which we shall discuss below, alleviates this discrepancy by postulating that the individual oscillators interact with each other.

By inspecting (20.14), we observe that this relation contains only one adjustable parameter, namely, the angular frequency, which we shall redesignate for this particular case by ω_E. By fitting (20.14) to experimental curves, the frequency of the phonon waves can be obtained. For copper, the angular frequency ω_E has been found in this way to be 2.5×10^{13} s^{-1}, which yields $\nu_E = 4 \times 10^{12}$ s^{-1}. It is customary to call frequencies up to 10^5 s^{-1} sound waves, frequencies between 10^5 and 10^9 s^{-1} ultrasonics, and frequencies above 10^9 s^{-1} thermal waves. The Einstein frequency is thus situated very appropriately in the thermal wave region.

Occasionally, the **Einstein temperature**, θ_E, is quoted, which is defined by equating the phonon energy

$$\hbar\omega_E = k_B\theta_E, \tag{20.15}$$

which yields

$$\theta_E = \frac{\hbar\omega_E}{k_B}. \tag{20.16}$$

Characteristic values for θ_E are between 200 K and 300 K. From the above-quoted ω_E value for copper, θ_E^{Cu} can be calculated to be 240 K.

20.2.2. Debye Model

We now refine the Einstein model by taking into account that the atoms in a crystal interact with each other. Consequently, the oscillators are thought to vibrate interdependently. We recall that the Einstein model considered only *one* frequency of vibration, ω_E. When interactions between the atoms occur, many more frequencies are thought to exist, which range from about the Einstein frequency down to frequencies of the acoustical modes of oscillation. We postulate that these vibrational modes are quantized (Fig. 20.2). The total displacement of a given atom in a crystal during the oscillation is found by summing up all vibrational modes. This has been done by Debye, who modified the Einstein equation (20.13) by replacing the $3N_0$ oscillators of a single frequency with the number of modes in a frequency interval, $d\omega$, and by summing up over all allowed frequencies. The total energy of vibration for the solid is then

$$E = \int E_{osc}D(\omega)\,d\omega, \tag{20.17}$$

where E_{osc} is the energy of one oscillator given in (20.12), and $D(\omega)$ is the density of modes given in (20.7). Inserting (20.7) and (20.12) into (20.17) yields

$$E = \frac{3V}{2\pi^2 v_s^3}\int_0^{\omega_D} \frac{\hbar\omega^3}{\exp\left(\dfrac{\hbar\omega}{k_B T}\right) - 1}\,d\omega. \tag{20.18}$$

The integration is performed between $\omega = 0$ and a cutoff frequency, called the Debye frequency, ω_D, (Section 19.4) which is determined by postulating that the total number of modes must be equal to the number of degrees of freedom.

The molar heat capacity, C_v, is obtained, as usual, by performing the derivative of (20.18) with respect to temperature. This yields

$$C_v = \frac{3V\hbar^2}{2\pi^2 v_s^3 k_B T^2}\int_0^{\omega_D} \frac{\omega^4 \exp\left(\dfrac{\hbar\omega}{k_B T}\right)}{\left(\exp\left(\dfrac{\hbar\omega}{k_B T}\right) - 1\right)^2}\,d\omega \tag{20.19}$$

or

$$C_v^{ph} = 9k_B N_0 \left(\frac{T}{\theta_D}\right)^3 \int_0^{\theta_D/T} \frac{x^4 e^x}{(e^x - 1)^2} \, dx, \qquad (20.20)$$

where

$$x = \frac{\hbar\omega}{k_B T} \qquad (20.21)$$

varies with the angular frequency, ω, and

$$\theta_D = \frac{\hbar\omega_D}{k_B} \qquad (20.22)$$

is called the **Debye temperature**. Values for θ_D can be obtained again by curve-fitting, particularly at low temperatures. They have been listed in Table 19.2. For low temperatures, i.e., for $T \ll \theta_D$, the upper limit of the integral in (20.20) can be approximated by infinity. Then (20.20) can be evaluated and it becomes

$$C_v = \frac{12\pi^4}{5} N_0 k_B \left(\frac{T}{\theta_D}\right)^3. \qquad (20.23)$$

From both equations, (20.20) as well as (20.23), it can be seen that C_v decreases proportionally to T^3 at low temperatures, which is quite in agreement with the experimental observations.

In summary, the main difference between the two theories is that the Debye model takes the low frequency modes into account, whereas the Einstein model does not. We have to realize, however, that the excitation of oscillators at low temperatures occurs only with a small probability, because at low temperatures only a few oscillators can be raised to the next higher level. This is a consequence of the fact that the energy difference between levels is comparatively large for the available small thermal energies, as schematically illustrated in Fig. 20.2.

It should be noted that even (20.20) is only an approximation, because the underlying model does not take into consideration the periodicity of the atoms in a crystal lattice. Thus, a refinement of the Debye model needs to utilize the actual density of modes function $D(\omega)$ for a given material. This has been done by scientists with good success. Equation (20.20) is, however, a fairly good approximation (see Fig. 18.2).

20.3. Electronic Contribution to the Heat Capacity

In the previous sections we have digressed considerably from the principal theme of this book, namely, the description of the *electronic* properties of

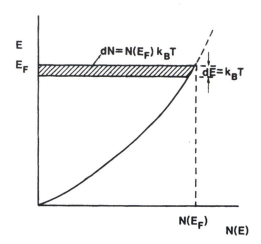

Figure 20.3. Population density as a function of energy for a metal. The electrons within the shaded area below E_F can be excited by a thermal energy $k_B T$.

materials. We now return to our main topic by discussing the contributions that the electrons provide to the specific heat. We will quickly see that this contribution is relatively small compared to that of the phonons.

First, we need to remember that only the kinetic energy of the *free* electrons can be raised with increasing temperature. Consequently, our present discussion is restricted to metals and alloys which have, as we know, partially filled bands and thus free electrons. Second, we need to remember that only those electrons which lie within an energy interval $k_B T$ of the Fermi energy can be excited in sufficient numbers into higher states, because only these electrons find empty energy states after their excitation. We know that the number, dN, of these excitable electrons depends on the population density of the metal under consideration (Section 6.4). In other words, dN is the product of the population density at the Fermi level, $N(E_F)$, and the energy interval $k_B T$, as indicated by the shaded area in Fig. 20.3.

We postulate that the electrons which are excited by thermal energy behave like a monatomic gas. We have already shown in (19.20) that the mean kinetic energy of gas molecules, or in the present case, the mean kinetic energy of the electrons above the Fermi energy is $\frac{3}{2} k_B T$. Thus, the thermal energy at a given temperature is

$$E_{\text{kin}} = \tfrac{3}{2} k_B T \, dN = \tfrac{3}{2} k_B T N(E_F) k_B T. \tag{20.24}$$

The heat capacity of the electrons is then, as usual,

$$C_v^{\text{el}} = \left(\frac{\partial E}{\partial T}\right)_v = 3k_B^2 T N(E_F). \tag{20.25}$$

We need now an expression for $N(E_F)$. We obtain this by combining (6.8) and (6.11) for $E < E_F$ (see Fig. 20.3 and Problem 9), which yields

$$N(E_F) = \frac{3N^*}{2E_F} \left(\frac{\text{electrons}}{\text{J}} \right), \tag{20.26}$$

where N^* is the number of electrons which have an energy equal to or smaller than E_F (Section 6.4). Inserting (20.26) into (20.25) yields

$$C_v^{el} = \frac{9}{2} \frac{N^* k_B^2 T}{E_F} \left(\frac{\text{J}}{\text{K}} \right). \tag{20.27}$$

So far, we assumed that the thermally excited electrons behave like a classical gas. In reality, the excited electrons must obey the Pauli principle. If this is taken into consideration properly, (20.27) changes slightly and reads

$$\boxed{C_v^{el} = \frac{\pi^2}{2} \frac{N^* k_B^2 T}{E_F} = \frac{\pi^2}{2} N^* k_B \frac{T}{T_F}.} \tag{20.28}$$

Let us assume now a monovalent metal in which we can reasonably assume one free electron per atom (see Part I). Then, N^* can be equated to the number of atoms per mole, N_0, and (20.28) becomes the heat capacity per mole

$$\boxed{C_v^{el} = \frac{\pi^2}{2} \frac{N_0 k_B^2 T}{E_F} = \frac{\pi^2}{2} N_0 k_B \frac{T}{T_F} \left(\frac{\text{J}}{\text{K} \cdot \text{mol}} \right).} \tag{20.29}$$

We see from (20.29) that C_v^{el} is a linear function of the temperature and is zero at $T = 0$ K, quite in agreement with the experimental observations, see Fig. 18.2. The room-temperature contribution of the electronic specific heat to the total specific heat is less than 1% (see Problem 3). There are, however, two temperature regions where the electronic specific heat plays an appreciable role. This is at very low temperatures, i.e., at $T < 5$ K (see Fig. 18.2). Second, we have learned in the previous sections that the lattice heat capacity levels off above the Debye temperature. Thus, the electron heat capacity can give at high temperatures a small contribution to the Dulong–Petit value.

An interesting aspect is added: Equation (20.25) may be rewritten in the following form:

$$C_v^{el} = \gamma T, \tag{20.30}$$

where

$$\gamma = 3k_B^2 N(E_F). \tag{20.31}$$

Furthermore, (20.20) can be rewritten as

$$C_v^{ph} = \beta \dot{T}^3. \tag{20.32}$$

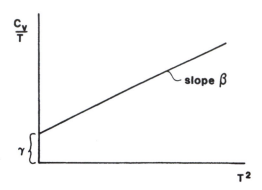

Figure 20.4. Schematic representation of an experimental plot of C_v/T versus T^2.

Below the Debye temperature, the heat capacity of metals is the sum of electron and phonon contributions, i.e.,

$$C_v^{tot} = C_v^{el} + C_v^{ph} = \gamma T + \beta T^3, \qquad (20.33)$$

which yields

$$\frac{C_v^{tot}}{T} = \gamma + \beta T^2. \qquad (20.34)$$

A plot of experimental values for C_v^{tot}/T versus T^2 provides the materials constants γ (intercept) and β (slope), see Fig. 20.4. Heat capacity measurements thus serve as a means to obtain the electron population density at the Fermi surface by using (20.31).

Some calculated and observed values for γ are given in Table 20.1. From the slight discrepancy between observed and free-electron γ-values, a **thermal effective mass** can be calculated, which is defined as

$$\frac{m_{th}^*}{m_0} = \frac{\gamma(\text{obs.})}{\gamma(\text{calc.})}. \qquad (20.35)$$

Table 20.1. Calculated and Observed Values for the Constant γ, see (20.31).

Substance	γ, observed $\left(\dfrac{J}{mol \cdot K^2}\right)$	γ, calculated $\left(\dfrac{J}{mol \cdot K^2}\right)$	$\dfrac{m_{th}^*}{m_0}$
Ag	0.646×10^{-3}	0.645×10^{-3}	1.0
Al	1.35×10^{-3}	0.912×10^{-3}	1.48
Au	0.729×10^{-3}	0.642×10^{-3}	1.14
Na	1.3×10^{-3}	0.992×10^{-3}	1.31
Fe	4.98×10^{-3}	—	—
Ni	7.02×10^{-3}	—	—

The deviation between the two γ-values is interpreted to stem from neglecting electron–phonon and electron–electron interactions.

Problems

1. How many electrons (in percent of the total number of electrons per mole) lie $k_B T$ (eV) below the Fermi energy? Take $E_F = 5$ eV and $T = 300$ K.

2. Calculate C_v at high temperatures (500 K) by using the quantum mechanical equation derived by Einstein. Assume an Einstein temperature of 250 K, and convince yourself that C_v approaches the classical value at high temperatures.

3. Calculate the electronic specific heat for $E_F = 5$ eV and $T = 300$ K. How does your result compare with the experimental value of 25 (J/mol K)?

4. Calculate the population density at the Fermi level for a metal whose electronic specific heat at 4 K was measured to be 8.37×10^{-3} (J/mol K).

5. Confirm that (20.14) reduces for large temperatures to the Dulong–Petit value.

6. At what temperature would the electronic contribution to C_v of silver eventually become identical to the Dulong–Petit value? (*Hint*: Use proper units for the heat capacity! Take $E_F = 5$ eV, $N_f = 10^{28}$ el/m^3.)

7. Show that for small temperatures (20.20) reduces to (20.23) and that for large temperatures (20.20) reduces to the Dulong–Petit value.

8. Derive (20.26) for $E < E_F$ as shown in Fig. 20.3 by combining (6.8) and (6.11) and eliminating the Planck constant.

9. *Computer problem.* Plot the Einstein equation (20.14) and the Debye equation (20.20) as a function of temperature by utilizing different values for ω_E, ω_D, and θ_D. Why can (20.23) not be used in the entire temperature range?

CHAPTER 21

Thermal Conduction

We stated in Chapter 19 that heat conduction can be described as the transfer of thermal energy from the hot to the cold part of a piece of material. We shall discuss now the mechanisms which are involved in this transfer of thermal energy.

We postulate that the heat transfer in solids may be provided by *free electrons* as well as by *phonons*. We understand immediately that in insulators, which do not contain any free electrons, the heat must be conducted exclusively by phonons. In metals and alloys, on the other hand, the heat conduction is dominated by electrons because of the large number of free electrons in metals. Thus, the phonon contribution is usually neglected in this case.

One particular point should be clarified right at the beginning. Electrons in metals travel in equal numbers from hot to cold and from cold to hot in order that the charge neutrality be maintained. Now, the electrons in the hot part of a metal possess and transfer a high energy. In contrast to this, the electrons in the cold end possess and transfer a lower energy. The heat transferred from hot to cold is thus proportional to the difference in the energies of the electrons.

The situation is quite different in phonon conductors. We know from Section 20.2.1 that the number of phonons is larger at the hot end than at the cold end. Thermal equilibrium thus involves in this case a net transfer of phonons from the hot into the cold part of a material.

21.1. Thermal Conduction in Metals and Alloys—Classical Approach

We now attempt to calculate the heat conductivity K (see (19.8)). The train of thought is borrowed from the kinetic theory of gases, because the same arguments hold true for electrons as for gas molecules.

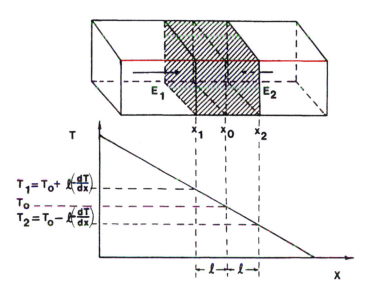

Figure 21.1. For the derivation of the heat conductivity in metals. Note that (dT/dx) is negative for the case shown in the graph.

Consider a bar of metal whose left side is hot and whose right side is cold (Fig. 21.1). Thus, a temperature gradient, dT/dx, exists in the x-direction of the bar. Consider also a volume at the center of the bar whose faces have the size of a unit area and whose length is $2l$, where l is the mean free path between two consecutive collisions between an electron and lattice atoms. We assume that at the distance l from the center, x_0, the average electron has had its last collision and has picked up the energy of this place. We calculate first the energy, E_1, per unit time and unit area, of the electrons that drift from the left into the above-mentioned sample volume. This energy E_1 equals the number of electrons, z, times the energy of one of the electrons. The latter is, according to (19.20), $\frac{3}{2}k_B T_1$, where T_1 can be taken from Fig. 21.1, and z is given in (19.12)

$$E_1 = z \cdot \frac{3}{2} k_B \left(T_0 + l \left(-\frac{dT}{dx} \right) \right) = \frac{n_v v}{6} \frac{3}{2} k_B \left(T_0 - l \frac{dT}{dx} \right). \tag{21.1}$$

The same number of electrons drift from right to left through the volume under consideration. These electrons, however, carry a lower energy, E_2, because of the lower temperature of the particles at the site of interaction. Thus,

$$E_2 = \frac{n_v v}{6} \cdot \frac{3}{2} k_B \left(T_0 + l \frac{dT}{dx} \right). \tag{21.2}$$

The excess thermal energy transferred per unit time through a unit area into the unit volume is therefore the heat flux

$$J_Q = E_1 - E_2 = -\frac{n_v v}{6} \frac{3}{2} k_B \left(2l \frac{dT}{dx} \right) = -\frac{n_v v}{2} k_B l \frac{dT}{dx}. \tag{21.3}$$

We compare (21.3) with (19.8):

$$J_Q = -K \frac{dT}{dx}.$$ (21.4)

Then, we obtain for the **heat conductivity of the electrons**

$$\boxed{K = \frac{n_v v k_B l}{2}.}$$ (21.5)

The heat conductivity is thus larger the more electrons, n_v, are involved, the larger their velocity, v, and the larger the mean free path between two consecutive electron–atom collisions, l. This result intuitively makes sense.

We now seek a connection between the heat conductivity and C_v^{el}. We know from (19.20) the kinetic energy of all n_v electrons per unit volume:

$$E = n_v \frac{3}{2} k_B T.$$ (21.6)

From this we obtain the heat capacity *per volume*,

$$C_v^{el} = \left(\frac{dE}{dT}\right)_v = n_v \frac{3}{2} k_B.$$ (21.7)

Combining (21.5) with (21.7) yields

$$K = \frac{1}{3} C_v^{el} v l.$$ (21.8)

All three variables contained in (21.8) are temperature-dependent, but while C_v^{el} increases with temperature, l and, to a small degree, also v, are decreasing. Thus, K should change very little with temperature, which is indeed experimentally observed. As mentioned in Section 19.5, the thermal conductivity decreases about 10^{-5} W/m · K per degree. K also changes at the melting point and when a change in atomic packing occurs.

21.2. Thermal Conduction in Metals and Alloys—Quantum Mechanical Considerations

The question arises as to what velocity the electrons (that participate in the heat conduction process) have. Further, do all the electrons participate in the heat conduction? We have raised a similar question in Section 20.3, see Fig. 20.3. We know from there that only those electrons which have an energy close to the Fermi energy, E_F, are able to participate in the conduction process. Thus, the velocity in (21.5) and (21.8) is essentially the Fermi velocity, v_F, which can be calculated with

$$E_F = \frac{1}{2} m v_F^2$$ (21.9)

if the Fermi energy is known (see Appendix).

Second, the number of participating electrons contained in (21.5) is proportional to the population density at the Fermi energy, $N(E_F)$, i.e., in first approximation, by the number of free electrons, N_f, per unit volume. Inserting the quantum mechanical expression for C_v^{el} (20.28),

$$C_v^{el} = \frac{\pi^2}{2} \frac{N_f k_B^2 T}{E_F} \left(\frac{J}{K \cdot m^3} \right),$$ (21.10)

into (21.8) yields

$$K = \frac{\pi^2 N_f k_B^2 T v_F l_F}{6 E_F},$$ (21.11)

which reduces with (21.9) and $l_F = \tau v_F$ (7.15a) (τ = relaxation time) to

$$\boxed{K = \frac{\pi^2 N_f k_B^2 T \tau}{3 m^*}.}$$ (21.12)

This is the result we were seeking. Again, the heat conductivity is larger the more *free* electrons are involved and the smaller the (effective) mass of the electrons.

Next, we return to a statement which we made in Chapter 18. We pointed out there that **Wiedemann and Franz** observed that good electrical conductors are also good thermal conductors. We are now in a position to compare the thermal conductivity (21.12) with the electrical conductivity (7.15),

$$\sigma = \frac{N_f e^2 \tau}{m^*}.$$ (21.13)

The ratio of K and σ (divided by T) is proportional to a constant called the **Lorentz number**, L, which is a function of two universal constants, k_B and e,

$$\frac{K}{\sigma T} = L = \frac{\pi^2 k_B^2}{3 e^2}.$$ (21.14)

The Lorentz number is calculated to be 2.443×10^{-8} $(J \cdot \Omega/K^2 \cdot s)$ (see Problem 2). Experiments for most metals confirm this number quite well.

21.3. Thermal Conduction in Dielectric Materials

Heat conduction in dielectric materials occurs by a flow of phonons. The hot end possesses more phonons than the cold end, causing a drift of phonons down a concentration gradient.

The thermal conductivity can be calculated similarly as in the previous section, which leads to the same equation as (21.8),

$$K = \tfrac{1}{3} C_v^{ph} v l.$$ (21.15)

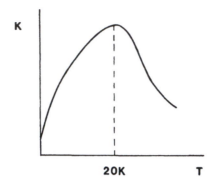

Figure 21.2. Schematic representation of the thermal conductivity in dielectric materials as a function of temperature.

In the present case, C_v^{ph} is the (lattice) heat capacity per unit volume of the phonons, v is the phonon velocity, and l is the phonon mean free path. A typical value for v is about 5×10^5 cm/s (sound velocity) with v being relatively temperature-independent. In contrast, the mean free path varies over several orders of magnitude, i.e., from about 10 nm at room temperature to 10^4 nm near 20 K. The drifting phonons interact on their path with lattice imperfections, external boundaries, and with other phonons. These interactions constitute a thermal resistivity, which is quite analogous to the electrical resistivity. Thus, we may treat the thermal resistance just as we did in Part II; i.e., in terms of interactions between particles (here phonons) and matter, or in terms of the scattering of phonon waves on lattice imperfections.

At low temperatures, where only a few phonons exist, the thermal conductivity depends mainly on the heat capacity, C_v^{ph}, which increases with the third power of increasing temperature according to (20.20) (see Fig. 21.2). At low temperatures, the phonons possess small energies, i.e., long wavelengths which are too long to be scattered by lattice imperfections. The mean free path, l, becomes thus a constant and is virtually identical to the dimensions of the material.

More effective are the phonon–phonon interactions, which are dominant at higher temperatures since, as we know, the phonon density increases with increasing T. Thus, the mean free path and, consequently, the thermal conductivity, decreases for temperatures above about 20 K (Fig. 21.2).

Another mechanism which impedes the flow of phonons at higher temperatures has been discovered. We explain this mechanism in quantum mechanical terms. When two phonons collide, a third phonon results in a proper manner to conserve momentum. Now, phonons (just like electrons) can be represented to travel in k-space. The same arguments, as discussed in Chapter 5, may then apply here. We need to consider Brillouin zones that represent the areas in which the phonon interactions occur. In the example of Fig. 21.3, the resultant vector $\mathbf{a}_1 + \mathbf{a}_2 = \mathbf{a}_3$ is shown to be outside the first

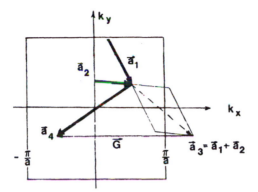

Figure 21.3. First Brillouin zone in a reciprocal square lattice. Two phonons $\mathbf{a}_1$ and $\mathbf{a}_2$ are shown to interact. In the example, the resultant vector $\mathbf{a}_3$ lies outside the first Brillouin zone.

Brillouin zone. We project this vector back to a corresponding place *inside* the first Brillouin zone by applying a similar vector relationship as in (5.34),

$$\mathbf{a}_1 + \mathbf{a}_2 = \mathbf{a}_3 + \mathbf{G}, \tag{21.16}$$

where $\mathbf{G}$ is again a translational vector, which has in the present case the modulus $-2\pi/a$ (see Fig. 21.3). As a consequence, the resultant phonon of vector $\mathbf{a}_4$ proceeds after the collision in a direction that is almost opposite to $\mathbf{a}_2$, which constitutes, of course, a resistance against the flow of phonons. This mechanism is called **umklapp process** (German for "flipping over" process).

Phonon collisions in which $\mathbf{a}_1$ and $\mathbf{a}_2$ are small, so that the resultant vector $\mathbf{a}_3$ stays inside the first Brillouin zone (i.e., $\mathbf{G} = 0$), are called *normal processes*. A normal process has no effect on the thermal resistance, since the resultant phonon proceeds essentially in the same direction.

Problems

1. Calculate the thermal conductivity for a metal, assuming $\tau = 3 \times 10^{-14}$ s, $T = 300$ K, and $N_f = 2.5 \times 10^{22}$ el/cm^3.

2. Calculate the Lorentz number from values of e and k_B. Show how you arrived at the correct units!

3. Calculate the mean free path of electrons in a metal, such as silver, at room temperature form heat capacity and heat conduction measurements. Take $E_F = 5$ eV, $K = 4.29 \times 10^2$ J/s · m · K, and $C_v^{el} = 1\%$ of the lattice heat capacity. (*Hint:* Remember that the heat capacity in (21.8) is given per unit volume!)

4. Why is the thermal conduction in dielectric materials two or three orders of magnitude smaller than in metals?

5. Why does the thermal conductivity span only 4 orders of magnitude, whereas the electrical conductivity spans nearly 25 orders of magnitude?

6. Is there a theoretical possibility of a thermal superconductor?

7. Discuss why the thermal conductivity of alloys is lower than that of the pure constituents.

CHAPTER 22

Thermal Expansion

The length, L, of a rod increases with increasing temperature. Experiments have shown that in a relatively wide temperature range the linear expansion, ΔL, is proportional to the increase in temperature, ΔT. The proportionality constant is called the **coefficient of linear expansion**, α_L. The observations can be summarized in

$$\frac{\Delta L}{L} = \alpha_L \Delta T. \tag{22.1}$$

Experimentally observed values for α_L are given in Table 22.1.

The expansion coefficient has been found to be proportional to the molar heat capacity, C_v, i.e., the temperature dependence of α_L is similar to the temperature dependence of C_v. As a consequence, the temperature dependence of α_L for dielectric materials follows closely the $C_v = f(T)$ relationship predicted by Debye and shown in Fig. 18.2. Specifically, α_L approaches a constant value for $T > \theta_D$ and vanishes as T^3 for $T \to 0$. The thermal expansion coefficient for metals, on the other hand, decreases at very small temperatures in proportion to T, and depends on the sum of the heat capacities of phonons and electrons in other temperature regions.

We turn now to a discussion of possible mechanisms that may explain thermal expansion from an atomistic point of view. We postulate, as in the previous chapters, that the lattice atoms absorb thermal energy by vibrating about their equilibrium position. In doing so, a given atom responds with increasing temperature and vibrational amplitude to the repulsive forces of the neighboring atoms. Let us consider for a moment two adjacent atoms only, and let us inspect their potential energy as a function of internuclear separation (Fig. 22.1). We understand that as two atoms move closer to each other, strong repulsive forces are experienced between them. As a conse-

Table 22.1. Linear Expansion Coefficients,
α_L, for Some Solids Measured at Room
Temperature.

Substance	α_L in 10^{-5} [K^{-1}]
Hard rubber	8.00
Lead	2.73
Aluminum	2.39
Brass	1.80
Copper	1.67
Iron	1.23
Glass (ordinary)	0.90
Glass (pyrex)	0.32
NaCl	0.16
Invar (Fe–36% Ni)	0.07
Quartz	0.05

quence, the potential energy curve rises steeply with decreasing r. On the
other hand, we know that two atoms also attract each other somewhat. This
results in a slight decrease in $U(r)$ with decreasing r.

Now, for small temperatures, a given atom may rest in its equilibrium
position, r_0, i.e., at the minimum of potential energy. If, however, the temperature is raised, the amplitude of the vibrating atom increases, too. Since
the amplitudes of the vibrating atom are symmetric about a median position
and since the potential curve is not symmetric, a given atom moves farther
apart from its neighbor, i.e., the average position of an atom moves to a
larger r, say, r_T, as shown in Fig. 22.1. In other words, the thermal expan-

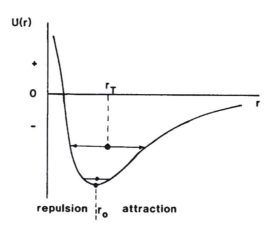

Figure 22.1. Schematic representation of the potential energy, $U(r)$, for two adjacent
atoms as a function of internuclear separation, r.

sion is a direct consequence of the asymmetry of the potential energy curve. The same arguments hold true if all atoms in a solid are considered.

A few substances are known to behave differently from that described above. They contract during a temperature increase. This happens, however, only within a narrow temperature region. For its explanation, we need to realize that longitudinal as well as transverse vibrational modes may be excited by thermal energy (see Section 20.2). The lattice is expected to contract if transverse modes predominate. Interestingly enough, only one known liquid substance, namely, water, behaves in a limited temperature range in this manner. Specifically, water has its largest density at 4°C. (Furthermore, the density of ice is smaller than the density of water at the freezing point.) As a consequence, water of 4°C sinks to the bottom of a lake during winter, while ice stays on top. This prevents the freezing of a lake at the bottom and thus enables aquatic life to survive during the winter. This exceptional behavior of water suggests that the laws of physics do not just "happen," but rather they were created by a superior being. I want to conclude my book with this thought.

Problems

1. Estimate the force that is exerted by the end of a 1 m long iron rod of 1 cm^2 cross section which is heated to 100°C.

2. Calculate the gap which has to be left between two 10 m long railroad tracks when they are installed at 0°C and if no compression is allowed at 40°C.

3. Explain some engineering applications of thermal expansion, such as the bimetal thermal switch, metal thermometer, etc.

4. What happens if a red-hot piece of glass is immersed in cold water? What happens if the same experiment is done with quartz?

5. Discuss thermal expansion in materials from an atomistic point of view.

Suggestions for Further Reading (Part V)

A.J. Dekker, *Solid State Physics*, Prentice-Hall, Englewood Cliffs, NJ (1957).
C. Kittel and H. Kroemer, *Thermal Physics*, 2nd ed., W.H. Freeman, San Francisco, CA (1980).
F.G. Klemens and T.K. Chu, eds., *Thermal Conductivity*, Vols. 1–17, Plenum Press, New York.
T.F. Lee, F.W. Sears, and D.L. Turcotte, *Statistical Thermodynamics*, Addison-Wesley, Reading, MA (1963).
J.M. Ziman, *Electrons and Phonons*, Oxford University Press, Oxford (1960).

APPENDICES

Periodic Disturbances

A *vibration* is a time-dependent *or* space-dependent periodic disturbance. We restrict our discussion to harmonic vibrations. In this case, the space-dependence of time is represented by a simple sine or cosine function or, equivalently, because of the Euler equations (see Appendix 2), by an exponential function. The use of exponential functions provides often a simpler mathematical treatment than using trigonometric functions. For this reason, exponential functions are usually preferred. We follow this practice.

A.1.1. Undamped Vibration

(a) Differential equation for *time-dependent periodicity*:

$$m\frac{d^2u}{dt^2} + \kappa u = 0 \qquad\qquad (A.1)$$

(m = mass, κ = retracting force parameter). A solution is

$$u = Ae^{i\omega t}, \qquad\qquad (A.2)$$

where

$$\omega = \sqrt{\frac{\kappa}{m}} = 2\pi v \qquad\qquad (A.3)$$

is the angular frequency and A is a constant called maximum amplitude.

(b) Differential equation for *space-dependent periodicity* (in one dimension):

$$a\frac{d^2u}{dx^2} + bu = 0. \qquad\qquad (A.4)$$

Solution:

$$u = Ae^{i\alpha x} + Be^{-i\alpha x}, \tag{A.5}$$

where A and B are constants, and

$$\alpha = \sqrt{\frac{b}{a}}. \tag{A.6}$$

A.1.2. Damped Vibration

(a) Differential equation for *time-dependent periodicity*:

$$m\frac{d^2u}{dt^2} + \gamma\frac{du}{dt} + \kappa u = 0 \tag{A.7}$$

(γ is the damping constant). The solution is

$$u = Ae^{-\beta t} \cdot e^{i(\omega_0 t - \phi)}, \tag{A.8}$$

where

$$\omega_0 = \sqrt{\frac{\kappa}{m} - \beta^2} \tag{A.9}$$

is the resonance frequency,

$$\beta = \frac{\gamma}{2m} \tag{A.10}$$

is the damping factor, and ϕ is the phase (angle) difference. In a damped vibration, the amplitude $Ae^{-\beta t}$ decreases exponentially.

(b) Differential equation for *space-dependent periodicity*:

$$\frac{d^2u}{dx^2} + D\frac{du}{dx} + Cu = 0. \tag{A.11}$$

Solution:

$$u = e^{-Dx/2}(Ae^{i\rho x} + Be^{-i\rho x}), \tag{A.12}$$

where

$$\rho = \sqrt{C - \frac{D^2}{4}} \tag{A.13}$$

and A, B, C, and D are constants.

A.1.3. Forced Vibration (Damped)

Differential equation for *time-dependent periodicity*:

$$m\frac{d^2u}{dt^2} + \gamma\frac{du}{dt} + \kappa u = K_0 e^{i\omega t}. \tag{A.14}$$

The right-hand side is the time-periodic excitation force. The solution consists of the start-up vibration and a steady-state part. The steady-state solution is

$$u = \frac{K_0}{\sqrt{m^2(\omega_0^2 - \omega^2)^2 + \gamma^2\omega^2}} e^{i(\omega t - \phi)}, \tag{A.15}$$

where

$$\omega_0 = \sqrt{\frac{\kappa}{m}} \tag{A.16}$$

is the resonance frequency of the undamped, free oscillation.

The tangent of the phase difference, ϕ, between the excitation force and the forced vibration is

$$\tan\phi = \frac{\gamma\omega}{m(\omega_0^2 - \omega^2)}. \tag{A.17}$$

A.1.4. Wave

A wave is a space- *and* time-dependent periodic disturbance. One distinguishes between *traveling waves*, which occur when the wave is not confined by boundary conditions, and *standing waves*, which are observed when a wave is reflected at a boundary and thus interacts with the oncoming wave. The wave motion in a vibrating string may serve as an example for the latter case. The simplest form of a traveling wave is a *harmonic wave*, which is expressed by a sine or cosine function, such as

$$u(t, x) = A\sin(kx - \omega t) \tag{A.18}$$

(when the wave is propagating in the positive x-direction), where

$$|\mathbf{k}| = \frac{2\pi}{\lambda} \tag{A.19}$$

is called the wave number vector. It has the unit of a reciprocal length. For convenience and because of the Euler equations (Appendix 2) one frequently uses instead

$$u(t, x) = A\exp(i(kx - \omega t)). \tag{A.20}$$

(a) The differential equation for the undamped wave is

$$v^2\nabla^2 u = \frac{\partial^2 u}{\partial t^2}, \tag{A.21}$$

where

$$\nabla^2 = \frac{\partial^2}{\partial x^2} + \frac{\partial^2}{\partial y^2} + \frac{\partial^2}{\partial z^2}. \tag{A.22}$$

The differential equation for a plane wave is

$$v^2 \frac{\partial^2 u}{\partial x^2} = \frac{\partial^2 u}{\partial t^2}, \qquad (A.23)$$

whose solution is

$$u(t, x) = e^{i\omega t}(Ae^{i\alpha x} + Be^{-i\alpha x}), \qquad (A.24)$$

or

$$u(t, x) = Ae^{i(\omega t + \alpha x)} + Be^{i(\omega t - \alpha x)}, \qquad (A.25)$$

or

$$u(t, x) = Ae^{i\omega(t + (x/v))} + Be^{i\omega(t - (x/v))}. \qquad (A.25a)$$

(b) The damped wave

$$v^2 \nabla^2 u = a \frac{\partial u}{\partial t} + b \frac{\partial^2 u}{\partial t^2} \qquad (A.26)$$

can be solved with

$$u(t, x, y, z) = Ae^{i(\omega t - k \cdot x)} \cdot e^{-\gamma x}. \qquad (A.27)$$

APPENDIX 2
Euler Equations

$$\cos \phi = \tfrac{1}{2}(e^{i\phi} + e^{-i\phi}), \tag{A.28}$$

$$\sin \phi = \tfrac{1}{2i}(e^{i\phi} - e^{-i\phi}), \tag{A.29}$$

$$\sinh \phi = \tfrac{1}{2}(e^{\phi} - e^{-\phi}) = \tfrac{1}{i} \cdot \sin i\phi, \tag{A.30}$$

$$\cosh \phi = \tfrac{1}{2}(e^{\phi} + e^{-\phi}) = \cos i\phi, \tag{A.31}$$

$$e^{i\phi} = \cos \phi + i \sin \phi, \tag{A.32}$$

$$e^{-i\phi} = \cos \phi - i \sin \phi. \tag{A.33}$$

Summary of Quantum Number Characteristics

The energy states of electrons are characterized by four quantum numbers. The main quantum number, n, determines the overall energy of the electrons, i.e., essentially the radius of the electron distribution. It can have any integral value. For example, the electron of a hydrogen atom in its ground state has $n = 1$.

The quantum number, l, is a measure of the angular momentum, L, of the electrons and is determined by $|\mathbf{L}| = \sqrt{l(l+1)}\hbar$, where l can assume any integral value between 0 and $n - 1$.

It is common to specify a given energy state by a symbol that utilizes the n- and l-values. States with $l = 0$ are called s-states; with $l = 1$, p-states; and with $l = 2$, d-states, etc. A $4d$-state, for example, is one with $n = 4$ and $l = 2$.

The possible orientations of the angular momentum vector with respect to an external magnetic field are again quantized and are given by the magnetic quantum number, m. Only m values between $+l$ and $-l$ are permitted.

The electrons of an atom fill the available states starting with the lowest state and obeying the Pauli principle, which requires that each state can be filled with only two electrons having opposite spin ($|\mathbf{s}| = \pm\frac{1}{2}$). Because of the just-mentioned multiplicity, the maximal number of electrons in the s-states is 2, in the p-states 6, in the d-states 10, and in the f-states 14.

The electron bands in solids are named by using the same nomenclature as above, i.e., a $3d$-level in the atomic state widens to a $3d$-band in a solid. The electron configurations of some isolated atoms are listed on the next page.

The designations s, p, d, and f are of an historical nature and are derived from certain early spectrographic observations. They stand for *sharp, principal, diffuse,* and *fundamental.*

Z	Element	K 1s	L 2s 2p	M 3s 3p 3d	N 4s 4p 4d 4f	O 5s 5p 5d 5f
1	H	1				
2	He	2				
3	Li	2	1			
4	Be	2	2			
5	B	2	2 1			
6	C	2	2 2			
7	N	2	2 3			
8	O	2	2 4			
9	F	2	2 5			
10	Ne	2	2 6			
11	Na	2	2 6	1		
12	Mg	2	2 6	2		
13	Al	2	2 6	2 1		
14	Si	2	2 6	2 2		
15	P	2	2 6	2 3		
16	S	2	2 6	2 4		
17	Cl	2	2 6	2 5		
18	Ar	2	2 6	2 6		
19	K	2	2 6	2 6	1	
20	Ca	2	2 6	2 6	2	
21	Sc	2	2 6	2 6 1	2	
22	Ti	2	2 6	2 6 2	2	
23	V	2	2 6	2 6 3	2	
24	Cr	2	2 6	2 6 5	1	
25	Mn	2	2 6	2 6 5	2	
26	Fe	2	2 6	2 6 6	2	
27	Co	2	2 6	2 6 7	2	
28	Ni	2	2 6	2 6 8	2	
29	Cu	2	2 6	2 6 10	1	
30	Zn	2	2 6	2 6 10	2	
31	Ga	2	2 6	2 6 10	2 1	
32	Ge	2	2 6	2 6 10	2 2	
33	As	2	2 6	2 6 10	2 3	
34	Se	2	2 6	2 6 10	2 4	
35	Br	2	2 6	2 6 10	2 5	
36	Kr	2	2 6	2 6 10	2 6	
37	Rb	2	2 6	2 6 10	2 6	1
38	Sr	2	2 6	2 6 10	2 6	2
39	Y	2	2 6	2 6 10	2 6 1	2
40	Zr	2	2 6	2 6 10	2 6 2	2
41	Nb	2	2 6	2 6 10	2 6 4	1
42	Mo	2	2 6	2 6 10	2 6 5	1
43	Tc	2	2 6	2 6 10	2 6 5	2

APPENDIX 4

Tables

The International System of Units (SI or mksA System)

In the SI unit system, essentially four base units—the meter, the kilogram (for the mass), the second, and the ampere—are *defined*. Further base units are the Kelvin, the mole (for the amount of substance), and the candela (for the luminous intensity). All other units are *derived units* as shown in the table below. Even though the use of the SI unit system is highly recommended, other unit systems are still widely used.

Quantity	Name	Symbol	Expression in terms of Other SI units	Expression in terms of SI base units
Force	Newton	N	—	$kg \cdot m/s^2$
Energy, work	Joule	J	$N \cdot m = V \cdot A \cdot s$	$kg \cdot m^2/s^2$
Pressure	Pascal	Pa	N/m^2	$kg/m \cdot s^2$
El. charge	Coulomb	C	J/V	$A \cdot s$
Power	Watt	W	J/s	$kg \cdot m^2/s^3$
El. potential	Volt	V	W/A	$kg \cdot m^2/A \cdot s^3$
El. resistance	Ohm	Ω	V/A	$kg \cdot m^2/A^2 \cdot s^3$
El. conductance	Siemens	S	A/V	$A^2 \cdot s^3/kg \cdot m^2$
Magn. flux	Weber	Wb	$V \cdot s$	$kg \cdot m^2/A \cdot s^2$
Magn. induction	Tesla	T	$Wb/m^2 = V \cdot s/m^2$	$kg/A \cdot s^2$
Inductance	Henry	H	Wb/A	$kg \cdot m^2/A^2 \cdot s^2$
Capacitance	Farad	F	C/V	$A^2 \cdot s^4/kg \cdot m^2$

Physical Constants (SI and cgs units)

Mass of electron (free electron mass; rest mass)	$m_0 \equiv m = 9.11 \times 10^{-31}$ (kg) $= 9.11 \times 10^{-28}$ (g)
Charge of electron	$e = 1.602 \times 10^{-19}$ (C) (SI-unit)
	$= 4.803 \times 10^{-10}$ (statcoul) $\equiv$ (cm$^{3/2} \cdot$ g$^{1/2}$/s) (el. static cgs units)
	$= 1.602 \times 10^{-20}$ (abcoul) $\equiv$ (g$^{1/2} \cdot$ cm$^{1/2}$) (el. magnetic cgs units)
Velocity of light in vacuum	$c = 2.998 \times 10^8$ (m/s) $= 2.998 \times 10^{10}$ (cm/s)
Planck constant	$h = 6.626 \times 10^{-34}$ (J $\cdot$ s) $= 6.626 \times 10^{-27}$ (g $\cdot$ cm^2/s)
	$= 4.136 \times 10^{-15}$ (eV $\cdot$ s)
	$\hbar = 1.054 \times 10^{-34}$ (J $\cdot$ s) $= 1.054 \times 10^{-27}$ (g $\cdot$ cm^2/s)
	$= 6.582 \times 10^{-16}$ (eV $\cdot$ s)
Avogadro constant	$N_0 = 6.022 \times 10^{23}$ (atoms/mol)
Boltzmann constant	$k_B = 1.381 \times 10^{-23}$ (J/K) $= 1.381 \times 10^{-16}$ (erg/K)
	$= 8.616 \times 10^{-5}$ (eV/K)
Bohr magneton	$\mu_B = 9.274 \times 10^{-24}$ (J/T) $\equiv$ (A $\cdot$ m^2)
	$= 9.274 \times 10^{-21} \left(\dfrac{\text{erg}}{\text{G}}\right) \equiv$ (g$^{1/2}$ cm$^{5/2}$/s)
Gas constant	$R = 8.314$ (J/mol $\cdot$ K) $= 1.986$ (cal/mol $\cdot$ K)
Permittivity of emtpy space (vacuum)	$\varepsilon_0 = 1/\mu_0 c^2 = 8.854 \times 10^{-12}$ (F/m) $\equiv$ (A $\cdot$ s/V $\cdot$ m)
	$\equiv \left(\dfrac{\text{A}^2\text{s}^4}{\text{kg m}^3}\right)$
Permeability of empty space (vacuum)	$\mu_0 = 4\pi \times 10^{-7} = 1.257 \times 10^{-6}$ (H/m) $\equiv$ (V $\cdot$ s/A $\cdot$ m)
	$\equiv$ (kg $\cdot$ m/A$^2 \cdot$ s^2)

Useful Conversions

1 (eV) $= 1.602 \times 10^{-12}$ (g $\cdot$ cm^2/s^2) $= 1.602 \times 10^{-19}$ (kg $\cdot$ m^2/s^2)

$\qquad = 1.602 \times 10^{-19}$ (J) $= 3.829 \times 10^{-20}$ (cal)

1 (J) $= 1 \left(\dfrac{\text{kg} \cdot \text{m}^2}{\text{s}^2}\right) = 10^7$ (erg) $= 10^7 \left(\dfrac{\text{g} \cdot \text{cm}^2}{\text{s}^2}\right) = 2.39 \times 10^{-1}$ (cal)

1 (Rydberg) $= 13.6$ (eV)

1 (1/Ωcm) $= 9 \times 10^{11}$ (1/s)

1 (1/Ωm) $= 9 \times 10^9$ (1/s)

1 (C) $= 1$ (A $\cdot$ s) $= 1$ (J/V)

1 (Å) $= 10^{-10}$ (m)

1 (torr) $\equiv$ 1 (mm Hg) $= 133.3$ (N/m^2) $\equiv 133.3$ (Pa)

1 (bar) $= 10^5$ (N/m^2) $\equiv 10^5$ (Pa)

1 (Pa) $= 10$ (dyn/cm^2)

1 cal $= 2.6118 \times 10^{19}$ (eV)

1 (horsepower) $= 746$ (W)

1 (KWH) $= 3.6$ (MJ)

1 (mm) (milli) $= 10^{-3}$ (m)	1 km (Kilo) $= 10^3$ m
1 (μm) (micro) $= 10^{-6}$ (m)	1 Mm (Mega) $= 10^6$ m
1 (nm) (nano) $= 10^{-9}$ (m)	1 Gm (Giga) $= 10^9$ m
1 (pm) (pico) $= 10^{-12}$ (m)	1 Tm (Tera) $= 10^{12}$ m
1 (fm) (femto) $= 10^{-15}$ (m)	1 Pm (Peta) $= 10^{15}$ m
1 (am) (atto) $= 10^{-18}$ (m)	1 Em (Exa) $= 10^{18}$ m

Electronic Properties of Some Metals

Material	Effective mass $\left(\dfrac{m^*}{m_0}\right)_{el}$	Effective mass $\left(\dfrac{m^*}{m_0}\right)_{opt}$	Fermi energy, E_F [eV]	Number of free electrons, N_{eff} $\left[\dfrac{electrons}{m^3}\right]$	Work function (photoelectric), ϕ [eV]	Resistivity, ρ [$\mu\Omega$ cm] at 20°C
Ag		0.95	5.5	6.1×10^{28}	4.7	1.59
Al	0.97	1.08	11.8	16.7×10^{28}	4.1	2.65
Au		1.04	5.5	5.65×10^{28}	4.8	2.35
Be	1.6		12.0		3.9	4.0
Ca	1.4		3.0		2.7	3.91
Cs			1.6		1.9	20.0
Cu	1.0	1.42	7.0	6.3×10^{28}	4.5	1.67
Fe	1.2				4.7	9.71
K	1.1		1.9		2.2	6.15
Li	1.2		4.7		2.3	8.55
Na	1.0		3.2		2.3	4.20
Ni	2.8				5.0	6.84
Zn	0.85		11.0	3×10^{28}	4.3	5.91

Electronic Properties of Some Semiconductors

	Material	Transition	Gap energy E_g [eV]		Room-temp. conductivity $\sigma \left[\frac{1}{\Omega \cdot m}\right]$	Mobility of electrons $\mu_e \left[\frac{m^2}{V \cdot s}\right]$	Mobility of holes $\mu_h \left[\frac{m^2}{V \cdot s}\right]$	Work function (photoelectric) ϕ [eV]	Effective mass ratio at 4 K $\frac{m_e^*}{m_0}$	$\frac{m_h^*}{m_0}$
			0 K	300 K						
Element	C (diamond)	I	5.48	5.47	10^{-12}	0.18	0.12	4.8	0.2	0.25
	Ge	I	0.74	0.66	2.2	0.39	0.19	4.6	1.64[a] / 0.08[b]	0.04[c] / 0.28[d]
	Si	I	1.17	1.12	9×10^{-4}	0.15	0.045	3.6	0.98[a] / 0.19[b]	0.16[c] / 0.49[d]
	Sn (gray)	D	0.09	0.08	10^6	0.14	0.12	4.4		0.3[d]
III–V	GaAs	D	1.52	1.42	10^{-6}	0.85	0.04	4.9	0.067	0.082
	InAs	D	0.42	0.36	10^4	3.30	0.046		0.023	0.40
	InSb	D	0.23	0.17		8.00	0.125		0.014	0.40
	GaP	I	2.34	2.26		0.01	0.007		0.82	0.60
	GaN	D	3.50	3.36		0.04	0.01		0.19	0.60
	InN	D		0.7						
IV–IV	α-SiC	I	3.03	2.99		0.04	0.005		0.60	1.00
II–VI	ZnO	D	3.42	3.35		0.02	0.018		0.27	
	CdSe	D	1.85	1.70		0.08			0.13	0.45
	ZnS	D	3.84	3.68		0.02	0.0005		0.40	
IV–VI	PbS	I	0.286	0.41		0.06	0.07		0.25	0.25

Transitions: D = direct
          I = indirect

[a] Longitudinal effective mass.
[b] Transverse effective mass.
[c] Light-hole effective mass.

Ionization Energies for Various Dopants in Semiconductors (Experimental)

Donor ionization energies are given from the donor levels to the bottom of the conduction band. Acceptor ionization energies are given from the top of the valence band to the acceptor levels.

| Semiconductor | Dopant | | Ionization energy (eV) |
	Type	Element	
Ge	Donors	Sb	0.0096
		P	0.012
		As	0.013
	Acceptors	B	0.01
		Al	0.01
		Ga	0.011
		In	0.011
Si	Donors	Sb	0.039
		P	0.045
		As	0.054
	Acceptors	B	0.045
		Al	0.067
		Ga	0.072
		In	0.16
GaAs	Donors	Si	0.0058
		Ge	0.006
		Sn	0.006
	Acceptors	Be	0.028
		Mg	0.028
		Zn	0.031

Physical Properties of Si and GaAs

	Si	GaAs
Lattice constant (Å)	5.431	5.654
Atoms (cm^{-3})	5.00×10^{22}	4.43×10^{22}
Band gap (eV) at 25°C	1.11	1.43
Temperature dependence of band gap ($eV°C^{-1}$)	-2.4×10^{-4}	-4.3×10^{-4}
Specific gravity ($g\ cm^{-3}$)	2.33	5.32
Dielectric constant	11.8	10.9
Electron lattice mobility ($cm^2\ V^{-1}\ s^{-1}$)	1.5×10^3	8.5×10^3
Hole lattice mobility ($cm^2\ V^{-1}\ s^{-1}$)	4.8×10^2	4×10^2
Number of intrinsic electrons (cm^{-3}) at 25°C	1.5×10^{10}	1.1×10^6
Coefficient of linear thermal expansion ($°C^{-1}$) at 25°C	2.33×10^{-6}	6.86×10^{-6}
Thermal conductivity ($W\ °C^{-1}\ m^{-1}$)	147	46

Optical Constants of Si and GaAs (from *Handbook of Optical Constants of Solids, Academic Press*, 1985)

		Si		GaAs	
E (eV)	λ (nm)	n	k	n	k
4.96	250	1.580	3.632	2.654	4.106
3.54	350	5.442	2.989	3.513	1.992
3.10	400	5.570	0.387	4.373	2.146
2.48	500	4.298	0.073	4.305	0.426
2.07	600	3.943	0.025	3.914	0.228
1.55	800	3.688	0.006	3.679	0.085
0.91	1370	3.5007	→0	3.3965	→0

Magnetic Units

Name	Symbol	em-cgs units	mks (SI) units	Conversions
Magnetic field strength	**H**	$Oe \equiv \dfrac{g^{1/2}}{cm^{1/2}\cdot s}$	$\dfrac{A}{m}$	$1\,\dfrac{A}{m} = \dfrac{4\pi}{10^3}$ Oe
Magnetic induction	**B**	$G \equiv \dfrac{g^{1/2}}{cm^{1/2}\cdot s}$	$\dfrac{Wb}{m^2} = \dfrac{kg}{s\cdot C} \equiv T$	$1\,T = 10^4\,G$
Magnetization	**M**	$\dfrac{Maxwell}{cm^2} = \dfrac{g^{1/2}}{cm^{1/2}\cdot s} = \dfrac{emu}{cm^3}$	$\dfrac{A}{m}$	$1\,\dfrac{A}{m} = \dfrac{4\pi}{10^3}\,\dfrac{Maxwells}{cm^2} = 10^{-3}\,\dfrac{emu}{cm^3}$
Magnetic flux	ϕ	$Maxwell \equiv \dfrac{cm^{3/2}\cdot g^{1/2}}{s}$	$Wb = \dfrac{kg\cdot m^2}{s\cdot C} = V\cdot s$	$1\,Wb = 10^8$ Maxwells
Susceptibility	χ	Unitless	Unitless	$\chi_{mks} = 4\pi\chi_{cgs}$
(Relative) permeability	μ	Unitless	Unitless	Same value
Energy product	BH	MGOe	$\dfrac{kJ}{m^3}$	$1\,\dfrac{kJ}{m^3} = \dfrac{4\pi}{10^2}$ MGOe
Magnetic moment	μ_m	emu*	Am^2	$1\,Am^2 = 10^3$ emu $1\,emu = 1\,\dfrac{erg}{Oe}$

*emu = electro magnetic unit (used in USA)

Conversions Between Various Unit Systems

SI	Electrostatic cgs (esu) units	Electromagnetic cgs (emu) units	emu-esu conversion
1 (C)	3×10^9 (statcoul) $\equiv \left(\dfrac{cm^{3/2} \cdot g^{1/2}}{s} \right)$	$\dfrac{1}{10}$ (abcoul) $\equiv (g^{1/2} \cdot cm^{1/2})$	1 (abcoul) = f (statcoul)
1 (V)	$\dfrac{1}{300}$ (statvolts) $\equiv \left(\dfrac{cm^{1/2} \cdot g^{1/2}}{s} \right)$	10^8 (abvolts) $\equiv \left(\dfrac{cm^{3/2} \cdot g^{1/2}}{s^2} \right)$	1 (abvolt) = $\dfrac{1}{f}$ (statvolts)
1 (A)	3×10^9 (statamps) $\equiv \left(\dfrac{cm^{3/2} \cdot g^{1/2}}{s^2} \right)$	$\dfrac{1}{10}$ (abamps) $\equiv \left(\dfrac{cm^{1/2} \cdot g^{1/2}}{s} \right)$	1 (abamp) = f (statamps)
1 (Ω)	$\dfrac{1}{9 \times 10^{11}}$ (statohms) $\equiv \left(\dfrac{s}{cm} \right)$	10^9 (abohms) $\equiv \left(\dfrac{cm}{s} \right)$	1 (abohm) = $\dfrac{1}{f^2}$ (statohms)

Note: The factor "f" in column four of this table has the value 3×10^{10} (s/cm).

Conversions from the Gaussian Unit System into the SI Unit System

The equations given in this book can be converted from the cgs (Gaussian) unit system into the SI (mks) system and vice versa by replacing the symbols in the respective equations with the symbols listed in the following table. Symbols which are not listed here remain unchanged. It is imperative that consistent sets of units are utilized.

Quantity	mks (SI)	cgs (Gaussian)
Magnetic induction	B	B/c
Magnetic flux	ϕ_B	ϕ_B/c
Magnetic field strength	H	$cH/4\pi$
Magnetization	M	cM
Magnetic dipole moment	μ_m	$c\mu_m$
Permittivity constant	ε_0	$1/4\pi$
Permeability constant	μ_0	$4\pi/c^2$
Electric displacement	D	$D/4\pi$

$\mu_0 = 4\pi \times 10^{-7} = 1.257 \times 10^{-6}$ (V·s/A·m) $\equiv$ (kg·m/C^2) $\equiv$ (H/m).
$\varepsilon_0 = 8.854 \times 10^{-12}$ (A·s/V·m) $\equiv$ (F/m).

Color Codes of Bands (Rings) on Commercial Resistors

First and second color band	Third color band	Fourth color band (Tolerances)
Black —0	Black —× 1	
Brown —1	Brown —× 10	Brown —1%
Red —2	Red —× 100	Red —2%
Orange—3	Orange—× 1,000 (1K)	Orange—3%
Yellow—4	Yellow—× 10 K	Yellow—4%
Green —5	Green —× 100 K	
Blue —6	Blue —× 1,000 K (1M)	
Violet —7	Violet —× 10 M	
Gray —8	Gray —× 100 M	
White —9	Gold — .1	Gold —5%
	Silver —× .01	Silver —10%
		None —20%

Periodic Table of the Elements

PAPERTECH

Editors: T. K. Varga, M.A.Sc. & C. Bello, M.A.Sc.

© Papertech Marketing Group Inc., 1994. All Rights Reserved.

APPENDIX 5

About Solving Problems

There are two types of exercises contained at the end of each chapter of this book; both of them are provided for the students to deepen their understanding of the material covered in the text. About 25% of the problems are concerned with conceptual reviews. These usually do not seem to be any major stumbling block to the reader. In contrast to this, however, the numerical problems are the ones which seem to provide some challenges. The goal of this section is to sketch a systematic approach for the solution of numerical problems and to give an actual example.

The first task is, of course, to find one or several equations which can be applied to the problem at hand. As a rule, however, the equations to be used are not yet provided in a form which lists the unknown variable on the left side of the equation and all the known variables plus a handful of constants on the right side. Thus, algebraic manipulations need to be applied until this goal has been achieved. (Under no circumstances should one insert numerical values immediately into the starting equations, in particular, if these variables are given in different unit systems.)

Once a final equation (containing the unknown quantity on the left side) has eventually been obtained, a unit check should be attempted by listing all known quantities in *one* unit system and inserting these units into the final equation. This provides a simple check on whether the algebraic manipulation was done correctly and in what unit the numerical result will turn out. Only then is a numerical calculation in place. At the end of each calculation the student should ask, "Does the result make sense?". A comparison with tabulated values in one of the appendices or with information given in the text can, most of the time, quickly answer this question. If the result seems to be off by several orders of magnitude, a recalculation should definitely be performed.

Example (Problem 2/1)

$$\left.\begin{array}{l} \lambda = \dfrac{h}{p}, \\[2mm] E = \dfrac{p^2}{2m}, \end{array}\right\} \quad \lambda = \dfrac{h}{\sqrt{2Em}}\left(\dfrac{\text{kg m}^2\text{ s s}}{\text{s}^2\text{ kg}^{1/2}\text{ m kg}^{1/2}}\right) \equiv (m),$$

$$E = 4\ (\text{eV}) = 4 \times 1.602 \times 10^{-19}\ (\text{J}) \equiv \left(\dfrac{\text{kg m}^2}{\text{s}^2}\right),$$

$$h = 6.626 \times 10^{-34}\ (\text{J s}),$$

$$m = 9.11 \times 10^{-31}\ (\text{kg}),$$

$$\lambda = 0.613 \times 10^{-9}\ (\text{m}) = 6.13\ (\text{Å}).$$

Solutions to Numerical Problems

Chapter 2

1. $\lambda = 6.13$ (Å).
2. $E = 4.18 \times 10^{-6}$ (eV).
4. $E = 2.07$ (eV).
5. $\lambda = 2.38 \times 10^{-24}$ (Å).

Chapter 4

6. $E = 13.6$ (eV).
10. $E_1 = 1.50 \times 10^{-18}$ (J) = 9.39 (eV) (zero-point energy)
$E_2 = 4 \times E_1$; $E_3 = 9 \times E_1$ etc.

Chapter 5

1. $L_1 \approx 14$ (eV); $L_2' \approx 8$ (eV); $L_2' - L_1 \approx 6$ (eV).
2. $E = 1.1$ (eV).
3. $\Delta E = h^2\pi^2/ma^2$; or $E_{111}/E_{100} = 3$.
5. (a) $X = 0 \rightsquigarrow E = 4C$; $X = \pi/a \rightsquigarrow E = 9C$ $(C = \pi^2h^2/2ma^2)$;
   (b) $X = 0 \rightsquigarrow E = 16C$; $X = \pi/a \rightsquigarrow E = 9C$.
6. (a) $X = 0 \rightsquigarrow E = 4C$; $X = 1 \rightsquigarrow E = 1C$ $(C = 2h^2\pi^2/ma^2)$;
   (b) $X = 0 \rightsquigarrow E = 2C$; $X = 1 \rightsquigarrow E = 5C$;
   (c) $X = 0 \rightsquigarrow E = 2C$; $X = 1 \rightsquigarrow E = 5C$.
7. $b_1 = (1/a)(\bar{1}11)$; $b_2 = (1/a)(1\bar{1}1)$; $b_3 = (1/a)(11\bar{1})$.
8. (a) $X = 0 \rightsquigarrow E = 0$; $X = 1 \rightsquigarrow E = \frac{1}{2}C$ $(C = 2h^2\pi^2/ma^2)$;
   (b) $X = 0 \rightsquigarrow E = 2C$; $X = 1 \rightsquigarrow E = 1\frac{1}{2}C$;
   (c) $X = 0 \rightsquigarrow E = 4C$; $X = 1 \rightsquigarrow E = 2\frac{1}{2}C$.

Chapter 6

1. $v_F = 1.38 \times 10^6$ (m/s).
3. $T = 290.5$ (K).
4. $E_F = 5.64$ (eV).

5. $Z(E) = 5.63 \times 10^{46}$ (electron states/J).
        $= 9.03 \times 10^{27}$ (electron states/eV).
6. For entire band: $N^*/V = 8.42 \times 10^{22}$ (1/cm^3).
7. $\eta = 2.5 \times 10^{23}$ (energy states).
8. (a) $N^* = 8.42 \times 10^{22}$ (electrons/cm^3);
   (b) $N_a = 8.49 \times 10^{22}$ (atoms/cm^3);
   (c) Not exactly one free electron per atom.
9. 0.88%.
10. $F(E) = \frac{1}{2}$.

12. (a)

n	1	2	3	4	5	6
error (%)	27	12	5	2	0.7	0.2

(b) $E = 5.103$ (eV).

Chapter 7

1. $N_f = 5.9 \times 10^{22}$ (electrons/cm^3).
2. See Fig. 7.9.
3. $\tau = 2.5 \times 10^{-14}$ (s); $l = 393$ (Å).
5. $N_f = 2.73 \times 10^{22}$ (electrons/cm^3) or 1.07 (electrons/atom).
7. $N(E) = 1.95 \times 10^{47}$ (electrons/J m^3) $\equiv 3.12 \times 10^{22}$ (electrons/eV cm^3). The joule is a relatively large energy unit for the present purpose.

Chapter 8

1. $N' = 9.77 \times 10^9$ (electrons/cm^3).

2.

T (K)	300	400	500	600	700
N_e (electrons/cm^3)	6.2×10^{-15}	2.4×10^{-6}	3.7×10^{-1}	1.1×10^3	3.5×10^5

3. $E_F = -E_g/2$ (using the bottom of the conduction band as the origin of the energy scale).
4. $T = 19,781$ (K) (!).
6. $E_g = 0.396$ (eV).
7. $E_F = -0.16$ eV; $\sigma = 31.2$ (1/Ω cm).
8. $(N_e)_{300°C} = 7.88 \times 10^{14}$ (electrons/cm^3);
   $(N_e)_{350°C} = 2.22 \times 10^{15}$ (electrons/cm^3).
   See also Fig. 8.9 (watch scale!).
9. (a) extrinsic $N_e = 1 \times 10^{13}$ (electrons/cm^3);
   (b) intrinsic $N_e = 9.95 \times 10^{10}$ (electrons/cm^3).
10. $E = 0.043$ (eV).

12.

Metal	Ag	Al	Au	Cu
ϕ_M (eV)	4.7	4.1	4.8	4.5

$\phi_M > \phi_S$,

$\phi_{Si} = 3.6$ (eV).

15. (a) $E_{light} > E_{gap}$;
    (b) $N = 1.6 \times 10^{14}$ (pairs/s).

16. $I_S = 2.97 \times 10^{-3}$ (A);
    $I_{net} = 3.2 \times 10^2$ (A).
20. $E = 2.58 \times 10^{-2}$ (eV).
22. 24 (nm).
23. $\eta = 0.97$.

Chapter 9

1. $\mu_{ion} = 3.32 \times 10^{-16}$ (m^2/Vs) (2 unit charges);
   $\mu_{S.C.} = 0.1$ (m^2/Vs).
2. $N_{ion} = 6.2 \times 10^{14}$ (sites/cm^3).
3. $Q = 0.83$ (eV).
4. $\sigma_{ion} = 1.35 \times 10^{-15}$ (1/Ω cm).

Chapter 10

2.	el. (DC)	opt. (AC)
ρ	1.67×10^{-6} (Ω cm)	3.85×10^{-3} (Ω cm)
σ	5.99×10^5 (Ω^{-1} cm^{-1})	2.6×10^2 (Ω^{-1} cm^{-1})

4. $Z = 27.8$ (nm).
5. $R_{Ag} = 98.88\%$; $R_{glass} = 5.19\%$.
6. $Z = 7.81$ (nm).
8. $T_2 = 94.3\%$.

Chapter 11

1. ν (s^{-1})	n	R (%)
1.43×10^{15}	0	100
1.44×10^{15}	0.1176	62
1.53×10^{15}	0.3556	23
2.0×10^{15}	0.6991	3.1
3.0×10^{15}	0.8791	0.4

2. $(\nu_1)_K = 1.03 \times 10^{15}$ (s^{-1});
   $(\nu_1)_{Li} = 1.92 \times 10^{15}$ (s^{-1}).
3. $(N_{eff})_{Na} = 1$; $(N_{eff})_K = 0.85$.
5. $R = 99.03\%$.
6. $\nu_1 = 2 \times 10^{15}$ (s^{-1}); $\nu_2 = 3.56 \times 10^{12}$ (s^{-1}).

7. λ (μm)	0.3	0.4	0.5	0.7	1	2	5
n_{calc}	1.599	1.460	1.416	1.384	1.369	1.359	1.356
ν (in 10^{14} s^{-1})	9.99	7.49	5.99	4.28	2.99	1.5	0.51

12. $N_{eff} = 5.49 \times 10^{22}$ (electrons/cm^3);
    $N_a = 5.86 \times 10^{22}$ (atoms/cm^3);
    $N_{eff}/N_a = 0.94$ (electrons/atom).
    600 nm is in the red part of the spectrum where the free-electron theory may be
    valid, see Fig. 11.1(a).

Chapter 12

3.	(a)	(b)	(c)	(d)	(e)
A	Metal (Ni) High R in IR (intraband tr.)	None	1.5 eV (weak) 3 eV (strong)	yes	partially filled bands
B	Semiconductor (GaAs) Low R in IR (no intraband transitions)	IR	1.5 eV (Band gap)	no	filled bands

4. 1.5 ($\Sigma_3 \rightarrow \Sigma_1$ and $W_2' \rightarrow W_1$).

Chapter 13

1. $n_2 - n_3 \approx 10^{-2}$.
2. $t = 2$ (μm).
3. (a) $\beta_T = 16.1°$;
   (b) $\beta_T = 41.8°$.
4. $N_{f_3} - N_{f_2} = 1.15 \times 10^{18}$ (cm^{-3}).
5. $\alpha = 1.6$ (cm^{-1}) $= 6.9$ (dB/cm).
7. $\lambda_{disk} = 503.2$ (nm); $\lambda/4 = 126$ nm.
8. $E = 1.4$ (eV); $\lambda = 886$ (nm)

Chapter 14

2. $H = 2.51 \times 10^2$ (Oe);
   $H = 2 \times 10^4$ (A/m).
4. Answers are in Table 14.1.

Chapter 15

1. $\chi = -70.9 \times 10^{-6}$. Note: $\chi_{SI} = 4\pi\chi_{cgs}$
6. $\zeta = 6.71 \times 10^{-3}$.
7. $\chi = 6.91 \times 10^{-5}$ (about one magnetic moment per atom).
8. FeO $\cdot$ Fe$_2$O$_3$, $\mu_m = 4$ μ_B;
   CoO $\cdot$ Fe$_2$O$_3$, $\mu_m = 3$ μ_B.
10. $H_M = 1.67 \times 10^9 \left(\dfrac{A}{m}\right)$.
11. Sure! (No tricks please.)

Chapter 16

1. $\chi_{para} = 5.41 \times 10^{-5}$ (Al?).
   (Diamagnetism *not* taken into consideration).
3. Fe: 7.8 out of 10; Co: 8.2 out of 10.
4. Ferro Fe: $\mu_m = 2.2$ μ_B; Co: $\mu_m = 1.8$ μ_B;
   Ferri Fe: $\mu_m = 4$ μ_B; Co: $\mu_m = 3$ μ_B;
   The number of Bohr magnetons for a single iron atom is zero. Ferromagnetism
   needs interaction with other atoms.

Chapter 17

1. $E = 10^{-3}$ (J).
2. For a synchrotron a steady magnetic field is used. No eddy current! High flux multiplication needed. Consult Table 17.1.
6. Joule heating in wires.

Chapter 19

1. $N_a = 2.4 \times 10^{11}$ (atoms/m^3).
2. $-J_Q$: Glass 3.8×10^3 (J/s m^2);
   Al 4.74×10^5 (J/s m^2);
   Wood 3.2×10^2 (J/s m^2).
3. $T = 1,142$ (K) $= 869$ (°C).
5. Proper heat dissipation is essential in semiconductor devices.
7. 1 BTU is the heat required to raise the temperature of one pound of water by one degree fahrenheit (!) (1 BTU $= 1055$ J)

Chapter 20

1. $\Delta N/N_{tot} = 0.566\%$.
2. $C_v = 24.4$ (J/K mol) $= 5.84$ (cal/K mol).
3. $C_v^{el} = 0.212$ (J/K mol).
4. $N(E_F) = 3.66 \times 10^{42}$ (energy states/mol J) $= 5.86 \times 10^{23}$ (energy states/mol eV); $N(E_F)$ per cubic centimeter is about one order of magnitude smaller.
6. $T = 2.1 \times 10^5$ (K).

Chapter 21

1. $K = 1.55 \times 10^2$ (J/s m K).
2. $L = 2.44 \times 10^{-8}$ (J Ω/K^2 s).
3. $l = 411$ (Å).

Chapter 22

1. $F = 2,600$ (N) (!).
2. $\Delta L = 4.9$ (mm).
4. Compare expansion coefficients (Table 22.1).

Index

Printed in the United States
134496LV00002B/19/P